The World in a Crucible: Laboratory Practice and Geological Theory at the Beginning of Geology

Sally Newcomb
13120 Two Farm Dr.
Silver Spring, Maryland 20902
USA

THE
GEOLOGICAL
SOCIETY
OF AMERICA®

Special Paper 449

3300 Penrose Place, P.O. Box 9140 ▪ Boulder, Colorado 80301-9140, USA

2009

Published by The Geological Society of America, Inc.
3300 Penrose Place, P.O. Box 9140, Boulder, Colorado 80301-9140, USA
www.geosociety.org

Printed in U.S.A.

GSA Books Science Editors: Marion E. Bickford and Donald I. Siegel

Library of Congress Cataloging-in-Publication Data

Newcomb, Sally.
The world in a crucible : laboratory practice and geological theory at the beginning of geology /
  Sally Newcomb.
     p. cm. — (Special paper ; 449)
  Includes bibliographical references and index.
  ISBN 978-0-8137-2449-2 (pbk.)
     1. Geology—History—19th century. 2. Petrology. 3. Geochemistry. I. Title.
QE11.N49 2009
550.9'034—dc22
                                                2008046585

Cover: This view of a mineralogist using his blowpipe is displayed in the University of Oklahoma History of Science Collections. From the collection of Everette Lee DeGolyer, founding donor of the University of Oklahoma History of Science Collections.

10 9 8 7 6 5 4 3 2 1

Dedication

To Bob
       Gail
          Rob
              Jesse
                  And small Mariah

They think they know those regions of the earth which never can be seen; and they judge of the great operations of the mineral kingdom, from having kindled a fire, and looked into the bottom of a little crucible.

—James Hutton, 1795 (v. 1, p. 251)

Contents

Preface

On several occasions I have asked groups of historians of geology when, in their opinion, experimentation began to contribute to geology. The number of people I asked is not large, but I believe it is a reasonably fair representation of opinion. The answers tend to fall into two groups. More than half immediately suggested the work of Sir James Hall. The others, while often noting his importance, mentioned less well-known, often later, people from specialties that related to their own work. A philosophically minded German colleague mentioned Hall, but then questioned what exactly was meant by "experiment." That is, of course, an important question whose answer for the purposes of this book lies somewhere between quantified and/or controlled observation, active "interference," and specific hypothesis testing. I will not attempt to impose twenty-first-century definitions of what constitutes an experiment, being far more interested in determining how important what the eighteenth- and nineteenth-century participants called "experiment" was to their perception of the science. The outcome of my informal enquiries was that it was evident that many historians of geology had given little thought to the question of experimental geology, and that in consequence they followed a historiographical "chestnut," or imposed their own experiences on the historical record.[1]

Early experimental topics mentioned by those colleagues and others included work with magnetism, earth density and tides, investigation of crater formation, the importance of microscopy and invention of achromatic lenses, the invention and use of the blowpipe, the work of chemists to clarify mineral composition, Nicolas Lémery's (1645–1715) "volcano," work on glacier ice flow with pressure effects, and rock mechanics. Those subjects easily span 200 years. Hatten Yoder's admirable *Timetable of petrology* (1993) included many of the investigatory milestones for that part of geology, as well as an excellent bibliography. Discussion in this book will be limited to the period prior to ca. 1840 and will concentrate mainly on experiments aimed at explaining or contributing to the understanding of rock origin.

Where does geology fit in the framework of science? Physics, chemistry, biology: These are the natural sciences generally agreed on by the end of the twentieth century. All of them, and their applied relatives in medicine, engineering, space and environmental sciences, are part of our intellectual landscape. But the questions continue. Where are neurology, physiology, and bacteriology? Shouldn't we mention the neurosciences—no mere offshoot of biology, but a unique blend of many of the sciences, testing hypotheses that are not in the domain of many of them, and whose annual meeting of the Society for Neuroscience draws in excess of 25,000 participants? When asked to define science, both scientists and philosophers of science agree that it is not a list of facts, but an activity that asks questions of nature with an accepted methodology. Different sciences are defined by their domains, those portions of the physical or biological worlds about which they ask questions. Inquiry methods and instruments are devised to provide answers. In his Carnegie

---

[1]It is important to state here that the concept of "experimentation" was nebulous during the period addressed in this book, which is mainly concerned with the time period from ca. 1760 until 1840, and most specifically 1780 until 1820. There were a number of activities that would not meet twentieth-first-century criteria for experiment, but that are included in the discussion. Where "observation" turns into "directed observation," which in turn may employ an instrument, is a wobbly frontier. It is safe to say that the investigators of this time period used any means they could think of to gain knowledge. And of course, some investigations, mesmerism and dowsing, for example, were discredited, as were many investigations trying to apply electricity and/or magnetism where they could not be applied. At the same time, when appropriate, multiple samples and purified reagents were used in analysis, and the value of repetition was recognized. One of the recognized activities of philosophical gatherings was the repetition of experiments or analyses. Today in geology and evolution and other sciences where we cannot test directly, the term "experiment" is applied. Hypotheses can still be made, and tested. For example, drilling deep into Earth is usually preceded by a hypothesis of what can be expected, often supported by geophysical data. The same logic holds for microscopy from the sixteenth to the twenty-first centuries. At first both microscopes and telescopes were used to increase vision, micro and macro. As the instruments improved, active structures could be identified or, in the case of mineralogy, growth features determined, which in turn could be related to formation conditions.

Figure 0.1. The title page of Thomas Burnet's (1635?–1715) seminal
The sacred theory of the Earth (1684).

evening lecture of 1997, Robert Hazen characterized scientific questions as of three kinds: (1) Questions of existence, or what's out there? (2) Questions of process, or how does it work? (3) Questions of origin, or where did it come from?[2] All of these kinds of questions were asked in early geology.

In some ways, the nineteenth century could be characterized as the century when the domains of biology, chemistry, physics, and geology were, if not first defined, at least refined. The more "mature" sciences of physics, astronomy, and chemistry, which encompassed mineralogy, were characterized first, followed by the often classificatory and field-based subjects of geology and biology (which had been preceded by botany and zoology). As the twentieth century advanced, boundaries blurred, admitting molecular biology, biochemistry and biophysics, geophysics, meteorology, the neurosciences, space science, and others. In truth, any excursion into the history of science shows that the divisions were rarely clear. At its inception, most who worked in the field of what we now call "geology" considered themselves natural philosophers who might use similar causal and/or quantitative methodologies to ask questions of agreed-upon importance about natural phenomena or materials, or natural historians concerned with classifying and describing natural objects.

---

[2]Robert Hazen, "Carnegie Science and the Endless Frontier" (lecture, Carnegie Institution, Washington, D.C., 7 May 1997).

The now-familiar term "geology" evolved in a series of steps. We have seen reference to the field change again in the late twentieth century into the appellations "geoscience," the "earth sciences," or "earth system science." In the eighteenth century, the word "geology" was just beginning to be commonly used as we now use it.[3] The word means, simply, "the study of the earth," but at the time that study existed as parts of areas that will be discussed in Chapter 1. The influential correspondent Jean André Deluc (1727–1817) used the word in 1778 because what he was describing concerned only Earth, and thus, "cosmology" was not appropriate (Rudwick, 2005, p. 135). The French Déodat de Dolomieu (1750–1801) had begun to use the term in reports in 1795 and 1797, in the latter year applying it to the title of a course he was teaching at the School of Mines in Paris (Rudwick, 2005, p. 338, 346). The Swiss geologist Horace Bénédict de Saussure (1740–1799) employed it in the fourth volume of his *Voyages dans les Alpes,* published in 1796 (p. 528). After mention of Saussure's remark that *géologie* was not made for sloths, and that it included both tiring and perilous journeys as well as intense work in the "cabinet," or laboratory, M.J.S. Rudwick remarked:

The comment was notable not only for Saussure's acceptance of "geology"—in the still novel use of the word—as a genuine empirical science, but also for his acknowledgment that indoor museum work was the necessary complement to outdoor fieldwork, though clearly his heart remained with the latter. (Rudwick, 2005, p. 346)

The unification of the varied parts of "geology" had begun, and the term, distanced from overarching Theories of the Earth, steadily gained acceptance at the beginning of the nineteenth century (Rudwick, 2005, p. 348).

How important experimentation, instrument-aided laboratory or field observation, or quantified field observation was to the early earth scientists remains a question, but it is the one that I seek to clarify in the present study.[4] The definition of what constituted experiment was not as well formulated as it is now. My concentrated effort to understand the place of deliberate experiment or, as some called it, "tampering with nature," in the beginnings of geology revealed a gap between what historians of science report about it, which has, in fact, been rather little, and what many participants said at the time. Moreover, there are many suggestions for the driving force that resulted in the emergence of geology as a science. Historians' views about this show a wide range of opinions. Each view betrays something of the historians' backgrounds and biases. Part of the variety is also due to the diversity of questions that geology attempts to answer, as well as that of the methods employed. Those methods may include, in addition to techniques unique to geology, procedures from biology, chemistry, physics, and statistical analysis, latter-day combinations of those, and computer analysis and modeling. A single historian or scientist will never be familiar with the entire range of possibilities. Consequently, for example, if the historian or scientist-historian is familiar with the methods of geophysics, less attention may be given to the chemistry or the implications of mineral composition.

I will not consider here the question of when a pre-science becomes a science, in what mode it can occur, nor its roots in what J. Schuster and G. Watchirs have discussed as natural philosophy (Schuster and Watchirs, 1990, p. 1–48). Also I will not attempt to fit the origins of all experimentation into a philosophical framework. Instead, this book will discuss experimentation and intentional investigations done to gain knowledge about the origins of earth materials and their behavior. This does not address all parts of what is now known as "geology," such as mountain chain or basin formation, explanations of surface geomorphology, the operation of the hydrologic cycle, or much about the age of Earth and its fossils. I will try to present, often in participants' words, what their investigations revealed to them about the origin and behavior of earth materials. I will concentrate on the impact of the experimental work on how people came to think of these problems, how geological thought was influenced by experimentation, and its role in the emergence of geology as a science. However, it will become clear that I believe that the objective activities with the broad designation of "experiment," whether in the laboratory or the field, whether instrument aided or not, constitute a sturdy "fifth leg" to the four components that Rudwick espouses (Rudwick, 2005, p. 59–132).

---

[3]That statement is accepted by most historians of geology. However, Vai and Cavazza (2006, p. 44) note that Ulisse Aldrovandi used it in that sense in his will, which was published in 1603.

[4]As explained in note 1, I will be using a less strict definition of "experiment" than is currently in use.

Acknowledgments

I owe many thanks to the reviewers, David Oldroyd, Ken Taylor, Bill Brice, and Davis Young, who saved me from many infelicities, and who improved the final book enormously. David has offered valuable advice about the project for a number of years. Kerry Magruder and his colleagues at the University of Oklahoma libraries supplied most of the images, working ahead of a deadline of their own. Stephen Brush taught me to "write a good story," and I have tried to follow his advice. Yildirim Dilek saw the possibilities of such a story and facilitated its progression. Sandra Herbert and Greg Good both read chapters, which benefited from their insights. And of course, much is owed to patient rare book librarians, especially David Corson and his staff at the Kroch Library at Cornell University, the late Peter Schmidt at the Bergakademie of Freiberg, and the staffs at the Library of Congress, the Dibner Library of the Smithsonian, the University of Maryland libraries, and at the USGS Library at Reston, Virginia. Robert Hull at the Othmer Library of the Chemical Heritage Foundation supplied images from the Roy G. Neville Historical Chemical Library as noted. Susanne Kittlinger at the Science and Picture Library of the Science Museum of London sent the image of the Wedgwood pyrometers in their collection. The completed project would never have seen the light of day without the help of my husband, Bob, who has been an enthusiastic solver of computer and printer glitches and has attempted to bring me from the eighteenth to the twenty-first century.

All of the portraits are from the Portraits Collection of the History of Science Collections, University of Oklahoma Libraries.

All of the other illustrations, with the three exceptions below, are courtesy of the History of Science Collections, University of Oklahoma libraries, © the Board of Regents of the University of Oklahoma, and Dr. Kerry Magruder.

Figures 1.1 and 3.5 appear courtesy of the Othmer Library of the Chemical Heritage Foundation, Robert Hull, digital librarian.

Figure 3.7 appears courtesy of the Science Museum, London.

The Geological Society of America
Special Paper 449
2009

The World in a Crucible: Laboratory Practice and Geological Theory at the Beginning of Geology

Sally Newcomb
13120 Two Farm Dr., Silver Spring, Maryland 20902, USA

ABSTRACT

In the late eighteenth and early nineteenth centuries, numerous ways of gaining knowledge about Earth combined to form the new discipline of geology. However, laboratory experimentation and/or instrument-aided observation, as applied to mineralogy and petrology, is little discussed by historians of geology, although primary literature of the time attests to its importance. It is my belief that it was information about the parameters of rock behavior gained in the laboratory that led to the convergence of emerging theories of petrogenesis, which had previously been only a matter of speculation and argument.

James Hutton's theory of igneous, or at least heat-mitigated, origin for nearly all rocks, was put forward in 1785, and more fully expressed in 1795. Historians of geology have not understood the lack of acceptance of the theory, when they have recognized it at all. From our vantage point, igneous origin, particularly of basalt, appears very obvious, and it is difficult to give credence to the once widely held belief that basalt (and other rocks now called "igneous") could have crystallized from solution. But in that earlier time, the appearance of basalt near volcanoes could be explained as solution-deposited rock simply altered by the somewhat ephemeral heat of the volcano. Rocks identical to basalt, called "whinstones" or "toadstones," were widely observed at great distances from active volcanoes, and apparently had no relation to them.

It is difficult to see why the origin of many rocks from solution would be accepted by those who were familiar with the field appearance and properties of minerals and rocks, while igneous origin was rejected. *The World in a Crucible* is an investigation of this paradox. The laboratory proved to be fundamental in determining parameters of rock and mineral behavior that didn't so much explain their origin, but gave insight into the possible as well as the impossible. This book is an investigation of the patient work over roughly 60 years by workers determined to solve the paradox of rock origin.

Newcomb, S., 2009, The World in a Crucible: Laboratory Practice and Geological Theory at the Beginning of Geology: Geological Society of America Special Paper 449, 185 p., doi: 10.1130/2009.2449. For permission to copy, contact editing@geosociety.org. ©2009 The Geological Society of America. All rights reserved.

CHAPTER 1

Origins of geology

The evolution of what came to be called "geology" is interesting. Geology is a science that draws from many disciplines, but yet maintains its integrity as a science in its own right. Rudwick noted that mineralogy, physical geology or geography, and geognosy,[1] were natural history sciences in that they "described and classified the diversity of the mineral world in atemporal terms." Earth physics described and investigated earth features in terms of natural processes in the framework of time. The latter was more the venue of natural philosophy (Rudwick, 2005, p. 290). Rudwick stated:

In Saussure's time, four fairly distinct sciences were concerned with the material objects and phenomena of the earth; or they could be described as four distinct sets of practical activities, each with its own characteristic genres of texts and pictures. (Rudwick, 2005, p. 59)

In the eclectic science of the day, practitioners often were active in more than one of the fields, and more or less the total of those areas of interest became the integrated science of geology. But certainly objects collected or questions engendered by the first three of the areas were taken back to the laboratory or the museum. Instruments were devised to make quantitative measurements. One could hardly say that measuring the angles of crystals or determining strike and dip would come under the rubric of "experiment." But at some point, observation or description leads to further investigations that could. Mineral color could simply be observed. To determine hardness requires further action with an "instrument," if a knife or nail could be so characterized. Chemical analysis of the mineral requires still more. One could argue that this constituted active "interference." Despite the simplicity of some of the procedures, I suggest that this sequence and its further development, ending with experiment as we would now define it, forms another main component of geological reasoning.

Just as no date can be given when alchemy and the practices of metal extraction and assaying, dyeing and distilling, medicine and medications, became chemistry, the same is true of the transition from the earlier categories to what is recognizably geology. It is generally accepted that the transition took place in the first decades of the nineteenth century. We note that in the past, the history of geology literature had mostly been dominated by accounts of the active and talented British practitioners, but there is an increasing literature addressing practices by the Germans, Italians, French, Swedish, etc.

One can categorize the precursors of geology in a somewhat different way, albeit with some of the same divisions. The interests of both philosophy and economics provided background, supplying tools and knowledge that were applied to conjectures about Earth, including:

1. alchemy;
2. observation of earth materials (rocks and minerals) and earth processes (folding and erosion), including mineralogy;
3. economics and technology, including mining, and the use of conceptual or experimental analogies from the use of glass, iron, and porcelain furnaces; and
4. social needs and connections.

None of those categories are exclusive, nor are they sequential as listed, nor do they include the same kinds of thought or knowledge. However, it is difficult to discuss any of them without including portions of the others. As noted above, I will often be concerned with what at the least could be called "extended observation," in other words, some sort of manipulation or measurement of materials as opposed to simple observation, as well as more formal experimentation. The first goes back to antiquity, so that even in the seventeenth century there was a prior history. In a book by P. Boccone (1633–1704), there are both reports of legendary and/or fanciful origins of minerals and fossils, and laboratory experiments done to clarify mechanisms of earth puzzles (Boccone, 1674). F.D. Adams's treatise (1938) gave a feel for the roots of information about earth materials from antiquity onward.

ALCHEMY

It may seem strange to begin a discussion about early geology with alchemy, represented allegorically in her laboratory (Fig. 1.1). However, there was a considerable wealth of knowledge about the behaviors of natural materials passed on from alchemy, and mineralogy and the mining tradition both informed and used that heritage. Cecil Schneer pointed out that early man used earth materials for adornment, light, hide tanning, metal working, preserving or fermenting food, embalming, making containers, and dyeing (Schneer, [1969] 1988, *Mind and matter*, p. 6–7).[2] Metal working became even more widespread, sophisticated, and necessary, highlighting the need to know about ores and their associations. Multhauf differentiated practical chemistry, meaning that of

---

[1]Geognosy came to be thought of basically as structural geology, derived from the insights of miners, which added the third dimension.

[2]Schneer provided a broad and informed tour of alchemy, metallurgy, and the growing craft tradition in the first chapters of his book.

Figure 1.1. An allegorical representation of Alchymya in her laboratory as shown in Gesner (1576). Courtesy of Othmer Library.

the craftsman, from alchemy with its theoretical basis, but noted that advancing research showed that the distinction was not clear (Multhauf, 1966, p. 14). The Greek, Arabic, and Latin worlds all had alchemical writings as did the Chinese. Alchemy itself changed in character during the eighteenth century, addressing more practical problems than transmutation of metals.

While alchemy has been dismissed as nothing more than misguided attempts to turn base metals into gold, many studies show it was the source of practical knowledge about materials and their transformations, as well as providing a theoretical basis for questions about matter (e.g., J. Read, [1957] 1961; Schneer, [1969] 1988; P.H. Smith, 1994; and Nobis *in* Fritscher and Henderson, 1998.) More recent studies further investigate chemistry's alchemical lineage. Newman (2000) traced the long-established use of the blowpipe and the balance in alchemy, as well as describing the use of test and experiment to determine the nature of metals. Principe (2000) discussed apparatus and reproducibility in alchemy. Fruton (2002) traced alchemical literature from the Greeks to the transition into chemistry.

Alchemy fostered a search for the ultimate nature and substance(s) of matter, application of "active principles" for medical intervention, and an inquiry into the nature of heat, which eventually took the form of phlogiston. Through the centuries, as now, there were charlatans who exploited their knowledge and others' credulity. However, those attempts resulted not only in increased understanding of the properties of many substances

and procedures for manipulating them, but also in equipment of many kinds in which to contain and process materials.

In one way, investigations of earth material properties in the seventeenth and eighteenth centuries seem straightforward. With some knowledge of the technical language used, a good dictionary and glossary, familiarity with laboratory and/or industrial procedures, and the aid of often-excellent illustrations of surviving apparatus, it is not difficult for us to understand those processes of fusion, oxidation or reduction, or the progress of serial analysis or synthesis reactions. However, through the eighteenth century, the ultimate nature of matter was unknown, though it was the subject of much argument. There are excellent discussions of the impact of changing matter theory on mineralogy and chemistry in Schneer ([1969] 1988), Melhado (1981), Laudan (1987), Oldroyd (1998), Siegfried (2002), and Fruton (2002). Melhado stated several problems with analysis in the seventeenth and eighteenth centuries, among them the question of whether the end materials of analysis had been originally present or were products of the fire often used for analysis, and whether they were truly end products or could be further broken down (Melhado, 1981, p. 51). These concerns were very much at issue in the important work of determining mineral composition. Antoine Lavoisier (1743–1794), John Dalton (1766–1844), Jöns Jakob Berzelius (1799–1848), Humphry Davy (1778–1829), and others advanced the understanding of matter. Oldroyd noted "the transition from alchemy to a theory of matter which suggested the hunt for specific constituent substances in objects" (Oldroyd, 1996, p. 40).

When Robert Hooke (1635–1703) spoke of manipulation of materials he used the term "Chymistry." He extended direct observation of one action to one of another kind by analogy, but at least in the lectures published in Drake (1996), he did not use obviously "alchemical" language. With some background in recent studies of alchemy, Robert Boyle's (1626–1691) work is not difficult to follow. Although he conjectured about gem formation by exhalations, his comments are generally rooted in what he saw, and he used both quantitative and qualitative measures to describe gems. While he spoke of medicinal stones and the virtues of gems, and while his knowledge was incomplete, he seemed to connect activity to either chemically (and/or physiologically) active ingredients, or to materials impregnated with such substances. Unsurprisingly, his language reflected the tradition of alchemical writing. For example, he spoke of virtues, exhalations, liquid principles, and lapidescent juices, but his reasoning can be followed (Boyle, [1672] 1972). John Read pointed out that Boyle's experiments were not random, but were focused on the solution of a particular problem such as combustion (J. Read, [1957] 1961, p. 114). Gain and loss of weight during heating was a commonly investigated problem, even in the seventeenth century. In the mid-eighteenth century, Johann Gottlob Lehmann (1719–1767), along with many others, seemingly transmitted some portion of alchemist belief in the influence of planets and the sun's rays on metal generation (Adams, [1938] 1954, p. 282–283).

In the process of trying to solve practical problems such as where to look for ores, direct observation eventually became more important than following alchemical writings. Like much else in mineralogy and ideas about rock origin, trusted knowledge in the form of reliable, testable information remained, and was augmented by new observations and instruments and the increased reliability of chemistry.

OBSERVATION

D.H. Hall has said:

By *observation* we mean, in its simplest form, the mere description of, or the measurement of parameters related to, phenomena of (in Francis Bacon's words) "uncontrolled nature". We can imagine a whole sequence of types of observation grading from the simplest through more sophisticated *designed observations* to true *experiments*. (1976, p. 61)

All aspects of Hall's description were employed by the earth-investigators at the end of the eighteenth century. The Earth is obviously hard to ignore. Despite often being a metaphor for stability and permanence, or perhaps because of that impression, when it does go into a paroxysm the impressions are vivid and unsettling. Earthquakes, volcanoes, landslides, and floods all give rise to fear, and then to an effort to understand their causes. Even quieter changes, such as sea-level change as indisputably recorded on the columns of the so-called temple (now identified as a marketplace) of Serapis near Naples, have given rise to wonder and speculation. As humankind slowly and variously retreated from belief that such destructive changes indicated the anger of god(s), there was increased effort to understand the phenomena in terms of known causes and effects. Early statements of, for example, the causes of volcanism seem fanciful to us. However, many of them, such as volcanic eruptions caused by water infiltration, heating, and expansion, are rooted at least in part in observation and truth.

An obvious advantage gained from observation of earth materials was that, if noted closely enough, one could find more of a useful substance, such as ores. Observation, even with a complete lack of interpretation, could still be useful. This held for minerals, ores, clay, rocks, and soils, as well as gemstones and native metals. Observation led naturally to classification, which was increasingly sophisticated in the emerging geology of the late eighteenth and early nineteenth centuries. Observation contributed to use of stone tools and ochre; working native iron, copper, and gold; the fascination with gems and precious stones; food preservation, sometimes with salts; winning metals from ores: all of which led to practical knowledge.

An observer from space would say that judging by current events, human understanding of earth processes is still most imperfect. In recent years, floods have ravaged China, India, Mexico, and Korea, with the loss of thousands of lives. Countries around the globe have been decimated by geological events. Hundreds of lives have been lost to floods and mudslides in California, and to volcanic eruptions in numerous places. The recent tsunamis in

Asia and earthquakes in Pakistan and elsewhere claimed many thousands of lives in 2004 and 2005. Our knowledge of rock mechanics and our arrays of seismic instruments rarely if ever predict earthquakes in a useful way. On an even larger scale, there is concern about global warming, with its effects on sea level around the world, and impacts of several kinds on agriculture and animal populations.

Nevertheless, observations of all of these earth processes from antiquity onward were a portion of the impetus that led to formal study of a science of Earth. There is a major genre of writings, from Pliny to the present, that is concerned with descriptions of earth processes, the most popular being the catastrophic. Volcanic eruptions and earthquakes were prominently featured. As observation admitted less of the supernatural to be a factor (the gods could do *anything*), and knowledge of materials, heat, and force increased, additional thought was given to more-plausible explanations. Recognition of predictable behavior discouraged supernatural explanation. Just going outside and looking led to important theory expansion. Some rock layers were clearly folded and distorted, and several modes of thinking could be adopted to explain this. It could be that originally horizontal layers of soft sediments were distorted by some later force. Or the layers might have been laid down on a tilted or uneven surface. On a smaller scale, mineralogists and chemists were subjecting minerals to laboratory procedures, which in turn led to increased understanding of necessary conditions for their natural occurrence. A major debate at the turn of the eighteenth to the nineteenth century, which continued for many years, was whether the forces currently observed were the same that had operated over geologic time, or whether they differed in magnitude, rate, or kind. Nearly every possible combination of magnitude and rate could be observed or postulated. The extent of geologic time was itself unresolved. As can be seen, unaided observation, whether of apparently rapid catastrophes or anomalies such as marine shells in layers on mountains, delineated many of the problems that were investigated in geology. The meaning and significance of some of those observations remained in question for many years.

TECHNOLOGY AND ECONOMICS

The history of technology, the technology being driven by practical economics, has always been a lively discipline.[3] For many of us, even if we don't work in technology ourselves, the *how* of industrial processes as various as metal smelting and working, glass production for normal and special uses, railroad building, fabrication and use of alloys for orthopedic purposes, silicon chip manufacture, etc., exercises a real fascination. In the seventeenth and eighteenth centuries, and into the present, the boundary between science and technology has been and is

---

[3]No matter what the technology, there had to be an economic benefit from it. Therefore, there was always an effort to find the richest ore, the most beneficial furnace temperature, or the process that produced the best or most durable product.

Figure 1.2. One of several representations of mining in Agricolae (1556).

famously indistinct. While science is normally credited with leading to useful technologies, the opposite is also often the case. Sometimes utility led to philosophy, rather than the other way around. As mentioned above, in the eighteenth century and onward, scientists did not hesitate to use resources provided by technology. Possibly better informed than current scientists about industrial processes, they applied analogies from those processes to scientific questions.

The glass, metal-smelting, porcelain, and ceramics industries supplied both equipment and procedures that could be applied to scientific questions. Analogies were made about appropriate combinations and conditions. Products and byproducts from forges and furnaces were compared to natural minerals and rocks. Some analogies went in what might be called the other direction, for example, when Josiah Wedgwood (1730–1795) named an attractive black ware that he'd produced "Basaltes" (Copeland, 1995, p. 9). Wedgwood developed his pyrometer, which will be discussed in later chapters, so that he could determine temperatures in his pottery kilns. It was the first successful method of quantifying the higher ranges of temperature fairly well and reproducibly. More unusual instruments, such as metallo- and hydroscopes were optimistically employed to learn more about metals and water underground.

Mining is and has been a constant contributor to geological and mineralogical knowledge. (See Fig. 1.2.) The importance of the early treatises on mining has been frequently noted (Agricola, [1556] 1950; Biringuccio, [1540] 1990; Ercker, [1580] 1951). Many twentieth-century authors have written about their importance, and that of mining itself, in the history of geology (D.H. Hall, 1976; T.M. Porter, 1981; Laudan, 1987; Beretta, 1998a, 1998b; Hamm, 1997; Vaccari, 1998b; Vaccari and Morello, 1998). Mining provided observations about minerals and their associations as well as rock layer structure and sequence. Ore manipulation and reduction, assaying, and development and uses of appropriate reagents contributed to geology through mineralogy, as well as being a prime mover in the change of chemistry to a quantitative science.[4] I shall discuss equipment and techniques from mining and assaying that were transferred from their original pragmatic application in technology to experiments that were designed to amplify or test geological theory. Even if the methods of technology did not lead directly to new ways of thinking about Earth, in some cases they served as a reality check to help filter out more fanciful suggestions.

SOCIAL NEEDS AND CONNECTIONS

It goes without saying that the emerging science of geology was quite influenced by the nation in which it was studied, by the wealth and power structure of that nation, which in turn depended on its climate and natural resources, and by the social relations among the scientists and their institutions. These relations are of major interest in the history of science, and many of the secondary references include treatment of one or more of them. Here we shall basically be concerned with the means by which those scientists "asked questions of nature" in the laboratory, and how the answers were interpreted.

---

4. See Oldroyd (1996, p. 68–71) for a good summary of this transition.

CHAPTER 2

Characters and how we know them

INTRODUCTION

A discussion of "external characters" may not be an obvious place to begin an exploration of laboratory work. Those observable features of minerals had been in use by mineralogists for many years by the end of the eighteenth century.[1] Surely such classificatory methods belong to the older science of geology-to-be or natural history, and refer as well to techniques that proved unworkable and less scientific compared to methods based on chemical composition and/or crystallographic principles. Putting aside the fact that mineralogy students still learn field identification by means of an updated version of these characters, a closer look at the criteria used and how they were applied will show both the developing state of specific knowledge as well as changing ideas about what qualified as useful. Quantitative measures increased, as did the use of instruments to augment human senses. With respect to conjectures about rock origin, the properties of minerals were what could be called the baseline. Any suggestion of rock origin had to be consonant with the properties of the minerals that made it up.

The various "characters" will be discussed in an effort to set the stage for a detailed look at the impact of laboratory and/or instrumental work in this period. Reporting character determination would seem to require only a simple recital, but it is evident that there was a long and rich tradition of methods to give objective knowledge of minerals, to define and describe them, to attempt to quantify their properties, and, more and more, to apply some sort of scale to the measurements. These methods were concerned with the physical, optical, electrical, and magnetic properties of matter. We shall consider the chemical properties in a later chapter. The lists of properties required for identification were remarkably similar from the eighteenth century to the present, and the order of the investigation was often the same. While recognizing that X-ray diffraction or fluorescence is now used commercially or to identify rare minerals or very small samples, it is still not uncommon for a geologist to follow an investigative sequence similar to what a Werner, Haüy, Kirwan, or Mohs would have done in order to begin characterizing a mineral or a rock formation. Color is a clue, but it is not definitive. Angles and hardness build the case. This is

usually sufficient for the common rock-forming minerals. But uncommon minerals pose a far greater problem.

It is unlikely that there was a time in human history when some elements of mineral characterization were not in use. Because of this, claims of "first" are more than usually suspect. Rock names, frequently based on mineral composition, also were in use very early. Naturally, the precise meaning of both rock and mineral names had to be standardized, refined, and changed, as information accumulated and as mineralogy became more scientific. In the eighteenth century the same name might have referred to different minerals or rocks in different regions or countries; or the same substance might have been given different names. This chapter will focus on methods and parameters rather than names, and will consider how properties were identified, regardless of how the particular character was classified.

The concept of external characters seems deceptively simple at first. What comes to mind are the obvious properties of color, luster, hardness, fracture, cleavage (or the shape of cleavage fragments), specific gravity, etc. A simple organizational scheme to describe work of the eighteenth century is, however, difficult to formulate. The same character might be classified as internal or external, the names of the classes could be changed, or the classification scheme itself might be changed. In this chapter, we will see what can be learned about the characters classified as external by most mineralogists, and how they were studied.

It is not uncommon for the historian of geology to first become acquainted with the importance of the external characters of minerals through reading Abraham Gottlob Werner's (1749–1817) *Von den äusserlichen Kennzeichen der Fossilien* ([1774] 1962). (See Fig. 2.1.) As Albert V. Carozzi has pointed out in his 1962 English translation, the question of external characters was a topic of lively interest at the time. Werner himself had translated the anatomist and mineralogist Johann Karl Gehler's (1732–1796) *De characteribus Fossilium externis* (of 1757) into German (Carozzi, *in* Werner, [1774] 1962, p. x). Franz Ritter von Kobell (1803–1882), the chief nineteenth-century historian of mineralogy, began his discussion of the history of systematic nomenclature from 1750 to 1800 with the Swedish Johann Gotschalk Wallerius's (1709–1785) treatment of external characters, a source with which Werner was familiar (Kobell, [1864] 1965, p. 155).

Descriptive and classificatory terms had been applied to minerals since antiquity. Two different, but related, organizational efforts can be seen, on two different levels. One was the desire to describe a mineral or rock in such a way that it would ultimately describe a specific substance and not something else

---

[1] A reminder: Rudwick noted that true laboratory experiment designed to mimic earth processes was on a far smaller scale than could be effective. However, the laboratory was most useful in classifying minerals. The museum, or cabinet, was also useful in that many specimens were collected and could be studied (Rudwick, 2005, p. 38).

Figure 2.1. Abraham Gottlob Werner of the Bergakademie, Freiberg.

1735, in the writings of Carl Linnaeus (1707–1778), minerals were just one type of a class of things we would call inorganic, the other members being stones and fossils (Oldroyd, 1974a, p. 508–509). This division continued roughly through the writings of Swedish chemist Axel Fredrik Cronstedt (1722–1765), with several variations.[3] The substances identified as belonging to the mineral kingdom were generally limited to salts, sulfurs (inflammables), and sometimes semi-metals, and metals.[4]

In terms of organizing principles, there was a general progression, particularly in the eighteenth century, from considerations of physical, or "external," characters to those of composition. There was overall acknowledgment that composition would be the best way, but for a long while it was felt to be insufficiently knowable. Oldroyd has emphasized that the use of Cronstedt's blowpipe yielded more of a "characterization" than an analysis (Oldroyd, 1974a, p. 511). A number of historians of mineralogy and geology has noted the increasing accuracy of wet, or "humid," analyses after their systematic introduction by the Swedish chemist, Torbern Olaf Bergman (1735–1784), whose methods were rapidly improved upon.[5] Composition, or the chemical or internal, character, was increasingly used for classification as methods engendered more confidence.[6]

The focus of this chapter will not be the various classification schemes or their importance, or how categories were ordered within the systems. Even today, in English, we begin almost unconsciously to produce a classification system when we distinguish a "rock" from a "stone."[7] What will be considered is what was actually known about these earth materials, when it was known, and the equipment or instruments used to extend human senses.

THE CHARACTERS

As noted above, any given character might be found in several of the major classification domains. There was not always a clear demarcation between physical and chemical characters, and several mineralogists of the time indicated their unease with some placements into one or another of the categories. For present purposes, external characters will generally be taken to

with a different constitution, and the description would identify it unambiguously. The other was that as knowledge developed, there was an increasing trend to order individual substances into groups. Individual characters were compared, and substances with similar characters were put into larger aggregations. This effort became especially important during the eighteenth century in what can be called the "Linnaean" period. Werner was admired at least as much, if not more, for his larger organizational scheme as for his assigned "characters." The meaning and use of this larger organization went beyond convenience or neatness. Just as characters of individual substances tell one something about the special nature of the substance described, a larger organizational scheme using similarities of individual properties can provide information about relationships between members of a group, their possible genesis, and what properties are definitive.[2]

For a long while there has been a perception that a mineral is anything that is not animal or vegetable. As pointed out by Beretta (1998a, p. 579), those three categories were in miner Georgius Agricola's (1494–1555) overall scheme, within which the first division was homogeneous bodies within mixtures. By

---

[2]Laudan discussed this dichotomy in her section "Identification vs. Classification" (Laudan, 1987, p. 81–83).

[3]The first edition of Cronstedt's *Försök til mineralogie, eller mineralrikets upställning* was published in Stockholm in 1758. There were multiple German, English, Russian, Italian, and French translations between 1760 and 1788 (Boklund, 1971, p. 474).

[4]Oldroyd (1974a) devised a useful synopsis of mineral classification schemes used from 1450 to 1850. His categories were organic, mechanical, natural history, chemical, crystallographical, and cosmogonies. There was a clustering of the natural history and chemical (by composition) systems between 1750 and 1850; there were fewer crystallographic systems.

[5]See Laudan (1987, p. 61–62) for a review.

[6]External characters were frequently called "physical characters." However, a specific character might be listed in one of several categories. I will generally use the designation of the primary author from that time period.

[7]A rock is generally understood as being an aggregate of one or more minerals. Presently, a stone is considered to be a rock that has been prepared in some way for construction. "Stone" also is used to refer to gemstones. The word "crystal" is commonly used to refer to a mineral with its faces intact, or else of good clarity.

be those that could be observed without altering the mineral's chemical composition.[8]

The lists of what were considered useful characters varied enormously. As early as the time of Aristotle and Theophrastus, minerals were identified by common properties, and the characteristics of useful and/or beautiful substances were well known. In 1546 in *De natura fossilium*, Agricola reviewed what some earlier authors had written. His first sentence was: "Mineral substances vary greatly in color, transparency, luster, brilliance, odor, taste, and other properties which are shown by their strength and weakness, shape, and form" (Agricola, [1546] 1955, p. 5). He continued with color, and mentioned variability of color within a given mineral type, and in multicolored minerals. He also discussed ease of melting and solubility, and difference in hardness as determined by scratching with steel.[9] Properties that could be determined by touch, as well as color, taste, and odor, were the best known. Miners knew the different reactions of stones to fire. As a person involved with mining, Agricola clearly knew the identities of useful minerals. Beretta (1998b) and other authors have pointed out that the practical operations of mining with its attendant technology probably had more influence on the early development of chemistry and geology than did alchemy (Beretta, 1993; Hall, 1976; Porter, 1981; Laudan, 1987; Rudwick, 2005). Werner said of Agricola: "[H]e was the first to introduce the real use of external characters in the description of minerals" (Werner, [1774] 1962, p. 7).

Long after Agricola, Wallerius ([1747] 1753) made use of color, how a mineral broke into fragments, how compact it was, what it felt like to touch, specific gravity, and relative hardness for identification. When appropriate, he added how its appearance changed in heat, or what reactions occurred with spirits of salt or niter, *aqua regia*, or other reagents.[10]

However, the definition of what characters were external varied with the mineralogist. In 1774, Werner identified the following as external:

Color	Cohesion	External form
External surface	External luster	Internal luster
Fracture	Form of fragments	Form of particles
Transparency	Streak	Stain
Hardness	Solidity	Flexibility
Adhesion to the tongue	Sound	Fluidity
Unctuosity	Coldness	Weight [density]
Smell	Taste	

(Werner, [1774] 1962, p. 23–108)

Werner also briefly mentioned internal or chemical, and physical and/or empirical characters. He preferred those characters he deemed "external" because all minerals had them in all samples, and they were subject to empirical inspection or investigation. Internal characters—meaning composition—could not be reliably determined, especially in small samples. Physical (for Werner, different from external) characters were not obvious to him and did not express essential differences. As an example, electrical activity was a physical character displayed by only a few minerals. Empirical characters were not necessarily present in different samples of the same mineral, and were not reliable. For instance, some copper or other metal ores might effloresce, and some might not (Werner, [1774] 1962, p. 4–5, p. 114–115). It is clear that he felt that if his external characters were used correctly, those in the chemical category were superfluous. For Werner, external characters were "thoroughly complete, reliably discriminative, best known, easiest to recognize, and most convenient to determine" (Werner, [1774] 1962, p. 4–5).

In 1794, the Irish chemist Richard Kirwan (1733–1812), follower of Wernerian geognosy, said: "[T]he properties which relate to our senses are called the *external*, and those that relate to chemical agents the *internal*, characters of bodies" (Kirwan, 1794a, v. 1, p. 23). He mentioned as external:

Color	Shape	Lustre
Transparency	Texture	Cohesion
Density	Taste	Smell
General feel	Color of streak	Adhesion to tongue and fingers
Absorption or diffusion in water		

Kirwan remarked that he had formerly thought that the only useful properties were the internal ones, but:

A more mature consideration has undeceived me. I perceived that names having been given to different mineral substances, sometimes by reason of their external properties, sometimes by reason of their internal, and the business of the mineralogist being to distinguish the subjects to which those names have been given, from each other, he must in some cases rely chiefly on the external, and in others chiefly on the internal; but if any ambiguity remain, the decision must rest with these. (Kirwan, 1794a, v. 1, p. 23–24)

He noted differences between the sorts of things included in lists of characters, such as Werner's (unusual) use of refrigerating powers, but said that he felt that, except for Werner and his immediate followers, the concept of character was often used in a vague sense (Kirwan, 1794a, v. 1, p. 26–27). German chemist Martin Heinrich Klaproth (1743–1817) (1801, p. 1) listed the study of minerals' *external characteristic marks* as one of five steps in learning about minerals. However, his work was concerned mainly with chemical analysis.

In 1816, the American chemist Parker Cleaveland (1780–1858) listed no fewer than 27 items under physical/external characters. These were:

---

[8]That understanding is serviceable until we get to the property of fusibility. If fusibility referred only to melting point it would still be within the "rules." But from the first quarter of the eighteenth century, and even before, in some technologies, admixtures were known to lower the melting point, and this behavior was used for mineral identification. However, because it goes well beyond identification, I will discuss fusibility in Chapter 3, which is on heat and its effects.

[9]Hardened iron, one variety of which is steel, had been known since antiquity to retain some carbon. The process was investigated in detail in the early eighteenth century by René-Antoine Ferchault de Réaumur (1683–1757) in his *L'art de convertir le fer forgé en acier, et l'art d'adoucir le fer fondu, ou de faire des ouvrages de fer fondu aussi finis que de fer forgé* (1722).

[10]Spirit of salt is hydrochloric acid, and spirit of niter is dilute nitric acid. *Aqua regia* is a mixture of the two, commonly with more nitric acid than hydrochloric.

Color	Smell	Hardness
Changeable colors	Taste	Fracture
Lustre	Adhesion to tongue	Frangibility
Transparency	Soil[11]	Shape of fragments
Refraction	Streak	Tenacity
Form	Distinct concretions	Magnetism
Surface	Flexibility	Electricity
Unctuosity	Sound	Phosphorescence
Coldness	Cohesion	Specific gravity

(Cleaveland, [1816] 1978, p. 57)

Cleaveland also had a section about chemical characters in which he discussed fusibility and various reactions. By this time, the study of chemical or internal properties was a constant feature of most methods.

In the following sections, I shall be particularly concerned with the major characters for which equipment or instrumentation was required for better determination, and generally will trace the determination of the major characters until the method stabilized. However, discussion of some, such as fusibility and composition, will be deferred until the chapters on heat and composition, respectively.

Hardness

Hardness was examined and used as a criterion for mineral characterization very early. In Albertus Magnus's *De mineralibus* ([ca. 1262] 1967), mention was made of various characters which included hardness, as had Aristotle much earlier. At first, hardness seems easy to discuss. It is determined by what either scratches or is scratched by one of a series of substances. A more formal comment is "The resistance that a smooth surface of a mineral offers to scratching is its hardness" (Hurlbut and Klein, 1977, p. 184). Hardness was progressively better described, although because it depends on several kinds of properties, there was no single means of expressing it at the end of the twentieth century (Newcomb, 2002).

It is unlikely that there was a time when comparative hardness was not used. Agricola included it among properties dependent on touch (Agricola, [1546] 1955, p. 9). Hard minerals were those that didn't take an impression. He did use a comparative scale and said:

Some minerals, too hard to take impressions, are soft enough to be scratched by iron, for example, marbles, almost all rocks and many stones known by special names. Other minerals are too hard to be scratched such as flint and almost all transparent gems. Minerals soft enough to be scratched by iron may be engraved and even turned in a lathe,... Some of the minerals which cannot be scratched with iron are very brittle such as flint while others such as the knots found in *schistos* and *basaltes* can be broken only with great effort....

Some gems are scratched by a file such as *topazius* (chrysolite) while others are not, such as lapis-lazuli and *carbunculus*. All gems can be engraved with emery except the diamond which can only be scratched by its own fragments. (Agricola, [1546] 1955, p. 11)

About quartz he said that it could be used for engraving, but it was softer than a file (Agricola, [1546] 1955, p. 120). Despite the ease of comparison, he seemed to mention hardness only somewhat incidentally in mineral descriptions.

Both Agricola, in the mid-sixteenth century, and Wallerius, in the mid-eighteenth, sometimes implied that hardness is linked to ease of fusibility. In 1747, Wallerius compared the hardness of rocks and minerals to each other. For one sort of quartz he said its hardness could be compared to that of a kind of pyrite. Petrosilex and quartz were described as "hard to very hard" and giving more or fewer sparks with steel (Wallerius, [1747] 1753, p. 193). He stated that diamond was the hardest of all the stones (Wallerius, [1747] 1753, p. 211). Hard steel was commonly used at this time in the pocket laboratory carried for field determination of minerals. It was used both for hardness testing and for producing sparks to light a spirit lamp or candle (Cronstedt, 1788, v. 2, p. 987).

Werner began his *Von den äusserlichen Kennzeichen der Fossilien* of 1774 with a review of the lists of external characters used by Agricola, Gehler, and Linnaeus (Werner, [1774] 1962, p. 8–13). Linnaeus had several divisions under hardness that included striking fire, capable of being scraped by a knife, hard, fragile, sectile, friable, brittle, flexible, malleable, staining, staining and marking, and various colors of streak. Werner valued the work of Wallerius, along with that of the German chemist and mineralogist Johann Friedrich Henckel (also spelled "Henkel") (1678–1744), the French naturalist and mineralogist Jacques Christophe Valmont de Bomare (1731–1807), and Cronstedt. He noted approvingly that Wallerius and the Berlin professor of mineralogy and mining sciences Carl Abraham Gerhard (1738–1821) had attempted to use both composition and external characters in order to decisively identify minerals, although their systems were not precise enough. Werner recognized that composition could either not be completely determined, or that two minerals of different external appearance could have quite similar chemical reactions (and presumably composition), so he focused on the external characters to make the final determination. However, he commented frequently on the possible arrangement of the integrant particles that would account for what was visible.[12]

Hardness was one of the fifteen categories Werner placed under the heading of the cohesion of solid minerals. In hardness, as in other characters, the one thing that seemed to distinguish Werner from his contemporaries was his reliance on what we might term "extreme tactile sensitivity," as when he said that

---

[11]Cleaveland differentiates between soil or stain, and streak. His definition for the latter is somewhat different than our current one, referring to the color of the powder produced by scratching the mineral with a hard instrument. We refer to the color of the material left behind on a plate of porcelain bisque, or streak plate. His soil or stain referred to the trace left behind when the mineral is rubbed on white paper (Cleaveland, [1816] 1978, p. 48).

[12]Werner defined integrant particles as those that could not be further divided without changing the character of the substance. He called larger, mechanical divisions aggregated parts. Constituent parts were those that were still combinations, but were subdivisions of the larger integrant particles. The simplest of all were the primitive constituent parts, simple units of matter (1774 [1962], p. 42 fn).

some minerals felt softer and more tender than others. He suggested that hardness differences could be confirmed by the use of instruments: "a *knife* in semihard and soft minerals, *steel* in hard minerals, and a *file* in very hard minerals" (Werner, [1774] 1962, p. 93–94). If a mineral struck sparks with steel but couldn't be scratched with a knife, he called it hard. He further differentiated hard minerals into those with three grades of response to being scratched with a file. Diamond, ruby, and emery could not be scratched. Some other minerals could be slightly scratched; and some could be easily scratched. The semihard minerals were those that did not spark, but could be scraped with a steel knife. If easily scraped with a knife, but not with a fingernail, he called them semisoft. Those scratched by a fingernail were soft. Werner supplied lists of minerals with each of the hardness grades, but concluded his discussion with a paragraph stating that the same mineral could have two levels of hardness close to each other (Werner, [1774] 1962, p. 94–95). Clearly Werner was intimately acquainted with the comparative hardnesses of a great many minerals but, at least in his published work, he did not assign numbers or standard minerals for different grades of hardness.

In 1784, Richard Kirwan, a skilled experimentalist, applied the comparative scale for hardness with considerable sophistication in his first book on mineralogy. Like many of his contemporaries, he demonstrated well-developed powers of observation. Kirwan noted that while many minerals struck fire with steel, their hardness could still be different. He used gradations of hardness to distinguish similar groups of minerals from each other, and also observed that color could be variable. As with Werner, final identification was made by considering an aggregation of characters.

Kirwan's numerical scale for hardness was in his 1794 second edition, but he must have disseminated it earlier. The editor of the 1788 English edition of Cronstedt's *An essay towards a system of mineralogy* inserted a numerical table of hardness and specific gravity that included some of Kirwan's numerical values. The table had been assembled by Bengt Andersson Quist (?–1799), director of an iron factory in Sweden (Cronstedt, 1788, v. 1, p. 225). The table included 32 minerals with the hardest listed first. Interestingly, although Kirwan gave the hardness of a diamond as 10, Quist called it 20. However, the numbers for the softer minerals at the bottom of the table supplied by Kirwan were Kirwan's own, which meant they were on a scale of 1 to 10. The last mineral was chalk, which Kirwan gave a 3. Almost all of Quist's numbers, from opal and crysolite upward, were greater than 10. Clearly, there could be and were significant differences between tables.

Not surprisingly, chemist Antoine François de Fourcroy (1755–1809) had a "chemical" view of mineral classification, but he still used hardness as one of the first indicators in identification. (See Fig. 2.2.) His first order of minerals included those that were insoluble in water and didn't burn. In the first class were those that sparked with steel, while the second class included those that did not give sparks (Fourcroy, v. 1, 1789). In a discussion of hardness he noted how variable it could be. Unlike harder stones, softer

Figure 2.2. The eminent chemist, Antoine François de Fourcroy.

stones like marble could take a good polish. Some stones, such as calcareous breccias that contained harder siliceous particles in a softer matrix, could sometimes give sparks (Fourcroy, 1789, v. 1, p. 256). While Fourcroy's discussion of hardness was typical, he added the use of a magnifying glass to look at the products of sparking (small pieces of iron detached from the steel) as collected on white paper (Fourcroy, 1789, v. 1, p. 255).

In 1792, the French naturalist Jean-Claude de La Métherie (1743–1817) did a second, expanded, edition of the earlier French translation of Bergman's *Manuel du minéralogiste* (1784a). In his extensive notes, La Métherie cited the Quist-Kirwan scale, but said that he himself had much extended the list of minerals whose hardness had been determined. He felt he needed many more intermediate steps than even Quist's total of 20 allowed, and proposed to use a maximum of 2000 instead. For a fixed point he settled on a glass plate from a particular glass-works. "Spath fluorique" could not scratch that glass, but zeolite could. Quist had given zeolite a value of 8, so 8 became the fixed point. In order to have more intervals, La Métherie gave it a value of 800. When the scale was thus expanded, diamond became 2000. However, he preferred to use a decimal table so, as in the conventions with specific gravity, he used 1000 as a base. Therefore, his fixed point, that of scratching glass, was 1000, and the hardness of diamond became 2500. After this explanation, La Métherie included a long list of mineral hardnesses in both scales. For example, serpentine was 620 or 750, while mica was 450 or 550.[13]

Kirwan was one of three people listed by the German mineralogist Friedrich Mohs (1773–1839) in his book of 1822–1824, translated by his assistant Wilhelm Karl Haidinger (1795–1871), as having used a hardness scale prior to that used by Mohs himself. The others were La Métherie and Jean Baptiste Louis de Romé de l'Isle (1736–1790) (Mohs, 1825, v. 1, p. 306). In his second edition of *Elements of mineralogy* (1794), Kirwan had also given numerical scales for luster, transparency, and shape of fragments. At first glance, Kirwan seemed to attribute a numerical hardness scale to Werner, but he was just listing Werner's categories. Kirwan's own list included a combination of minerals and the instruments used to test it. It is worth reviewing this list to see how the idea of comparative hardness was evolving:

I distinguish the various degrees of hardness by figures.

   3. denotes the hardness of chalk.[14]
   4. a superior hardness, but yet what yields to the nail.
   5. that which will not yield to the nail, but easily and without grittiness to the knife.
   6. that which yields more difficultly to the knife.
   7. that which scarcely yields to the knife.
   8. that which cannot be scraped with a knife, but does not give fire with steel.
   9. that which gives a few feeble sparks with steel, as basalt.
  10. that which gives plentiful lively sparks, as flint.

The superior degrees are discovered by observing the order in which stones cut each other. Under this head also, *flexibility* and *elasticity* may be noticed. (Kirwan, 1794a, v. 1, p. 38)

Kirwan's list augmented Werner's sequence with numerical designations. And yes, he really did start with number 3, not number 1. He did not explain or justify this, as far as I have been able to determine. Kirwan employed his hardness numbers when discussing minerals and rocks in the rest of his book, but frequently used a range such as 9–10, or 7–9. This had the value of requiring fewer words than noting, say, "many sparks with steel."

Writing at the same time, Johann Gottfried Schmeisser (1767–1837) numbered his sequence of hardness. His list went from hard to soft, and was somewhat different from those of his contemporaries:

Hardness according to the different Degrees.

1. Such substances, whose surfaces cannot be scratched by another Mineral, as the Diamond.
2. Scratched only by diamond,   strike fire
3. _____ _____ rock crystal,   with
4. _____ _____ a hard knife,   steel
5. Scraped by a knife,
6. Scratching tin, or lead,   strike not fire
7. Suffering impressions by the nails,   with steel

(Schmeisser, 1795, v. 1, p. 30–31)

Later in his book, Schmeisser didn't use the numbers, but characterized the hardness of topaz by saying, "It scratches hyacinth and rock-crystal; though its surface can be scratched by diamond, saphire [*sic*], and ruby" (Schmeisser, 1795, v. 1, p. 62). He also mentioned minerals that strike sparks with steel or which cut glass. Like Mohs's later effort, Schmeisser's list seems to be a comparative ordering rather than the assignment of a numerical value. Table 2.1 compares some of the schemes, indicating some of the minerals. Robert Jameson (1774–1854)'s minerals are put in increasing order of hardness.

In describing the properties of minerals according to their hardness, the famous crystallographer René Just Haüy (1743–1842) said that he combined the effect of striking the mineral with steel along with that of rubbing a corner of one body on the surface of another. For the latter, he used three objects of different hardnesses: a piece of clear quartz, a plate of plain glass, and a plate of clear calcium carbonate (Haüy, 1801, v. 1, p. 268). Haüy arranged a table of minerals in this order: substances that scratch quartz, or give sparks; those that scratch glass, divided into those that commonly, or those that sometimes, give sparks with steel; and those that scratch calcium carbonate, or do not give sparks. He followed a particular sequence when testing solids for hardness, as recorded in his tables at the end of the fifth volume of this work.

Numerical designation was slow to win favor. The German chemistry professor Johann Ludwig Georg Meinecke (1781–1823) used just four designations for hardness in his 1808 textbook: hard, semihard, soft, and very soft. The hard items gave sparks with steel. Semihard items gave fewer sparks. Soft materials could be scratched with a knife, and a very soft mineral could be scratched with a fingernail. However, since Meinecke's (1808) book was designed for use in schools and private instruction, it was not as elaborate as one written for the specialist, and perhaps did not represent then-current procedures accurately.

The Scottish antiquary and historian John Pinkerton (1758–1826) gave us an extraordinary amount of opinion in his two volumes titled *Petralogy, a treatise on rocks* (1811). He didn't think much of numerical indicators of hardness. In his reaction to the work of John Murray (d. 1820), a Scottish chemist, Pinkerton wrote:

Murray, in his excellent *System of Chemistry*, has justly observed that it is difficult to attach precise ideas to arbitrary numbers. Every reader must have observed, that he passes without reflection the ciphers 1, 2, 3, &c. when applied to Hardness, Specific Gravity, Lustre, or Transparency. It therefore seemed more advantageous to employ terms derived from the substances themselves, which, though only relative and recollective, yet convey ideas more clear, and, so to speak, more tangible than barren ciphers. (Pinkerton, 1811, v. 1, p. xix)

Pinkerton argued that new words were necessary because he felt the science of mineralogy was completely new (Pinkerton, 1811, v. 1, p. xix). But he wrote:

[W]hile some recent authors of mineralogy pollute the classical language of our fathers with an inundation of barbarous German words, derived from the vulgar dialects of illiterate miners, who of course first observed the distinctions between mineral bodies; it became the more

---

[13]La Métherie, editor's notes, *in* Bergman (1792, v. 1, p. 340–345).

[14]David Oldroyd has pointed out that harder minerals could be numbered 11, 12, etc., but that if softer minerals than chalk were found their numbers would be negative (D.R. Oldroyd, 2008, personal commun.).

TABLE 2.1. MINERALS LISTED ON VARIOUS HARDNESS SCALES

Kirwan	Quist	La Métherie	Jameson	Mohs
10 Diamond	20 Diamond	2500 Diamond	1 Diamond	10 Diamond
	16 Sapphire	2100 Saphir	2 Sapphire	9 Corundum
			3 Zircon	
	15 Topaz	1800 Topaz	4 Topaz	8 Topaz
	12 Garnet	1500 Garnet	5 Precious garnet	
			6 Spinel	
			7 Beryl	
			8 Saussurite	
	10 Schorl, opal	1300 Rock crystal	9 Rock-crystal	7 Quartz
	tourmaline, quartz	1100 Schorl	10 Chrysoprase	
		1000 Scratches glass	11 Felspar	6 Felspar
		900 Lazuli	12 Prehnite	
			13 Actynolite	
	8 Zeolyte	800 Zeolite	14 Arragonite	
	7 Fluor		15 Fluor-spar	
		620–750 Serpentine	16 Apatite	5 Apatite
			17 Calcareous-spar	4 Fluorite
		550 Mica	18 Witherite	3 Cleavable calc-haloide
	5 Gypsum	500 Gypsum	19 Gypsum	
			20 Talc	2 Gypsum-haloide
3 Chalk	3 Chalk		21 Chalk	1 Common talc

Note: These values came from the following references: Kirwan (1794a, v. 1, p. 38); Quist, *in* Cronstedt (1788, v. 1, p. 224–225); La Métherie, *in* Bergman (1792, v. 2, p. 343–345); Jameson (1817, p. 260); Mohs (1820, p. xvi). Not all members of a sequence are shown.

an object of ambition to treat this difficult subject with such a degree of classical purity, as not to disgust the eye of taste, contemn [*sic*] the discussions of grammar, or vitiate the eternal tenor of our language. (Pinkerton, 1811, v. 1, p. xx)

In pursuit of this noble ideal, he proposed using a series of familiar words to represent grades of hardness. While other authors might have used the dreaded ciphers to denote hardness and weight, Pinkerton wrote:

As Chalk, Gypsum, Marble, Basalt, Felspar, Rock Crystal, and Corundum, form various stages of hardness, at distances of 200 or more in the common tables, they have been chosen to express the relative hardness of other substances, by the following terms: *Cretic, Gypsic, Marmoric, Basaltic, Felsparic, Crystalic, Corundic.* (Pinkerton, 1811, v. 1, p. xx)

He made hardness one of the first tests because, he said, most people have a lens to examine texture, and a knife handy for hardness. After quibbling about the order in which people recorded the characteristics, he quoted at length from Werner (1774) on hardness, and included his footnote on the instruments that should be available, namely, the knife, steel, and a file, and provided the name of Werner's supplier for them. As Pinkerton got to the business of describing rocks and minerals, he followed his prescription for denoting hardness, and, not surprisingly, called the hardness of basalt "basaltic" (Pinkerton, 1811, v. 1, p. 17). For feldspar he commented: "Hardness, of course, felsparic" (Pinkerton, 1811, v. 1, p. 157). In his mineral descriptions, quartz minerals were "crystallic," marble was "marmoric," and compact limestone had a hardness of "gypsic to marmoric." While it might appear that he was introducing a

few barbarous terms of his own, it could also be considered as a move in the direction of Mohs's scale which had, admittedly, those meaningless ciphers, but which used specific minerals. Pinkerton's intent, however, was not quite the same, nor would his results be so reproducible, due to the sometimes different hardnesses of substances such as basalt, which could have a variable composition. In his second volume, Pinkerton used numerical designations without comment when quoting mineral characterizations by Kirwan.

As the nineteenth century advanced, hardness remained an important property. Parker Cleaveland did not give it the prime position that some other mineralogists did. His arrangement of minerals was basically chemical, and hardness was listed prominently only if it was a distinguishing character. He essentially followed Werner's hardness criteria, and added a somewhat expanded list of minerals or objects to scratch or be scratched by. In 1817, the Scottish Wernerian Robert Jameson listed 21 minerals from hard to soft. Diamond was given as 1, chalk was 21, and feldspar was in the middle at 11. By inserting other minerals between the ten in Mohs's scale, introduced in 1812 and which we use now, the relative distances between any two minerals were somewhat altered (Jameson, 1817, p. 260).[15] The English Robert Bakewell's (1768–1843) discussion of hardness in 1813, 1819, and 1833 was typical for the period, and emphasized the need for care in determining which mineral was scratched. It is interesting that although these books were written after Mohs's introduction of his scale, the authors did not use his numbers.

---

[15]The reference list includes only English translations of Mohs in 1820 and 1825 to which I had access. Mohs first published his *Versuch einer Elementar-Methode zur naturhistorischen Bestimmung und Erkennung der Fossilien* in 1812 in Vienna.

Mohs is best, and now almost exclusively, known as the originator of the Mohs's scale of hardness. He studied with Werner in Freiberg some time after 1798, then visited and studied in Great Britain, before being commissioned to catalog the mineral collection of J.F. von der Null in Vienna. Mohs was then called to Graz where he worked on the mineral collection at the Johanneum, and became professor of mineralogy there in 1812, after which he became professor of mineralogy at Freiberg in 1818 after Werner's death in 1817 (Burke, 1974). Mohs evolved his hardness scale as he worked to catalog those large mineral collections, partly because he felt there were inadequacies in Werner's method. To classify, he started with the crystal system, then went to hardness and specific gravity which "must be tried with proper accuracy, and expressed in numbers" (Mohs, 1820, p. xx). He proposed his scale in 1812, but according to Burke it was not adopted widely until the 1820 translation of his *Of the classes, orders, genera, and species; Or, the characteristics of the natural history system of mineralogy* (Burke, 1974, p. 448). Mohs said:

The degrees of HARDNESS, or (if not constant) their limits, are expressed by numbers; and the letter H. designs hardness in general (as Sp.Gr. does specific gravity).... These numbers refer to the following scale:

The number
1. denotes the degree of hardness of a variety of PRISMATIC TALC-MICA, known by the name of common talc.
2. of a variety of PRISMATOIDAL GYPSUM-HALOIDE, of imperfect cleavage, and not perfectly transparent. Varieties perfectly transparent and crystallised, are commonly too soft.
3. of a cleavable variety of RHOMBOHEDRAL CALC-HALOIDE;
4. of OCTAHEDRAL FLUOR-HALOIDE [fluorite];
5. of RHOMBOHEDRAL FLUOR-HALOIDE [apatite];
6. of PRISMATIC FELD-SPAR;
7. of RHOMBOHEDRAL QUARZ [sic];
8. of PRISMATIC TOPAZ;
9. of RHOMBOHEDRAL CORUNDUM;
10. of OCTAHEDRAL DIAMOND.

(Mohs, 1820, p. xvi)

This was a relative, nonlinear scale, but it eventually came to be seen as so practical that it is still in use, and utilizes the common substances used in the eighteenth century, namely, the fingernail (~2.5), a cupro-nickel penny (at ~3), a knife blade (~5.5), and window glass (also ~5.5). A file is ~6.5.

In 1825, Mohs spoke of the importance of being accurate with hardness and the advantage of being able to divide his intervals into tenths, although he rarely needed to go to a division of less than 0.5. Mohs usually gave a range rather than a single number for hardness, since it was known that different faces of the same crystal could show somewhat different hardnesses.

Recognition of Mohs's influence in codifying hardness comes not from his being first with a numerical scale—which he wasn't—but from the use of common, or reasonably common, minerals in a series that could be applied in a way that had been familiar for centuries. He stated clearly that the difference in hardness between any two successive minerals on the

scale was not the same, but that more convenient minerals were not available. He also stressed that the faces tested should be smooth and even. Cleavage faces were best, if available. Edges and corners used for scratching were equally important, and Mohs discussed the difficulty of standardizing them. Scratching alone wasn't enough. He wrote:

But if we take several specimens of one and the same mineral, and pass them over a fine file, we shall find that an equal force will everywhere produce an equal effect, provided that the parts of the mineral in contact with the file be of a similar size, so that the one does not present to the file a very sharp corner, while the other is applied to it by a broad face. It is necessary also that *the force applied in this experiment, be always the least possible*. (Mohs, 1825, v. 1, p. 304, italics as in original)

It was necessary to practice to get consistent results, and to follow a standardized procedure.

By 1836, the Scottish chemist Thomas Thomson (1773–1852) declared hardness to be one of three properties necessary for mineral identification, the others being specific gravity and primary crystalline form. He mentioned Kirwan as being the first to use convenient numbers to represent hardness. About Mohs he remarked: "Mohs has adopted the same plan, though he has neglected to mention the source whence he derived his first idea of it" (Thomson, 1836, vol. 1, p. 10). He then gave Mohs's list, and reiterated the usefulness of careful use of the file.

Instruments and Equipment Used for Hardness Determination

Instruments, or standards, were used in mineral identification very early. In much of the literature, the properties of the preferred "knives," "steel," etc., seemed to be of such well-known character that they were not given specific descriptions, as was a glass plate. The standard aids, besides the minerals themselves, were files and pieces of hard steel, the latter being of dual purpose. The fingernail was used to differentiate soft minerals. A magnifying glass or microscope might be used to inspect the sparks that resulted from striking hard minerals with steel (Fourcroy, 1789, v. 1, p. 254–255). The glass plate was added somewhat later.

There was a gradual progression toward the more precise determination of hardness, with more specific methods and measures applied. In 1819, Bakewell listed his equipment for determining hardness:

Crystallized quartz, window-glass, fluor spar [sic], and calcareous spar, are used for this purpose. A knife with two blades, one of copper, the other of steel, and a small well-tempered file, may also be employed. According to the resistance which minerals yield to these instruments, they are called by Werner extremely hard, hard, rather hard, and soft; but these terms being indefinite, it is much better to state the precise degree of hardness as proved by trial. (Bakewell, 1819, p. 41)

When he used Mohs's scale in 1836, Thomson refined its use. Rather than apply the corner of a mineral to see if it would scratch another, he preferred to use a file on both, with the same force, and compare the scratches made (Thomson, 1836, v. 1, p. 10).

A list of things other than minerals used to determine hardness from the eighteenth century until now is as follows, with approximate Mohs's scale numbers:

fingernail:	a little greater than 2
copper coin or knife blade:	~3
brass pin:	a little over 3
steel of pocket knife:	a little over 5
window glass:	5.5
steel file:	6.5
steel for sparks	

Some mineral identification schemes that were carried into the twentieth century used hardness as their first division into groups, with increasing sophistication of instruments to determine it (Smith, 1998, p. 303–304).

Streak

Although it can be useful in specific cases, streak has not been as important over time as some of the other mineral properties such as hardness. It could be called a property that is transitional between hardness and color, a fact that was recognized almost from the beginning. We now think of streak simply as the color of a powdered mineral, the powder being produced by rubbing the mineral against a piece of unglazed porcelain, which has a hardness of ~7. A mineral of greater hardness does not leave a streak, or is listed as white.

Knowledge of streak must have evolved both from observations of mineral powder colors and the long tradition of the touchstone. In a footnote in their edition of Agricola, the Hoovers credit Theophrastus with the first description of the touchstone (Agricola [1556], 1950, p. 252). Touch needles were used in conjunction with the touchstone. By the time of Agricola, the use of touch needles was common, and there were instructions for their production and use (see Fig. 2.3). There might be sets of about

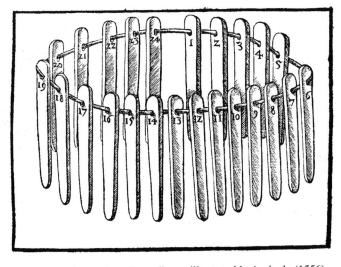

Figure 2.3. A set of touch needles as illustrated in Agricola (1556).

30-some needles in a series of known compositions, fabricated with carefully weighed amounts of gold, silver, and sometimes copper. The unknown metal, suspected of being in some part gold, was rubbed on the touchstone (the flint), then that streak was compared with a streak from the appropriate touch needle. Those adept in their use could be very accurate in determining the relative composition of the unknown metals. This method was sometimes preferred to fire assay because it was much simpler, used much less material, and was less destructive (Agricola, [1556] 1950, p. 252–260).

In 1546, Agricola also discussed colors seen after rubbing a mineral on flint (Agricola, [1546] 1955, p. 87). We recognize the minerals he described as mainly being ores of iron, which are notorious for having streaks of a different color from that of the mineral. The early use of flint for streak color is interesting, as it was not used merely as a touchstone. Agricola's book provides a detailed discussion of the various minerals that give a precisely colored streak.

The 1753 edition of Wallerius's *Minéralogie* spoke of the colors of powders produced by scraping a mineral with a knife. In 1774, Werner noted that the powder produced when a mineral was "scraped or rubbed with a knife or other hard substance" could be either the same as or a different color from the solid mineral (Werner, [1774] 1962, p. 92). He gave examples of minerals in which the two colors were different. The method was not standardized and as he viewed it, didn't take account of the fact that substances used to do the rubbing might themselves have different hardnesses. Bergman mentioned that color varied with particle size, so that if a stone was powdered it might, using the criterion of powder color, be placed in a class different from that of a large piece (Bergman, 1784a, p. 8).

In the 1788 English edition of Cronstedt's *An essay towards a system of mineralogy*, von Engeström's "Dry Laboratory" included a piece of black flint used as a touchstone. When the touchstone was rubbed by a metal sample and left a streak that didn't dissolve in *aqua fortis*, the metal was gold.[16] In 1795, Schmeisser did not include anything that would serve as a streak plate in his quite detailed list of instruments required by a mineralogist, while in 1808, Meinecke said that streak was the color of powder produced when a mineral was scratched with a steel engraving pencil. He noted it might be the same as or different from that of the original mineral color (Meinecke, 1808, p. 12).

In 1820, Mohs used investigations of streak, specific gravity, and hardness in combination to begin dividing minerals into classes. He pointed out that it was important to distinguish between the color of a mineral and that of its streak. The streak might be the powder produced by rubbing or, if the sample was a malleable metal, it appeared as the greater luster shown where scratched. While it was probably employed earlier, Mohs's use was the earliest example I've seen of what he called "a plate of porcelain biscuit," or unglazed porcelain, on which to rub the minerals. Powder was made this way or by scratching with

---

[16]*Aqua fortis* is concentrated nitric acid.

a file. He listed classes of minerals where the mineral and its powder were the same color and those where they were different (Mohs, 1820, p. xx; 1825, v. 1, p. 296).

The use of streak color evolved over time. Kirwan stated that little explanation was needed for it. I would have agreed until I saw the numerous ways in which it was employed even in his time, quite apart from the changes over the next two hundred years.

Color and Related Properties

Color is one of the most obvious characteristics of minerals but one of the least trustworthy for purposes of identification. It is a rare professor of mineralogy who has not produced an atypically colored mineral to use in an identification test. This variability, and the difficulties to which it could give rise, were common knowledge. For example, under the heading of crystalline quartz there can be eleven or more named sorts, depending on color and/or inclusions. The colors can include pink, violet, dark or light yellow, brown, or milky. Some inclusions might be mica, hematite, limonite, tourmaline, or rutile. However, a few minerals such as malachite and sulfur have distinctive colors that do not vary much. Some minerals show iridescence or opalescence, and some change color on exposure to air so that a fresh surface must be obtained.

Agricola's first division of minerals was by color, listing minerals of the same color together. He noted also that specimens of the same mineral could have either mixed colors or two separate colors. While this may seem simplistic, Agricola applied other tests, such as hardness, fusibility, etc., as well as knowledge of location and associations, to assist him in his mineral identifications. However, many minerals and precious stones came from locations far from him, and positive identification could be difficult despite extensive investigations of several properties. On the other hand, Wallerius ([1747] 1753) didn't discuss color much. He noted it, but was more concerned with reactions or other properties. He didn't list as many colors for quartz as did later workers, and judging by some of the colors he did list, he may have misidentified some minerals. He did record that the colors of precious stones were due to mixtures of mineral substances. For example, Wallerius thought that the red of rubies was produced by iron; copper made the blue of sapphires; cobalt could also give blue; and copper and iron mixed made emeralds green (Wallerius, [1747] 1753, p. 231).

Werner remarked: "Among the common generic characters of minerals, *color* is the first which is discovered by our senses" (Werner, [1774] 1962, p. 23). He felt it was a good aid to identification, along with other properties, and mentioned some of the dramatically colored ores and salts. Color was less to be trusted in earths and stones, but was a help. He named eight principal colors: white, gray, black, blue, green, yellow, red, and brown (Werner, [1774] 1962, p. 25). His discussion continued meticulously, and he considered mixed colors as well as their intensity. Further, Werner listed subdivisions under each principal color.

While he numbered the different possibilities, 7 under white, 5 under black, 13 under red, etc., he used the appropriate word or word combination in his mineral descriptions, which include no numerical data. He defined and used such precise wording for all varieties of all characters that, with a dictionary, his descriptions can be followed and verified even today.

Kirwan listed seven colors for calces of iron "according to the degree of their dephlogistication" (Kirwan, 1784, p. 15). In other words, he recognized what we would now call the different oxidation states of iron. He listed four colors for quartz, five for feltspar (his spelling), and seven for mica. He too recognized the source of some colors: "The coloured transparent crystals derive their tinge generally from metallic particles in exceeding small proportion" (Kirwan, 1784, p. 104). Mineralogists routinely referred to this fact, which was particularly well stated by Haüy:

Dans un certain nombre de ces corps, et en particulier dans les substances terreuses et acidifères les couleurs sont dués aux molécules d'un principe étranger, qui est souvent le fer, et quelquefois le chrome ou le manganèse, dissemine entre les molécules propres du corps coloré. (Haüy, 1801, v. 1, p. 223–224)[17]

Kirwan described colors with words and combinations of words. In 1794, he named color as the first of his external characters. His five shades of gray might be difficult for us to distinguish according to his description, and in general it is rather difficult to determine the precise color meant by some of the expressions.

Fourcroy (1789) basically observed that color could be fixed or variable, and thus could be of more or less help. Sometimes his varieties of a single mineral were listed by color, sometimes by other descriptive terms. He was far less detailed and precise with color terms than Kirwan or Werner, as was La Métherie (1795). Haüy discussed minerals such as fluorspar that might have several colors or some that might have mixed colors. In one table he listed color of the mass of the mineral, as well as that of the powder and small fragments (Haüy, 1801, v. 5). Pinkerton gave a list of characters and then said: "to which colour is sometimes added, though the most vague and insignificant of all the characteristics" (Pinkerton, 1811, v. 1, p. xix). Like Werner, he generally used the principal color names with little embellishment.

Color did not figure largely in the French chemist Charles-Louis Cadet's (1769–1821) discussion of mineral characteristics (1803), but he remarked that one could distinguish several minerals of the same family by their different colors, which sometimes were uniform, or in patches or bands. In 1816, Cleaveland made a statement that showed how far understanding of what Kirwan had called the "degree of dephlogistication" had advanced. About reasons for color Cleaveland remarked:

---

[17]"In a certain number of these bodies, and in particular in earthy and acid substances, the colors are due to molecules of a foreign element, which is often iron, and sometimes chrome or manganese, disseminated between the characteristic molecules of the colored body" (my translation).

First, in many minerals the coloring matter is both accidental and variable; and arises from the presence of metallic oxides, particularly those of iron and manganese. Now these oxides may exist in different proportions, or with different degrees of oxidation; either of which would produce a variation in the color, or at least in the shade of the color of different varieties, belonging to the same species....

But, secondly, the color sometimes depends on the nature of the mineral, and is produced by light reflected from its essential, component parts. Here it is a character of very considerable value. (Cleaveland, 1816 [1978], p. 37–38)

He followed this statement with paragraphs of description that included shades of color, and a section on tarnished surface colors.

When writing about often transparent precious stones, Haüy's discussion of color was concerned with Newton's theories of light as related to his own observations (Haüy, 1817). In his table of stones there was a column for color, but he titled it "Accidens de lumière." Mohs (1825) began his section, "Of the optical properties of minerals," by saying:

The consideration of the natural-historical properties in general, presupposes the presence of light. Yet all of them do not *depend* upon its presence. This is the case, however, with the properties to be considered in this chapter. We cannot maintain that a mineral possess in the dark that same colour, lustre, or transparency, which it exhibits when observed under the influence of light. This character, therefore, is sufficient for distinguishing the optical properties from all other properties of the minerals. (Mohs, 1825, v. 1, p. 274)

He divided colors into metallic and nonmetallic, and referred back to Werner's eight principal colors of white, gray, black, blue, green, yellow, red, and brown. Each of these had a metallic representation and a number of different nonmetallic possibilities. Interestingly, Mohs considered that composition was of little value in natural history because it only told what the simple mineral was and he did not think that specific characteristics could be derived from it (Mohs, 1825, v. 1, p. 278).

From this point, the use of color in the natural history sense did not change appreciably in visual identification of minerals. One still sees colors, such as brown, gray, green, or yellow, given as possible variations for a particular mineral. However, current use now includes the physics of reflected, refracted, and transmitted light, as well as behavior in polarized light. The latter became part of the mineralogists' arsenal in the first decades of the nineteenth century.

Transparency and Luster

To Albertus Magnus it appeared that transparent stones were made of water (Magnus, [~1262] 1967, p. 14). Wallerius and others of his time described stones as opaque, semitransparent, and transparent, sometimes with color notations. Werner said:

A mineral is *perfectly transparent* when its integrant particles are arranged in such a manner that all the interstices between them line up in perfectly straight directions, thus giving rays of light completely free passage. On the contrary, a mineral is *opaque* when these interstices lie confusedly one among another. (Werner, [1774] 1962, p. 88)

It was thought that the minerals must have been dissolved, or the constituents couldn't have been arranged into the necessary positions. Werner related degree of transparency of the solid mineral to how well it had been dissolved, the least dissolved making the less transparent part of the crystal, which he felt would be near the surface from which the crystal had grown. He divided transparency into five degrees: transparent, semitransparent, translucent, translucent at the edges, and opaque (Werner, [1774] 1962, p. 89). He then noted that some transparency is *doubling*, that is, two images could be seen through that kind of mineral.[18]

In 1794, Kirwan tried to quantify transparency, saying:

Mr. Werner distinguishes several *degrees*, which I note by figures.

4. denotes that degree which allows objects to be clearly distinguished.
3. that which suffers objects to be perceived, but not distinctly.
2. that which transmits light, but does not permit objects to be discerned.
1. that which transmits light only at the edges.
0. denotes perfect opacity.

When a fossil possesses these qualities in various degrees, I place the figure denoting the most usual degree first, and the least common degrees follow in their order. (Kirwan, 1794a, v. 1, p. 32)

Kirwan followed his explanation with footnotes giving the degrees of transparency in German, and used the numbers for transparency in his mineral descriptions. Schmeisser also used numbers to designate transparency, also in four divisions, with 1 representing transparency, and 4, opacity. His description of each grade was slightly more complete or, alternatively, assumed less knowledge on the part of his readers (Schmeisser, 1795, v. 1, p. 30). He noticed that some mineral crystals are more transparent in one orientation than another. It is not hard to see how mineralogists, then and now, could accuse each other of misidentification when the characters could be so variable.

In 1808, Meinecke used a short paragraph to discuss the grades of transparency: "transparent, half-transparent, translucent, translucent on the edges, opaque."[19] Mohs's list of 1825 was the same, but numbered one to five (Mohs, 1825, v. 1, p. 297). The whole question of light transmission became much more technical in the nineteenth century.

Luster is another of those qualities that has seemingly always been used to describe minerals. Agricola ([1546] 1955) cautioned that although many minerals have natural luster, they may also have been polished or otherwise have had their luster enhanced.

---

[18] Werner only mentioned this property with respect to calcite, for which it had long been known, but it didn't take long for it to be demonstrated in other minerals.

[19] His words were "durchsichtig," "halbdurchsichtig," "durchscheinend," "den Kanten durchscheinend," "undurchsichtig" (Meinecke, 1808, p. 11). Interestingly, in 1834 Karl V. Presl used exactly the same list as Meinecke had, numbered them 1 through 5, but used the words in his mineral descriptions (Presl, 1834, p. 86).

Not content to simply liken luster to the appearance of familiar surfaces (vitreous, resinous), Werner divided it into external and internal and defined it as follows:

Luster refers to the relation which a mineral displays with respect to the reflection of light, which is caused partly by the smoothness of the surface—or at least of the aggregated parts which constitute the surface—and partly by its compactness. Smoothness is the cause of the different degrees of luster, and compactness, its different kinds. (Werner, [1774] 1962, p. 71)

For intensity, he listed "splendent," "shining," "weakly shining," "glimmering," and "dull," with explanations of each. His "kinds" were more familiar, but less descriptive, and consisted of "common" and "metallic." Werner's attempt to be more precise was not always followed by his contemporaries, who generally used the familiar terms such as metallic, possibly semimetallic, vitreous, pearly, waxy, greasy, or resinous.

Kirwan again used numbers to designate degrees of luster:

4. denotes the strongest, and is perceived at a certain distance, such as that of diamonds and polished metals.
3. denotes a weaker, such as that of crystals or metals not much polished.
2. denotes a still weaker, as that of silk, or still less glossy.
1. when only a few particles reflect any lustre, or it is exceeding [*sic*] weak.
0. dull, reflecting no lustre at all.

(Kirwan, 1794a, v. 1, p. 32)

For one kind of garnet, he said its luster was 3–4. Kirwan's is the only example I know of where numbers were used to designate luster. There was also no statement of the significance of a difference in numbers (does a "4" shine twice as much as a "2"?). However, mineralogists have continued to distinguish between degrees and kinds of luster, and thus our vocabulary for it today is virtually identical to that of the late eighteenth century.

Specific Gravity

Differences in relative masses of minerals were readily determinable, and specific gravity was a property familiar to those who worked in natural philosophy. We have all had the experience of picking something up that was unexpectedly light or heavy. I have not tried to trace the use of specific gravity through the voluminous literature, interesting though that exercise might be. It is commonly claimed that Archimedes used specific gravity to determine whether a crown was gold (Buchanan, 1998, p. 49), and coins were checked for their gold content by its use. Boyle was well versed in what was called "Hydrostaticks" and made use of specific gravity measurements. Interested in marble, he wrote:

And accordingly having weigh'd a piece of white Marble in Air and Water, I found it to be in weight to an equal bulk of that Liquor very near 2 and 72/100 to 1, or, (that the proportion with very little errour may be the better remembered,) as two and seven tenths to one. (Boyle, [1672] 1972, p. 127)

In Boyle's time, as now, specific gravity was defined as the mass of a substance compared to the mass of an equal volume of water. For the most reproducible results, this should be done at the temperature at which water is of maximum density, 4 °C, but various temperatures were specified in the eighteenth century. The numerical value could be listed with as many as four decimal places or determined simply by the "heft" of similarly sized pieces of minerals.

By the end of the eighteenth century, as accurate weight measurements became more important in science, there were improved balances for determining specific gravity. The history of balances in general is a long one and began in antiquity. It is summarized in Chapter 6 prior to the discussion on mineral reactions in Chapter 7.[20] Since specific gravity was one of the properties first used in mineral identification, hydrostatical balances were of great interest. Buchanan (1998, p. 49–50) has written a short, somewhat incomplete, review of them, and Lundgren (1990, p. 257) discussed specific gravity in the context of early chemistry.

Wallerius routinely included specific gravity in his mineral descriptions, generally immediately after color and hardness. He followed the normal procedure of assigning the given volume of water a value of 1000, then giving the weight of the same volume of mineral in proportion. For one kind of quartz his value was 2600:1000 (Wallerius, [1747] 1753, p. 194).

The 1757 English edition of Henckel's *Pyritoliga, or, a history of the pyrites* said:

As I myself was without a good hydrostatical balance, I in this case applied to the celebrated Dr. Meuder, who readily gave me his assistance here, and not only examined the *pyrites* and its concomitant matters in the most accurate manner possible, and with repeated care and diligence, but also many others, and those the principal, both dense and fluid bodies, as laid down in the following tables; to which he has subjoined a set of uncommon remarks and principles, and peculiar ways of managing the hydrostatical balance. (Henckel, 1757, p. 359)

Interestingly, while Meuder started with amber at ~1 as expected, he had rather unlikely values for other substances, such as 1098 for gold and 300 for rectified spirit of wine (Henckel, 1757, p. 359–360). His technique cannot have been very satisfactory.

The so-called hydrostatical balances were designed to enable one to ascertain the weight of a substance in air and then in water. Weights were placed on one pan of a beam balance to counterbalance the weight of the sample, which was hung by a thin thread below the second pan. After recording the weight of the sample in air, it was immersed in distilled water at a constant temperature, which could be different for different purposes or conditions, but which was constant for the series of weighings. The weight of the immersed sample was recorded, and the second weight was subtracted from the first. Since the volume of water has a 1:1 ratio to its mass, that difference represented the mass and volume

---

[20]For a quick reference, see D.F. Crawforth-Hitchins, *in* Bud and Warner (1998, p. 47–49).

of a volume of water that was equal to the volume of the sample. The weight of the sample in air divided by that difference gave the specific gravity, which as noted above was the mass of the substance divided by the mass of the same volume of water.

In 1774, after a discussion of what "specific weight" (his term) meant, Werner cautioned that the procedure must be done using pure distilled water. He then declared that unlike physicists, mineralogists didn't have the required apparatus or instruments for specific gravity determinations. Most mineral collections would not allow such use of specimens, and the determinations would take too much time. He also thought that remembering numbers would be too difficult. He preferred to lift minerals in his hand for their "heft," and assigned five degrees: floating, light, rather heavy, heavy, and extremely heavy. Using numbers reluctantly, he said:

the weight of water is divided into 1,000 parts, and the specific weight of the body to be examined is then determined by calculating the number of these divided parts to which it is equal. (Werner, [1774] 1962, p. 105)

He had a range of values for each of his five degrees, using 1000, not 1 (as we do now), as his number for comparison.[21] This was one of several numerical systems for expressing specific gravity proposed at the time.

In 1785, the British chemist William Nicholson (1753–1815) published notice of his new, easy, and portable hydrostatical balance in the second volume of the *Memoirs of the Philosophical Society of Manchester*.[22] It won immediate favor with practicing mineralogists, as evidenced by frequent favorable citations. Unlike the regular balance for this purpose by which a substance was weighed first in air and then immersed in water, while suspended by a hook, the Nicholson balance consisted of a funnel attached to a tube, which in turn was vertically attached to a bulb and then to another receptacle. The tube was marked at the level to which it sank in distilled water when the funnel was loaded with 1000 grains.[23] The funnel was then loaded with the unknown mineral specimen until it sank to the same level. That substance was then loaded into the lower receptacle until the tube recorded the same level, which indicated the amount of water displaced by the sample. Dividing the weight in air by the second number gives the specific gravity. Nicholson added a note that the water itself and its temperature should be the same for all experiments, and a thermometer was supplied with the balance. 60 °F was suggested as a normal room temperature.

Apparently Lavoisier didn't use the Nicholson balance for he spoke of suspending a substance in water from the balance pan by a fine wire. However, Lavoisier stated that it was rare for a chemist to need to determine the specific gravity of a solid. For him, it was a more useful determination for liquids, as it helped determine their purity and concentration (Lavoisier, 1789, v. 2, p. 337). He gave detailed instructions for liquids, and described the instru-

ments he used. Some of the tables at the end of the 1789 edition of *Traité elémentaire de chimie* differ numerically from those of the 1790 English edition. English units were inserted by the translator to account for different weight standards in the two countries.

Schmeisser summarized specific gravity as an external character of minerals in 1795 and mentioned further methods and precautions, among them that whatever liquid was used to immerse it, the sample should not react with it. Some of his values were: copper, 9000; lead, 11,000; mercury, 14,000; Brasil topaz, 3564; shoerl, 3000; quartz, 2653; Brasil turmaline [*sic*], 31,500. The latter is an interesting number. Other authors place the specific gravity at ~3, so Schmeisser was off by a factor of 10. My first thought was that this had been a printing or calculation error, but Schmeisser listed the value for another kind of tourmaline at 30,541. In trying to see what could possibly have made his specimens so dense (heavier than gold!), I looked at the analysis he gave for the second specimen, credited to Klaproth. It was: "55 siliceous, 25 argillaceous, 0,9 calcareous earth, and 0,9 iron, 0,01 manganese" (Schmeisser, 1795, v. 1, p. 87). There is nothing in it to account for a very high specific gravity. Note also that the analysis only totals ~82%.[24]

By 1801, the form of the Nicholson balance referred to by Haüy had changed slightly and was called by him an *affleurer aréomètre*, or a leveling hydrometer (Fig. 2.4). Haüy's discussion of the balance was less obscure than those of some of his compatriots. In still another conjecture, I wonder whether that was because he performed the operations himself, rather than having a colleague or assistant do them. It is useful to remember that numerical measurement was in its infancy. The values Haüy reported for specific gravity, sometimes to four or five decimal places, were not unusual. The concepts of decimal calculations and significant figures were only just beginning to be disseminated.[25] Haüy included instructions on how to adjust his aréomètre for various circumstances, such as the adherence of some water to it, allowing differences in water temperature, and how to determine specific gravity for substances that absorbed water.

Meinecke dealt swiftly with the question of specific gravity. He first quoted Werner's terms, followed by a brief comment on how numerical values were obtained. He specified that the water temperature should be at 61 °F, or 14 °R (Réaumur). He gave values for a few substances, but only in whole numbers. No masses or volumes were given, and his calculations were not provided.

The Nicholson balance, as described by Cleaveland ([1816] 1978), was virtually identical to that of Haüy in 1801. As in Haüy, there were instructions for determining specific gravities for materials more and less dense than water. Some of his calculations were hard to follow, and in some cases I could not obtain

---

[21]For an explanation, see Murray (1806–1807, v. 3, p. 24–25).

[22]As reported in Cronstedt (1788, v. 2, p. 1007).

[23]The grain was a small unit of mass that was slightly different in different countries. A Paris grain was about 0.053 of the new metric system's grams.

[24]I believe neither anomaly (for specific gravity or low total percent composition) revealed ignorance or carelessness. It might be explained by the difficulty in producing and correcting a handwritten manuscript. The errors might have many sources: lab practice, instrument malfunction, misread instruments, miscalculation, transcribing/recording, or typesetting.

[25]Trying to ascertain the meaning of numerical data is part of the constant effort to understand the science of the past in the context of its own time. It is not easy to figure out what units an instrument was calibrated in, or to know calculation conventions.

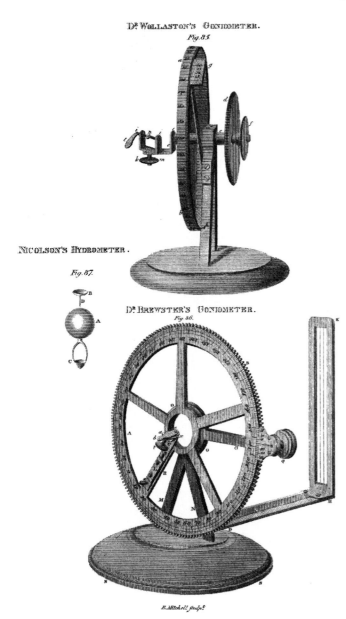

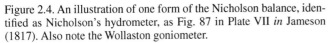

Figure 2.4. An illustration of one form of the Nicholson balance, identified as Nicholson's hydrometer, as Fig. 87 in Plate VII *in* Jameson (1817). Also note the Wollaston goniometer.

his results from his data. In his section on mineral properties, Cleaveland frequently cited values reported by others.

In 1817, Haüy spoke approvingly of the use of specific gravity measurements for distinguishing precious stones that might resemble each other in color or hardness. The Nicholson balance he discussed was identical to the one he used in 1801. He pointed out the importance of using water of the same density in different determinations, with constant temperature, namely, 14 ° Réaumur or 17.5 °centigrade (Haüy, 1817, p. 89). He recommended the use of distilled or rain water, but said filtered water might suffice. An experiment on a red topaz was used to illustrate his method and calculations.

The units used in these physical quantities are not always so obvious. The metric system had been introduced in France only toward the end of the eighteenth century, and one is likely to see unfamiliar unit names because the changeover was incomplete. As noted earlier, the English system was not only different from the French, there were two English systems, troy and avoirdupois. The use of significant figures does not follow modern conventions. In some cases, authors explained the source of the figures and calculations, but often they would not.[26] However, much agreement can be seen in the values for specific gravities during this period, and evidently the determination helped a great deal in mineral identification.

Refraction

Refractivity had been carefully investigated by the end of the eighteenth century. It would be interesting to have a good idea of when the property of double refraction was noticed for the first time. It was probably in antiquity. Double refraction in calcite was reported in a publication by Erasmus Bartholinus (1625–1698) (Burke, 1966, p. 67). It was routine toward the middle of the eighteenth century to mention this property in the case of Iceland spar. Wallerius began his discussion of "Cristal de' Islande" with "Spatum dilucidum objecta duplicans" and gave the terms used by Agricola, Pliny, and others for it (Wallerius, [1747] 1753, p. 117). The mineral's distinctive property, along with transparency, shape, and hardness, allowed easy identification of calcite.

The property of double refraction was soon noted in crystals other than calcite. Not all mineralogists had access to suitably transparent specimens, so it was slowly reported in other crystals over time as people looked for the phenomenon. All transparent, birefringent crystals, essentially those in any system but the cubic, show the property. Haüy stated that light rays passing through media of different densities produced refraction (1801, v. 1, p. 229). When a ray could take two different routes it gave double refraction. Haüy wrote: "Il seroit difficile de trouver un caractère plus saillant que celui qui se tire de la double réfraction, puisqu'il tient à l'essence même des minéraux dans lesquels il existe" (Haüy, 1801, vol. 1, p. 234).[27] He used zircon as an example. The relatively minor character of refractivity, or double refractivity, could at that time be definitive in the identification of otherwise similar minerals.

Magnetism and Electricity

Like refractivity, magnetism and electricity (see Fig. 2.5) have a long history and were discussed in antiquity. Magnetite, or lodestone, was well known for its property of attracting iron,

[26]For background on measuring systems and use of numbers including significant figures, see an excellent chapter by Heilbron, "Measure of Enlightenment," *in* Frängsmyr, Heilbron, and Rider (1990).

[27]"It would be difficult to find a more remarkable character than that of double refraction, since it is connected with the essential nature of those minerals in which it exists" (my translation).

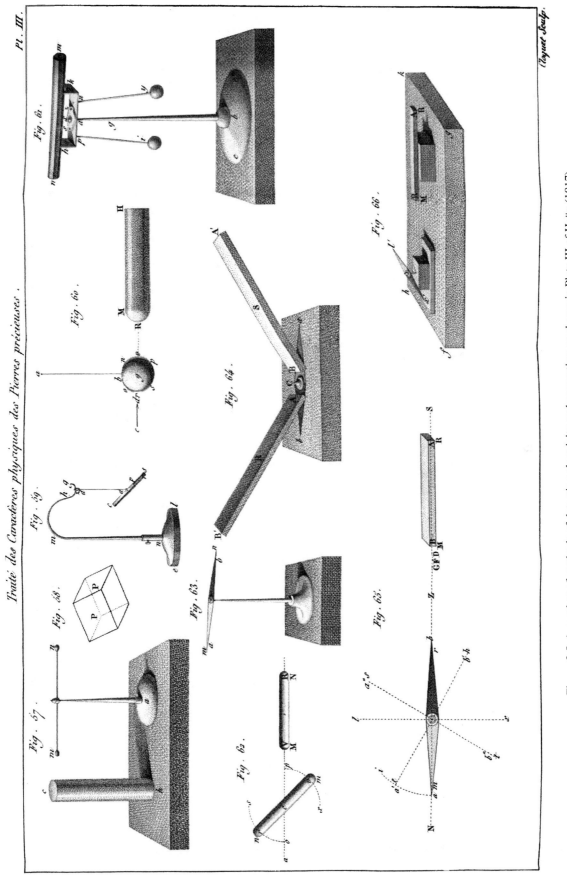

Figure 2.5. A number of methods of detecting electricity and magnetism as shown in Plate III of Haüy (1817).

and was in use for this purpose in the mining culture described by Agricola. Its ability to make nonmagnetic iron magnetic was also noted. It was recognized that not all iron ores were magnetic. Werner (1774) and Schmeisser (1795) included a magnet in their instrument lists for mineralogists. Von Engeström included a magnet, which he called an artificial lodestone, in his portable dry laboratory (Cronstedt, 1788). Mineralogists discussed advantageous ways of using magnets in the laboratory. In field reports, it was common to observe the response of rocks, particularly volcanics, as well as minerals, to the magnet, and then to make the comment that the iron content must be high.

Early mineralogists also commented on the electrical phenomena associated with some minerals.[28] To Werner, electricity was a physical character or property that was shown by observation that a mineral attracted small pieces of paper after being rubbed (Werner, [1774] 1962, p. 115). Schmeisser (1795) included what he called an electrometer in his instrument list.[29] He noted that electricity could be produced by rubbing or by heating, and referred to the work of Romé de l'Isle and La Métherie who showed that most transparent schorls (schoerls) "are electric when heated to about 200° Fahrenheit" (Schmeisser, 1795, v. 1, p. 77). Schmeisser expanded this when speaking of "turmalin" (tourmaline), saying "It is electric, when heated to 200° Fahr. at this heat attracting light bodies by one end, and repelling them by the other" (Schmeisser, 1795, v. 1, p. 79). This unusual mineral could become charged just by being moved from a cool to a warm room.

In 1801, Haüy wrote that electricity could be excited in a body by heat, rubbing, or by communication with another body already electrified. He credited "Francklin" (Benjamin Franklin) with designating the positive (*vitrée*) and negative (*résineuse*) electric fluids, and gave examples of minerals that showed each.[30] This, combined with the method used to excite the electricity, could be helpful in identification. Haüy's instrument to demonstrate electrification was a familiar one. Two pith balls were suspended from a copper needle on a pivot. When a charged mineral was brought near he could judge something of the attractive or repulsive force by the distance the balls moved. Science students still use this simple instrument in the twenty-first century. In his treatise on precious stones, published in 1817, Haüy had a chapter on electricity produced by rubbing and one on electricity produced by heat.[31] He discussed several pieces of apparatus, based on the same principle as that above, and experiments on various minerals. This went far beyond mineral identification to investigations of both electricity and matter in general.

The effort to measure electricity and magnetism, along with the other "imponderables" of light and heat, is one of the great stories in the history of the science at the end of the eighteenth century and the beginning of the nineteenth (see Heilbron, 1979). The treatment here of how it was applied in mineralogy is necessarily brief. For our story, we see that many of the people who worked on mineral identification were well acquainted with the broader issues of science in their day, and tried to apply the new, more quantitative, methods originated by Coulomb and others to the electrical and magnetic examination of minerals. During this period, measuring instruments of many kinds became much more reliable. This fascinating story is told in a number of books and need not be recounted here.[32] Electricity and magnetism will be revisited in Chapter 8 with respect to larger phenomena than single minerals.

Cleavage, Fracture, and Fragments

This section is concerned with how a crystal breaks, and into what it breaks. Mineralogically, cleavage is defined as the tendency of a crystal to break cleanly along particular planes. We now recognize that this breaking is parallel to molecular planes, and it is generally parallel to a face. It can be classified according to direction. Both halite (NaCl) and calcite ($CaCO_3$) exhibit perfect cleavage, clean and marked parting planes. While both substances are colorless, transparent, relatively soft minerals, and both have cleavage, the cleavage angles immediately identify cubic halite and rhombohedral calcite. When there are no planes of weakness in a crystal where fracture occurs preferentially due to weaker or longer bonds, the break is curved and is called conchoidal from its resemblance to a shell shape.[33] The fragments produced by fracture can also be classified. These properties of minerals were well observed by the mid-eighteenth century.

There were many efforts to classify types of cleavage and the kinds of cleavage fragments, as well as the kind of fracture or break that occurred. In the course of the acute observation needed, it is not surprising that conjectures were made about the particles that made up matter, their shape, and the forces that held them together. Twentieth-century books tend to limit the terms for fracture to a few. A typical list now includes conchoidal, fibrous or splintery, hackly, or uneven.

In 1774, Werner said of fracture: "By it is understood the form of the internal surface of a solid material" (Werner, [1774] 1962, p. 75). He listed compact, fibrous, striated, and foliated as his main fracture types, and he subdivided each of those into as many as five different kinds, some of which were also subdivided. Under "form of fragments," Werner listed cubical, rhomboidal, pyramidal, cuneiform, scaly, tabular, and irregular,

---

[28]Until early in the nineteenth century, with the advent of Volta's "pile," investigations of electricity were concerned only with static electricity.

[29]We know the electrometer as a calibrated device that shows a potential difference by measuring the electrostatic force between two charged bodies. I suspect Schmeisser meant what we call an electroscope, which detects the presence and sign of static electricity.

[30]Franklin observed that when rubbed, glass and amber (a resin) acquired opposite charges, which attracted each other. The jury is out on why he chose positive for glass and negative for amber.

[31]These are Chap. 5: "Durée de l'électricité acquise à l'aide du frottement," and Chap. 6: "Électricité produite par la chaleur."

[32]See, for example, Frängsmyr et al., 1990, *The quantifying spirit in the 18th century*; Heilbron, 1993, *Weighing imponderables and other quantitative science around 1800*; and appropriate articles in Bud and Warner, eds., 1998, *Instruments of science: An historical encyclopedia*.

[33]Glass and flint break this way.

for each of which he gave mineral examples (Werner, [1774] 1962, p. 84). As expected with Werner, careful observation was required, and his next section on characters suggested that it would not be easy to follow Werner's methods. It is titled "Form of distinct particles," which to me is difficult to distinguish from "Form of fragments."

Kirwan used Werner's four kinds of fracture, but added slaty. He also subdivided those major kinds. For fragments, he was concerned with both their shape and their sharpness. His shapes were:

> most frequently *indeterminate*, but sometimes *cubical*, *rhomboidal*, or *pyramidal*, or *trapezoidal*, although these forms are often disguised; also the *long splintery*, the *broad splintery*, and the *tabular* which consist of plates that grow thinner and sharp at the extremities. (Kirwan, 1794a, v. 1, p. 36)

If none of the other types were listed, Kirwan said it would belong in his indeterminate class. He also classified the sharpness of fragments numerically, from 4 for sharpness like glass to 0 for blunt. This numerical system did not exactly take the mineralogical world by storm, although when authors used Kirwan's description of a particular mineral they sometimes included it. It should be remarked that Kirwan did not distinguish between rock and mineral cleavage.

The use of fracture and cleavage as diagnostic characters became rather routine by the end of the eighteenth century. Haüy and others dealt with them briefly, by comparison with the longer discussions from earlier authors. During this period, however, these properties, and those variously termed "tenacity," "cohesiveness," "ductility," "particle appearance," "friability," etc., all entered into discussions of how external properties might reveal internal structure or composition.

Form

Many mineralogy books contain more about form, or shape, as a character than any of the others, and the varieties were illustrated in eighteenth- and nineteenth-century texts with beautifully drawn plates (see Fig. 2.6). It was here that many ideas about internal structure and external form were voiced. The earlier ideas about a few different particles of sharp or smooth shapes that caused observed properties had evolved to the recognition that external and internal form must be related, but external form was not necessarily a direct result of the shapes of the internal particles. The perfect shapes of quartz, calcite, pyrite, fluorite, and other minerals were obvious. The tendency of calcite and halite to break into cleavage fragments that are the same shape as the original large crystal was part of the impetus for study. The various crystal shapes were minutely described in one of the major methods of mineral classification, for which Romé de l'Isle and Haüy were credited as initiators, along with Bergman.

Werner said: "The *external form* of a solid mineral is nothing else but the form of the natural outline which its individuals have" (Werner, [1774] 1962, p. 44).[34] For him, that natural form was the result of the mineral growing by itself, its natural outline. There were many ways that natural forms could be altered, and they were subject to the vagaries of solvents, solutions, precipitation events, and the sort of space or surface on which they were deposited. Since, however, the same external form was not present in each sample, Werner did not consider it possible to use it to tell the essential character of the mineral (Werner, [1774] 1962, p. 45). As one would expect from someone who worked with Werner's thoroughness, in the next 25 pages he described the regular figures in great detail, as well as the galaxy of forms, accumulations, and concretions that might occur. Because of that variety, the forms were part of Werner's classification, but were not considered as trustworthy as his external characters.

At about the same time, during the decades of the seventies and eighties in the eighteenth century, others were adding to ideas about crystal formation. In his influential *Dictionnaire de chymie* of 1766, Pierre Joseph Macquer (1718–1784) adopted the concept of the polyhedral integrant molecule as the smallest unit of composition (Mauskopf, 1976, p. 8). Those molecules aggregated to form larger, visible, crystals, of the same form as the integrant molecules. However, crystals of the same chemical composition may have several outward forms, a problem that Macquer did not address.

Romé de l'Isle attempted to explain the several forms that crystals of the same substance may assume. He postulated a primitive form for each substance, dependent on the shape of the integrant molecule, the form of which could be combined in different ways to result in the final external shape (Romé de l'Isle, 1772; Mauskopf, 1976). Having noticed the constancy of interfacial angles, as Nicholas Steno and Hooke had, and now having access to a goniometer, he attempted to produce the various shapes, sometimes by means of supposed beveled and truncated edges or angles. He later gave equivocal statements about the geometry of the integrant molecules (Mauskopf, 1976).

Haüy brought further order to the descriptions of external form, and the attempt to relate it to composition. He first used the term *molécules constituantes* to describe the smallest unit that retained the properties of the substance, but later used *molécules intégrantes*. These were of a particular shape and size for each substance. The shapes might combine in different ways to produce the different external shapes that the same substance might display. Crystals had a primitive form, or nucleus, that could be revealed by cleavage, upon which layers of more integrant

---

[34]The definition of an "individual" was refined as knowledge of chemical composition increased, and the possibility of fixed or variable composition in compounds with the same elements was recognized. Romé de l'Isle and Haüy needed the geometrical unit of the *molécule intégrant*, so that "each species is characterized by its geometrical as well as its chemical type" (Hooykas, 1958, p. 312). Chemical individuals were those having the same constituents, which might be in different proportions, but the group belonged to the same species. Thus, Werner's definition of "individual" was amplified as fixed, fixed with different proportions, and variable, such as the copper sulfides. Compounds were recognized in conjunction with the work of John Dalton and Claude Louis Berthollet. See Hooykas (1958) for a complete discussion.

26

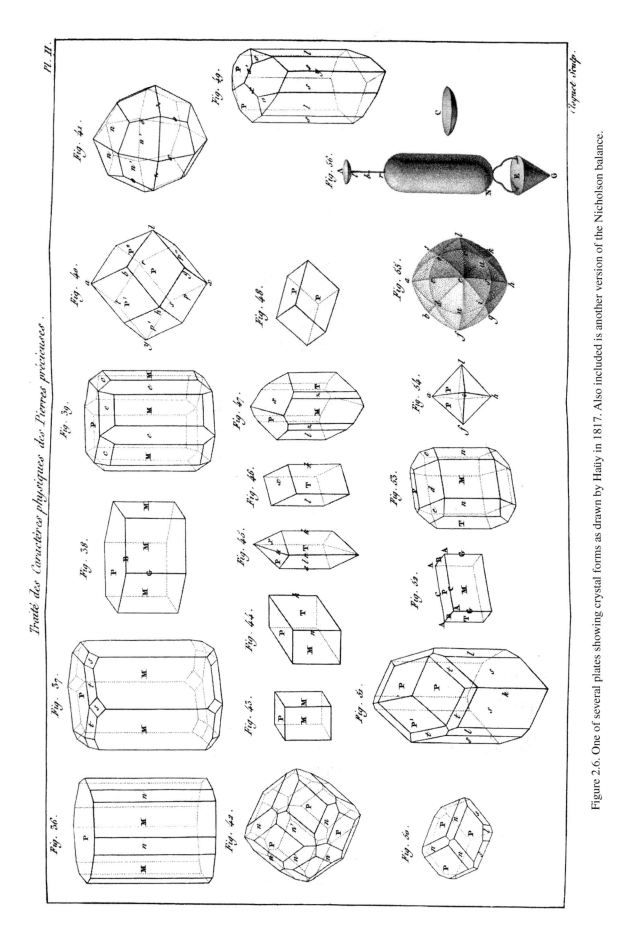

Figure 2.6. One of several plates showing crystal forms as drawn by Haüy in 1817. Also included is another version of the Nicholson balance.

molecules were deposited. The pattern of deposition determined the final shape (Mauskopf, 1976, p. 12).

The relation of shape to composition was an ongoing concern, and there is much of importance here in the contribution of mineral description and classification to both chemistry and matter theory. However, I am chiefly interested in the instrumentation that improved the knowledge of shape during this fecund period in the sciences. And here the most important instrumental innovation was the goniometer. It appeared and evolved during the late eighteenth and early nineteenth centuries, although development didn't stop there.

The contact goniometer is a simple instrument, still in use, which enables a quick determination of the angles between crystal faces. The constancy of interfacial angles in crystals of the same kind is sometimes called Steno's Law, named for Nicholas Steno (1636–1686), because he recognized that the angles between similar faces of different quartz crystals were the same. He did not extend this observation to the crystals of other minerals, however. And while others noted this constancy in some crystals in the intervening century, the general statement of the law, that interfacial angles are the same in all crystals of the same substance, is credited to Romé de l'Isle in 1783 (Burke, 1966, p. 67 and 69; Burchard, 1998, p. 520–521). Romé de l'Isle's priority was contested by the student he assigned to make clay models and to design an instrument to measure the correct angles (Burke, 1966, p. 70; Birembaut, 1970, p. 61).

This origin of the contact goniometer (Fig. 2.7) is generally agreed upon, although there are interesting hints of precursors.[35] Romé de l'Isle wished to include drawings of many more crystals in the second edition of his *Essai de cristallographie* than had been in the first edition in 1772, and hired an engraver to make and draw them (Burchard, 1998, p. 521). The engraver requested models of the crystals, and in order to make them with the correct angles, Romé de l'Isle asked two of his students to make a measuring instrument for them. This instrument was designed by Arnould Carangeot (1742–1806), and consisted of a semicircular arc (a protractor), graduated in degrees, whose base served as one measuring arm. The second, moveable arm was fixed in the center of the arc's base. If two faces of the crystal were made to lie flush with the two arms, the angle between them, or their internal angle, could be read from the arc. With some justice, Carangeot claimed discovery of the law of constancy of angles, but he was never accorded that recognition (Burke, 1966, p. 70; Burchard, 1998, p. 521). Carangeot later designed an improvement to the goniometer that allowed the measurement of angles of groups of crystals.

Early in the nineteenth century, Robert Bancks (1796–1834) designed a goniometer that placed the measuring arms slightly above the diameter of the semicircle; the arms were slotted so that they could be moved to accommodate different sizes of crys-

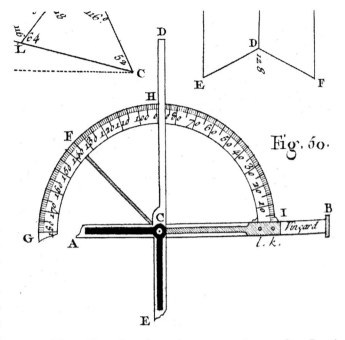

Figure 2.7. An illustration of an early contact goniometer, from Romé de l'Isle (1783, v. 4, Fig. 50).

tals; and the arms could be fixed with a screw after the crystal was removed—all of which led to more accurate measurement. Full- and quarter-circle contact goniometers were devised, as were means of increasing the precision of the measurement, which was generally between fifteen minutes and one degree (Burchard, 1998, p. 524). There were also stationary instruments that left the hands free to manipulate samples. We will briefly discuss a different kind of goniometer that was soon invented. However, the simple contact goniometer is in use today to demonstrate the constancy of angles to students in the laboratory and the field, and to aid in mineral identification.

Figure 2.4 shows the Wollaston goniometer. An important advance came with the introduction of the reflecting goniometer by the gifted English chemist William Hyde Wollaston (1766–1828) in 1809. His instrument used the law of reflection from a plane mirror, which was that the angles of incidence and reflection of a light beam are equal. The crystal was mounted so that the edge between faces coincided with and was parallel to the axis of the instrument. To measure the angle between those faces, a light ray impinging on a face was directed through a telescope to the eye. The crystal was rotated so that light reflected from the next face was visible. The angle that the crystal face was rotated through to throw beams from adjacent faces to the same spot was the angle between the faces. Measurements could be made even for very small and/or imperfect crystals whose adjacent faces did not meet cleanly. The reflecting surfaces had to be smooth. Measurements of great accuracy were attained with this instrument and its successors, which were used by most mineralogists.

Schmeisser mentioned the contact goniometer as the means to measure "the degree of the side or end angles" of a crystal, and

---

[35]Much of the following discussion comes from an excellent and comprehensive monograph by Ulrich Burchard (1998). He has traced the precursors and development of contact and reflecting goniometers from the seventeen century until X-ray diffraction came into use.

included an illustration (Schmeisser, 1795, v. 1, p. 26). Cleave-land discussed both kinds of goniometer at length. For his contact goniometer, which he attributed to "Carangeau," both arms rotated and could be shortened by sliding them in slots to avoid protruding parts of the sample. He also spoke about the convenience of the reflecting goniometer of Wollaston, particularly where crystals were not regular. This was followed in Cleaveland's book by eleven pages of discussion supported by five pages of plates, which illustrated the crystal shapes.

In 1817, Jameson described the use of contact and reflecting goniometers, including a beautifully drawn illustration of the latter. However, he also spoke approvingly of Werner's method of "ocular inspection" without the need of an instrument (Jameson, 1817, p. 115). Jameson described angles at lateral edges, terminal edges, and the summit (top) angle, as well as outward forms. It is interesting that this method persisted after the introduction of both simple and more complex goniometers. Also in 1817, Haüy only mentioned the contact, and not the reflecting, goniometer when he began his discussion of the crystal shapes of precious stones (Haüy, 1817, p. 14–15).

To be sure of identification, Mohs was typical of his time when in 1820 he said:

Having advanced in this manner to the character of the species, [after hardness and specific gravity] it will in some instances be necessary, and in all cases advisable, for the sake of certainty, to have recourse to the dimensions of the form. This is particularly necessary, if the genus to which the mineral belongs contains several species, having forms of the same system, as is the case in the genus *Augite-Spar*. This determination of the dimensions of the forms may be effected by the common gonyometer [*sic*], the differences in the angles being in general so great, that they cannot be easily missed, even by application of this instrument. (Mohs, 1820, p. xx–xxi)

As the nineteenth century progressed, both kinds of goniometers were in use, and both, particularly the reflecting goniometer, were under nearly constant development to improve ease of use and accuracy.

CONCLUSION

Minerals have been collected since antiquity, some for their properties or derivatives, and some for adornment. The cabinet, or collection, of minerals was generally associated with the natural historical approach to mineralogy: that of classification. Some collections belonged to institutions that might have been denoted museums. Cabinets of curiosities, which included minerals, were a mark of the educated gentleman, and prominent scientists such as Mohs were hired to identify and classify collections in the eighteenth and into the nineteenth centuries. For the scientist, there was not only the desire to possess as many kinds of minerals as possible, but also to have type specimens to compare with newly collected minerals. The collections also were used for instruction, not just for young students, but for practicing mineralogists, as has been admirably set out in Sweet's article about Robert Jameson's visit to Ireland (Sweet, 1967). Not everyone,

individual or institution, could afford high-quality specimens, and in any case there was disagreement about what really was a good example of a particular type. One of the five categories in the Leskean cabinet that Jameson observed in Dublin was called the Characteristic Collection, which had been made according to the external characters (Sweet, 1967, p. 122). To assign a mineral to a type there had to be substantial agreement in those characters, but there were numerous examples of disagreement among well-known mineralogists about identification of a specific mineral and its classification into a particular group. Because of the meticulous attention paid to characters, as more travel, observation, and collecting was done, new minerals were described frequently, adding rapidly to the list of about 1000 that Beretta estimated were known at the beginning of the eighteenth century (Beretta, 1998b, p. 583).

Thus, there was not only an effort to assign minerals to known divisions, but also to develop a means of classifying new minerals in the correct association. Chemical analysis had seemed to hold hope of definite classification. More accurate analysis toward the end of the eighteenth century both helped and hindered. It could differentiate minerals by composition, but also revealed that, contrary to expectation, the same compound could have different forms, such as calcite and aragonite, or different compounds could have the same form. Partly because of the unreliability of chemistry, Romé de l'Isle defined mineral types by hardness, crystal form, and specific weight. There was a period of approximately forty years when the questions of definite and indefinite composition, and isomorphism and polymorphism were being worked out.

Werner, with his connections to the practical mining industry, was more interested in the properties of minerals than in pursuing matter theory. This explains the eventual popularity both of better measurement methods such as the goniometer or the use of numerical scales, and also the extremely detailed description of multiple shades of different colors.

Measurement is defined as ascertaining the dimensions, extent, or quantity of something by comparison with an appropriate scale. Scales can be absolute or relative, merely comparative, as in Kirwan's hardness, or mathematically more rigorous as in the Richter scale in powers of ten measuring energy for earthquake magnitude. Several scales have been mentioned in this chapter.[36] The hardness scale was simply an ordering of soft to hard in Mohs's scale. Hardness is a particularly interesting "character" to follow into the very late twentieth century (Newcomb, 2002). It proved very difficult to come up with a number that could represent absolute hardness, because various methods of determining hardness depend on different properties of the material tested, and the numbers are not consonant. Other numerical scales, when adopted, had even less relation to a property that could be empirically tested. There was a long period of development of numerical

[36]Notice that the word "scale" is not used to refer to the scientific instrument, the balance, which was so denoted at the time because of the balancing operation of the sample with known weights.

expressions and calculations that accurately reflected properties of interest. Thermometer scales are a different matter, and will be discussed in the chapter on heat (Chapter 3). The concept of using a number to express a property more than survived, however, as we have empirical-mathematical scales such as those for pH and the Richter scale for earthquakes in constant use.

The metric system, whose initiation, determination, and adoption are another of those great stories in the history of science, was newly introduced in France at the end of the eighteenth century, became indispensable in science, and eventually provided international standards for mass, length, and volume. The use and applications of numerical data increased enormously at the end of the eighteenth century. It is tempting, not to say almost automatic, to interpret those numbers, whether of hardness, weight, or specific gravity, in the way we currently do, but two factors are immediately evident, and hinge on our current use of significant figures. First, we must make an effort to know how "good" the numbers are, if generated by an instrument such as a balance that was increasingly used in science, moving from its earlier place in mining technology and alchemy. If, in calculating specific gravity, a mass of 2.445 is given, does that really reflect the accuracy of the balance? Often the eighteenth-century authors did record to what mass their balances were accurate. Second, it wasn't until after the metric system was introduced and strongly advocated, that decimal calculations became common. It became recognized very early in the nineteenth century that many decimal places were artifacts of the calculations themselves and should be dropped (Heilbron, 1990, p. 241). Recorded numbers should reflect the accuracy of the measurement instrument, and not artifacts of calculation.

In this chapter, we have surveyed various methods of obtaining knowledge of external characters in order to identify minerals, the building blocks of rocks. We will now proceed to methods applied to both minerals and rocks to go beyond identity and to ask questions about conditions of origin and the history of Earth itself.

CHAPTER 3

Heat matters: Fusion

*A thorough knowledge of the effects produced by fire upon fossils is
of the greatest importance in the cultivation of many arts.*
—Bergman (1784–1791, v. 3, p. 222)[1]

INTRODUCTION

Heat and its effects were central to geology as it emerged as a distinct branch of science toward the end of the eighteenth century. Although not all people interested in the origin and actions of earth processes were Plutonists or Vulcanists, heat was clearly both a cause of some phenomena such as volcanoes and a useful variable to apply in the investigation of mineral and rock behavior. Investigators recognized that many of the earth materials seen on field trips were the ingredients for industrial processes. Furnaces, forges, and smelters are part of some of the oldest technologies in human history, having been in use for at least four millennia. Heat was used to work native metals such as copper, gold, silver, and the occasional native iron mass, as well as to win metals from ores or to react an ore with a native metal, as in the making of brass. By the middle of the eighteenth century, furnaces had long been part of the laboratory, used by alchemists and metal assayers. The natural philosophers of the eighteenth century also made use of metal smelting, glassworks, and porcelain furnaces in their investigations. Because of this long history, the behavior of numerous kinds of materials under fusion conditions was known, and many of these materials were minerals. Subjecting a substance to heat was one of the first things done when trying to discover its properties, and analogies were made between the products of nature and those of furnaces. The English physician to Charles II, Thomas Sherley (1638–1678), declared: "[T]he Art of Pyrotechny is the only true means of informing the mind with Truth, and acquainting it with realities" (Sherley, [1671] 1978, p. 15). Debates about the nature of heat were most prevalent during the eighteenth century, but those philosophical searchings had little immediate effect on the engines (literally) of technology.

On a large scale, heat and geological phenomena had also been linked for millennia by way of volcanoes, for which there obviously had to be some kind of natural heat source. Differences in land elevation linked to volcanic eruptions were also noted. An early analogy was made between the expansion noted in mate-

rials when heated artificially, and the influence of heat on the expansion of rocks in situ. The correlation between the increasing temperature of Earth and depth was being documented as well. As in other parts of this story, it is virtually impossible to establish a "first" for such ideas. Apart from instances of volcanism and deformation, heat also figured in the debate about the origin of rocks, and whether they were formed by solution or fusion. Laboratory-scale experiments were designed to test both theories.

Questions about heat permeated the sciences of the late eighteenth century. While it was acknowledged that there was no satisfactory theoretical understanding of the nature of heat, its effects were applied unhesitatingly to questions of all kinds. The clarification of the nature of heat from that time and throughout the nineteenth century is another classic tale in the history of science.[2] Thus, at the end of the eighteenth century, heat, as caloric, was the subject of much investigation by those working in several branches of mathematical and empirical science, while the true character of combustion was just being unraveled in the work of Lavoisier and others. Acceptance of a heat-dependent theory of earth processes, such as that of the Scottish geologist James Hutton (1726–1797), was hampered, partly due to the lack of understanding of the nature of heat.

Among those who studied Earth and its materials, heat was frequently discussed on both large and small scales. For the Freiberg, Saxony, metallurgist Christlieb Ehregott Gellert (1713–1795), fire was the principal one of the six chemical agents, "for without its assistance no chymical operation can be performed" (Gellert, 1776, p. 72). Its properties were *light*, and the *expansion* it caused in bodies. The expansion was supposedly caused by fire forcing interior parts of bodies toward the surface, which meant that those parts must be in motion, as must the fire itself, and the more fire the more the motion, and thus the greater the expansion (Gellert, 1776, p. 73). Donovan (1978) noted the linkage between Hutton's theory of Earth and his knowledge of the Scottish chemist Joseph Black's (1728–1799) clarification of the calcination of limestone. Hutton suggested that limestone

---

[1]However, Bergman is better known for his introduction of a systematic method of wet analysis.

[2]Stephen Brush has written extensively about the history of the understanding of heat and its actions. Several references are: *The kind of motion we call heat* (1976); and the three volumes of his *Kinetic theory* (1965–1966, v. 1 and 2; and 1972, v. 3).

Figure 3.1. One of many depictions of furnaces and their collateral equipment in Agricola (1556).

as heat sources. Next is a discussion of containers for the materials being heated. The following section is concerned with that important instrument, the blowpipe, the use of which (for mineral characterization) began in earnest in the last decades of the eighteenth century and continued through to the mid-twentieth century. The next section concerns the instruments available to quantify heat, namely, pyrometers and thermometers. There is then a section about the temperatures attained by the various methods. Chapters 9 and 10 will discuss specific experiments and their influence on questions of rock origin.

FURNACES

Les fourneaux de fonderie des forges, sont les instruments avec lesquels l'art peut approcher le plus près des opérations de la Nature pour imiter les produits des volcans.
—Grignon (1775, p. 477)

Les forges sont des laboratoires immenses dont le travail en grand fourniroit tous les jours sujet à des découvertes intéressantes.
—Grignon (1775, p. 17)[3]

Because heat was the first choice of agent when one wanted either to find out something about a substance or to alter it, a great many ways were devised to supply it. By the time of the great metallurgical treatises of the sixteenth century,[4] furnaces (Fig. 3.1) were well understood and carefully fabricated to deliver the appropriate amount of heat for specific products and processes. Anderson pointed out that each of the different chemical processes of "fusion, cupellation, calcination, reverberation, cementation, distillation, digestion (and the bain marie), sublimation, fierce distillation, and distillation per ascendum" (Anderson, 2000, p. 17) required a different kind of furnace.

At the end of the eighteenth century, scientists had a wide choice for delivering heat for fusions or reactions.[5] They might use an industrial furnace itself, putting samples in smelting, glasshouse (glassworks), or porcelain furnaces. Pottery kilns were also used. On a smaller scale, there were different kinds of furnaces specially made for the laboratory, where they were considered to be at the heart of operations. Still smaller furnaces were designed for traveling laboratories. Alcohol lamps might be used while traveling or in the laboratory, as might lamps with fuels such as horse fat. Finally, water and sand baths were used for lower or for sustained heating. Details of furnace construction will generally not be described here in detail, but references are given.

strata could be consolidated by heat if they were subject to great pressure that retained the "fixed air" usually given off by limestone when heated. After an examination of the two entities that caused fluidity, namely, fire and water, Hutton ([1788] 1973; [1795] 1972) concluded that heat had to be the cause because he considered that there was no solvent that could have held all rock material in solution, as suggested by some German theorists. There was no way for the solvent to be removed if it had formerly held Earth in solution, and nowhere for it to go.

Heat also provided an explanation for the expansion that was thought to have deformed rock layers. Murray objected to portions of the Scottish mathematician John Playfair's (1748–1819) (1802) defense of Hutton's theory by considerations of the behavior of heat. Murray argued that the distribution of heat throughout Earth must approach equilibrium, since heat from a central source would eventually spread to contiguous layers, and the heat thus would be distributed equally over the globe. In this condition of equilibrium, Hutton's effects could not occur, as they required large temperature differences (Playfair, [1802] 1964, p. 92–96; Murray, 1815, p. 411–412). Despite these arguments, questions of rock origin could be tested by heating without knowing Earth-wide effects, and without knowing the true source and nature of heat and the mathematical laws of its operations.

This chapter is concerned with the equipment and procedures used to investigate and change the materials that make up Earth by means of heat. The first section covers industrial, laboratory, and portable furnaces; their construction; their fuels; and how they were controlled. Sand and water baths, several heating lamps, and burning mirrors and glasses will be briefly considered

---

[3]"The smelting furnaces of foundries are the instruments with which art can most nearly approach the operations of nature in order to imitate volcanic products" (my translation).

"Forges are immense laboratories whose work will supply interesting discoveries on a large scale every day" (my translation).

[4]See V. Biringuccio ([1540] 1990), G. Agricola ([1556] 1950), and L. Ercker ([1580] 1951). C.S. Smith discussed this literature in the Introduction to this translation of the *Pirotechnia* (Biringuccio, [1540] 1990).

[5]The words "science" and "scientist" will be used here for brevity, although the terms did not come into general use until later in the nineteenth century.

Materials

No matter what the use, all furnaces, regardless of how sophisticated, must be made of materials that are not fused by the fire they are intended to contain. The first furnace was probably a hollow in the earth with rocks stacked around it. Natural clefts in rocks were sometimes used, so that walls only had to be constructed on one or two sides. Some ores could be smelted by piling the ore in a heap, surrounding it with more ore or rocks, covering it with charcoal, and subjecting it to blast from a bellows (Biringuccio, [1540] 1990). A simple furnace might be made of several bricks with iron bars for support, while larger furnaces were principally made of bricks.

The Italian metallurgist Vannoccio Biringuccio (1480–1539) wrote: "The practice of making bricks is so well known that it seems a shame to write at length of it here" (Biringuccio, [1540] 1990, p. 400). The process is no longer so generally familiar. First, all lumps and stones were removed from clay, which was then pressed into boxes or molds after it was coated with dry sand so it wouldn't stick to the molds. After drying in the sun, the future bricks were fired in a furnace for seven or eight days and then allowed to cool slowly (Biringuccio, [1540] 1990, p. 401). However, it was noted that bricks might react with the materials being smelted, or deteriorate when they were cleaned of slag. Biringuccio also explained how to make mortar and plaster of paris to hold the bricks in place.

At the end of the eighteenth century, Pierre Loysel (1751–1813), the French official and glassmaker, listed three sorts of bricks that might be employed in making a *four de fusion* for glassmaking. He suggested one kind (*briques molles*), difficult to fabricate, for use when a good grade of glass was to be made in open pots. Ordinary bricks would suffice when common glass was to be made in closed pots (Darcet et al., 1791).[6] After a furnace was constructed, the bricks were frequently covered with lute—a mixture of clay and water that sometimes included ingredients such as sand to give it body.

The use of stone was recommended to construct furnaces that would be subjected to great heat, such as blast furnaces. The stone had to be of a kind that would not crack or burst. Therefore talc, flintstone, sandstone, and some volcanic rocks could not be used. Stone was harder than bricks to work and to shape into the desired configuration, but it could be employed to make the entire furnace, or as Biringuccio did in one of the endless variations, employed to heat refractory materials. He made a receptacle "in the form of a little cradle" of either stones or coals and powdered clay, baked it with good charcoal, and circled it with rocks, in order to further heat a "matte" (the result of a first smelting) of copper (Biringuccio, [1540] 1990, p. 170–171). Later, at the end of the eighteenth century after the discovery of the very refractory chromite, the chromite was either itself shaped into bricks, or powdered, mixed with silica, and then fashioned into bricks

for furnace construction.[7] Most furnaces had a relatively short life, and needed frequent repairs and rebuilding.

Smaller assay furnaces might be made of brick, clay, or iron. Agricola described in detail a furnace made of shaped and joined iron bars with an iron plate for a stand, the whole covered with iron plates and smeared with lute to protect it from the fire. The clay furnace was to be the same dimensions as those made of iron, and the clay was supported with iron wires. Agricola noted that small brick furnaces could be made more quickly but had to stay in the same place, while clay and iron furnaces could be moved as needed. As mentioned above, a simple furnace could be made by putting an iron plate across several bricks, leaving the front open (Agricola, [1556] 1950, p. 225–226).

Materials for constructing furnaces changed little in the next 300 years, with a very few exceptions. Graphite, or black lead as it was called, was used for small furnaces, as discussed in the section on portable furnaces below. Johann Heinrich Pott (1692–1777), the mineral chemist who experimented with a vast variety of mineral mixtures in a search for the procedure for making true (hard paste) porcelain, mentioned that his furnace was nearly the same as the one the alchemist Johann Joachim Becher (1635–1682) had described nearly a century earlier. It was made of iron plates covered with a lute of white clay mixed with equal parts of the same clay that had been calcined, and beef blood to hold the lute together (Pott, [1746] 1753, p. 421). Sometimes small hooks were riveted on the inside of the furnace to support a heavier coat of lute. The lute might also serve to hold elements of the furnace in place (Gellert, 1776, p. 147). Other recipes for lute included a mixture of clay and pounded charcoal that might be covered with a second coat of pipe clay (Schmeisser, 1795, p. 371), a mixture of slaked lime and linseed oil (Cronstedt, 1788, v. 2, p. 995), or a combination of quicklime and egg white (Henry, 1814, v. 1, p. 345). Lavoisier included a chapter on making lutes both for heat resistance and to seal vessels in chemical investigations (Lavoisier, [1790] 1965, p. 407–413).

Muffles were refractory pieces shaped like a roofing tile and inserted in furnaces to protect the sample, which was placed underneath the muffle. A muffle might be inserted in any furnace, particularly the smaller ones, as needed. The variation in muffles consisted of the number and shape of openings in the sides, designed to facilitate heat and gas flow over the sample. They were made of the same materials as the furnaces and sample containers, generally clay.

Fuel

The fuel used in furnaces was determined by availability, suitability, and economics. There was also a body of practical knowledge about the interaction of some fuels with particular ores, so the choice for smelting ores sometimes depended on the kind of product desired. Small laboratory heating lamps used oil

---

[6]Loysel's results were discussed at length in v. 9 of the *Annales de chimie* by d'Arcet (Darcet), de Fourcroy, and Berthollet (1791).

[7]Chromite, the most common ore of chromium, is a double oxide, $FeO \cdot Cr_2O_3$, with a great deal of substitution of other +2 and +3 metals.

or alcohol (spirit of wine). The latter was recommended because it was inexpensive, had no odor when burned, didn't leave a residue, and didn't smoke (Cronstedt, 1788, v. 2, p. 997). A fuel that must have displayed the very antithesis of those virtues, melted horse fat, was recommended for a lamp used in conjunction with a blowpipe (Smeaton, 1966, p. 86).

Wood and charcoal were commonly used because of their wide availability and the relative ease of obtaining and transporting them. Other fuels were not sought until the forests were depleted. Biringuccio included a chapter titled "Concerning the properties and differences of charcoals and the customary methods of making them" in which he discussed wood sources and the properties of charcoal made from different kinds of trees (Biringuccio, [1540] 1990, p. 173). He contrasted the action of wood and charcoal in the furnace, stated differences in charcoal made from various kinds of wood, and gave directions for the best way to make it. It is clear from Agricola's instructions for several processes that charcoal was used for both fuel and/or a reducing agent, as has continued until now in metal smelting, although this dual function was not understood until the more advanced chemistry of the later eighteenth century (Smith, 1967, p. 146). Use of wood persisted when a flame was desired, while charcoal remained in constant use.

Coal had been in use since antiquity and the different kinds such as bituminous or anthracite were well described, although names varied across times, locations, and languages. The combined descriptions of appearance and behavior allow us to make fairly informed assumptions about the sort that was actually used. Some coal had such a high sulfur content that it interfered with iron smelting, both because the product might have too much sulfur in it, and also because the fuel's large residue could interfere with heating. In such a case, the ironmaster and archaeologist of Bayard, Pierre Clement Grignon (1723–1794), recommended what he called the "English method" of preparing coke from the coal (Grignon, 1775, p. 102). Essentially, coke was produced from coal by heating it enough to drive off the most volatile constituents, a process which involved numerous complicated chemical reactions, but which did not, however, need to be understood.

Not surprisingly, many other combustible substances were used when convenient, including cork, sawdust, or cinders. The latter was identified as "common cinders, taken from the fire when the coal has just ceased to blaze, sifted from the dust, and broken into very small pieces" (Henry, 1814, v. 1, p. 336). There were also mixtures of various fuels, such as a little charcoal with coal. Glass furnaces might be fired with coal or wood. Peat was not very satisfactory, although it was sometimes used. Even the heat of a lamp furnace could be varied by use of different fuels or sizes of wicks. The temperature of a furnace could be controlled by selection of an appropriate fuel, its amount, and the operation of bellows and draft holes. The relation of reaction speed to particle size was also clearly known, as the size of fuel pieces was often specified. Coal gas, driven off by heating coal in the absence of air, and leaving a residue of coke, was used in laboratories after ~1820 (Anderson, 1998, p. 251).

Heating Methods

Industrial Furnaces

Metal-working furnaces: Blast and wind. Furnaces for industrial metal production, or assaying on a smaller scale, while of disparate sizes, materials, shapes, and modes of operation, operated on a few well-understood principles, known from at least the middle of the sixteenth century. All were designed to answer the requirements of different ores and desired products. There was frequently a preliminary step, that of roasting the ore. This might be, in our language, for either physical or chemical purposes, to facilitate breaking up the ore, or to drive off volatiles that might contaminate or alter the product. A source of oxygen (air) for combustion of the fuel was necessary, and the fuel itself. All furnaces included means of controlling the heat as well.

Blast and wind furnaces were applied to many tasks. A wind furnace used a natural draft, while the blast was a forced stream of air usually supplied by bellows. Conditions were altered according to the kind of ore to be treated and its contaminants. Biringuccio gave excellent directions for making a blast furnace. It was necessary to choose a site conveniently located by supplies of wood, ore, and water power, the latter being necessary to work the bellows that supplied the blast. The housing, walls, connectors, and water wheel needed to be built first. He suggested rock such as black flintstone, or one that was half talc, for furnace walls (Biringuccio, [1540] 1990, p. 146). Both he and Agricola gave such precise measurements and directions for furnace construction that they could be followed today, and the same general principles were followed for centuries. With such a background of experience, it is not surprising that smaller furnaces were well-understood tools in the laboratory.

Grignon went well beyond the reports expected of an ironmaster. Not surprisingly, he wrote about furnace construction, what could go wrong with furnaces, and how they should be built. He observed smelting processes closely, and the resultant products and slags. Grignon quoted Romé de l'Isle, who had suggested that metal crystals could only originate by the mediation of water. Grignon then announced that he would use facts and knowledge he had acquired to show how regular crystals of metals, minerals, and vitreous substances could be formed in his furnaces without water (Grignon, 1775, p. 475). Some of the crystals he saw in his furnaces resulted from long heating and longer cooling.

Reverberatory furnaces. Biringuccio is also credited with having given the first good description of a reverberatory furnace (Smith, 1956); see Figure 3.2. It is obvious the technology was not then new, as the description occupied an entire chapter with numerous drawings and suggestions for various arrangements (Biringuccio, [1540] 1990, p. 281–288). Steel was made in this kind of furnace as early as 1614 (Wolf, [1952] 1961, v. 2, p. 633). Basically, in a reverberatory furnace, the material to be melted was separated from the fuel, but the configuration of the furnace was such that the flame or reflected radiant heat reached the

Figure 3.2. Biringuccio (1540) showed one example of a reverberatory furnace.

material. William Henry (1774–1836), the chemist and a member of both the Wernerian Natural History Society of Edinburgh and the Geological Society of London, who published a book on experimental chemistry, explained:

In this furnace, the fuel is contained in an anterior fire-place; and the substance, to be submitted to the action of heat, is placed on the floor of another chamber, situated between the front one and the chimney. The flame of the fuel passes into the second compartment; by the form of which it is concentrated upon the substance exposed to heat, which is not confined in a separate vessel or crucible, but placed on the floor of the furnace. When reduced to a state of fusion, the melted mass is allowed to flow out through a tap-hole. (Henry, 1814, v. 1, p. 339)

Reverberatory furnaces, of different sizes and designs, were generally used when great heat was required and the sample needed to be protected. In at least one example, this kind of industrial furnace crossed over into the service of science when Sir James Hall (1761–1832), conducting experiments in relation to James Hutton's geological theory, placed his samples of whinstone and lava "into the great reverberating furnace at Mr. Barker's iron foundery" (Hall, 1805b, p. 47). Lavoisier and other chemists adapted smaller reverberatory furnaces for laboratory needs. He illustrated this type of furnace and discussed its use (Lavoisier, [1790] 1965, p. 463–466).

Glassworks Furnaces

La verrerie est peut-être de tous les arts, celui qu'on peut soumettre le plus rigoureusement à des principes déterminés par la physique, celui par conséquent qui peut parvenir à la plus grande précision, mais il demandoit un qui fût également familiarisé avec tous ses procédés & instruit en physique.

—Darcet et al. (1791, p. 113)[8]

The above quote about Loysel, by Darcet, Fourcroy, and Berthollet—leading lights in French chemistry at the end of the eighteenth century—again demonstrated the close connection of technology and science. The technology of glass making was improved by application of science (*physique*) as expected. Continuing the traditions exemplified by Pott, materials were tested for fusibility and/or resistance to fusion. Loysel investigated not only the materials with which to make the furnace, but also the shape of all its parts, the fuel, the heat developed, and the interaction of heat and the containers. It was required that mixtures to make glass would fuse, while the opposite was desirable for containers.

The furnaces used by glassmakers, in what were called glassworks, were surmounted by a dome or gable so that the heat, rising from below, was also deflected downward, thus surrounding the melting glass with heated air. The clay pots that contained the glass mixture were set on benches that might be at different levels if different temperatures were desired. In a glassworks, there would also be an annealing furnace in which fired glass was placed to cool slowly, a furnace to fire the clay pots in which the glass mixture was melted, and sometimes one for drying the wood used as fuel (Polak, 1975, p. 14–15).

From the detailed reports of Loysel's investigations with glass, and of Josiah Wedgwood's in his pottery furnaces, it could be argued that they used and produced scientific knowledge.[9] Additionally, availability of their technology advanced science for others. James Keir (1755–1820), who managed a glass factory for several years, was an associate of Wedgwood, Sir James Hall, and the chemist Joseph Priestley (1733–1804) (Scott, 1975, p. 277). Keir noticed and drew the shapes discernible in a large mass of glass that had been left in a glassworks pot while the

---

[8]"Perhaps of all the arts, glassworking is the one that can be submitted the most rigorously to physical principles, consequently the one that can succeed with the greatest precision, but it requires an observer who is equally familiar with all its procedures and instructed in natural philosophy" (my translation).

[9]There have been questions about whether those applications could be termed "science," or whether they fell more within technology. See Schofield (1959) for discussion of this issue and references. J. Uglow's magnificent *Lunar men* (2002) detailed this period and how interwoven the science and technology were.

fire slowly went out. This observation was later referred to by geologists thinking about rock origins. After precise description, Keir suggested that inferences could be drawn about basalt. Hall referred to Keir's (1776) observations as well as to a similar occurrence at another glass house when he began his work on rock fusion. Gregory Watt (1777–1804), the son of James Watt, was inspired by the work of Keir and Hall, and used an iron foundry reverberatory furnace to fuse a large mass of the basalt called Rowley Rag (Watt, 1804a). On slow cooling from a vitreous state, the mass showed forms that he interpreted as the sorts of figures seen in stony basalt.

Porcelain and pottery furnaces. Baked wares made of clays and minerals could be classified into nine different orders depending upon their hardness and translucency, which were the combined result of the materials they were made of and the temperatures at which they were fired (Jacquemart, 1873).[10] They are of interest here for many reasons: Their history and surviving samples tell us about mineral supplies and the technology applied to them; we can judge the furnace heat and fuel supplies required; this work supplied the containers in which high-temperature work was done; finally, studies done with furnaces and related materials made possible analogies to the natural world.

Pott had said proudly of his smaller furnace that he could fuse mixtures that could not be melted in the furnaces for glass or porcelain making, even when they were fired for several days (Pott, [1746] 1753, p. 421). However, the French chemist Jean Darcet (1725–1801) doubted Pott's claim. He thought Pott's clay-covered iron furnace too small in diameter and thickness, and with its use of coal rather than wood, plus the lack of good evidence that he actually used bellows for blast, believed that it could not produce the effects claimed for it (Darcet, 1766, p. 7–9). Darcet used a porcelain furnace that he felt was far superior. It could be slowly brought up to temperature and maintained thus for several days. His crucibles were carefully made and covered to prevent the entrance of foreign matter (Darcet, 1766, v. 1, p. 7–11). Besides his many fusions of minerals and mixtures, Darcet fused rocks including basalt, and showed that diamonds burnt and disappeared when subjected to the heat of the porcelain furnace.

Klaproth continued the use of a porcelain furnace for experiment, in his case that of the Royal Porcelain manufactory at Berlin, into which he put his covered charcoal crucibles. He noted that Pott, Gellert, and Darcet had not allowed for the effect of clay crucibles interacting with their samples to give erroneous fusion results. However, enclosing his samples in charcoal meant that any iron was reduced, as Klaproth noted (1801).

Hazards. The use of industrial furnaces had another kind of impact on subsequent laboratory science. Many of the procedures, materials, and reactions were dangerous, or used or produced toxic substances. Life spans of everyone, but particularly of mine and smelter workers, were often short, and descriptions of some methods, for example, of roasting sulfide ores, raise questions about how long a person could work in such an atmosphere. Many accounts of laboratory work in the eighteenth century raise doubts about whether any lessons about toxicity had been learned from earlier work.[11] However, there are a few comments that indicate knowledge of the problems. When one particular furnace and design was used to separate arsenic from associated ores, workmen retrieving the arsenic were advised to eat bacon and then bind their faces as protection against the arsenic. In a slightly later design, workers had less exposure to the fumes, but were still to bind their faces when stirring or removing the arsenic (Henckel, 1757, p. 341–342). There must have been many such primitive precautions of doubtful efficacy that transferred into the laboratory, although dangerous situations remained.

Those examples show how industrial furnaces of various kinds were used rather frequently by natural philosophers, and in turn, work by industrialists supplied information back to them. The latter was often in the form of advice as to how to obtain replicable temperature conditions of operation, or information about the characteristics of minerals, ores, and earths.

Laboratory Furnaces

Small furnaces were constructed for many purposes, so it is not surprising that a variety of them moved into the laboratory. Until the increasingly accurate and successful analysis "in the wet way" was instituted toward the end of the eighteenth century, the "dry" method of fusion had to suffice to give most information about minerals and mixtures. Historian of science Jon Eklund has said: "If fire was the central instrument of the chemist, then the operational heart of the chemical laboratory was the furnace (or furnaces) and the chimney" (Eklund, 1975, p. 6). Laboratory furnaces were frequently illustrated in books, along with detailed directions about their materials, fuels, and regulation.

Portrayals of alchemists' laboratories nearly always included one or more furnaces. By the time of Becher, who was associated with the transition from alchemy to chemistry, the function of the laboratory was changing but furnaces were still a necessity (P.H. Smith, 1994). They could be constructed so that minor alterations allowed them to serve for more than one function. Valmont de Bomare (1731–1807) discussed the furnace alterations needed to extract zinc from its ores, in an instance where it is difficult to categorize the use as either industrial or laboratory scale (Valmont de Bomare, 1762, v. 2, p. 67–68). The arrangements for that and other reactions worked with the variables of air exposure and amount of blast.

---

[10]Jacquemart used the classification given by the French geologist and director of the pottery at Sèvres, Alexander Brongniart (1770–1847). Within the class of soft paste he listed unglazed baked clay, lustrous pottery with a thin glaze, glazed pottery, and enameled pottery. In his second class of hard paste opaque pottery were fine earthenware or pipe clay and stoneware. Within hard paste translucent pottery were soft, or English pottery, and French pottery. Each was produced in a different way.

[11]Found in virtually any chemistry book of the time, examples abound of operations undertaken with strong acids and bases and combustible or explosive mixtures.

Besides his quantitative work with alkalis, Joseph Black was recognized for his expertise with furnaces, some of the first to be designed specifically for the needs of the emerging science of chemistry. He designed a laboratory reverberatory furnace, and was consulted about an air furnace (Anderson, 1978, p. 24). His best-known furnace, modeled on Becher's, will be discussed with it in the section on portable furnaces.

Gellert stated that chemical apparatus in general was the equipment that allowed chemical agents such as fire, air, water, earth, and menstrua (solvents) to work. The furnace was the agent for fire (Gellert, 1776, p. 143). His furnaces were varied to provide different degrees and durations of heat. The basic furnace had a space for ashes, then a grate upon which the fuel and the samples were placed. There could be one or more chambers above those, depending on use. He discussed blast and draft (wind) furnaces, and gave precise directions for the kind of assay furnace that the analytical chemist and assayist Johann Andreas Cramer (1710–1777) used. Muffles could be inserted to protect the sample. An "athanor" or tower furnace could be used on its own or in conjunction with another furnace as shown by Lavoisier. (See Fig. 3.3.) There were many arrangements possible with the athanor, but usually it consisted of a tall top cylinder that was filled with charcoal so that as the fuel at the bottom was slowly consumed, it was replenished by more falling from above. Heat, particularly of modest degree, could be maintained for a very long time. Athanors were often adapted for distillation or for use with water or sand baths. Since furnaces were relatively expensive to build and maintain, the configuration could be altered to fit the purpose at hand. Gellert discussed most of the other sorts of furnaces as well.

Not surprisingly, Lavoisier's *Elements of chemistry* (1790) includes a chapter titled "Of the instruments necessary for operating upon bodies in very high temperatures," of which the second section concerns furnaces. He discussed a general sort of furnace that provided moderate heat for melting substances with relatively low melting points such as lead and tin, for evaporation, and for heating sand baths. It was fired by charcoal, and the heat produced could be regulated by a stream of air. The reverberatory furnace was more useful in his view, but those sold by potters in Paris had to be modified by increasing the size of the airholes so that the heat supply could be increased (Lavoisier, [1790] 1965).

When Robert Jameson was in Ireland in 1797, he visited Richard Kirwan. Apparently one of his two small laboratory rooms was dominated by a "pretty large brick assay furnace" and the other by what was called a sand range, a large area for heating by means of sand baths. The bellows for fusion were outside in a separate small building (Sweet, 1967, p. 110). Despite his Neptunist stance and objections to Hutton's theory, Kirwan clearly believed in the efficacy of fire, which he put to good use in his analytical work (Kirwan, 1784, 1794a).

Schmeisser described his wind furnace for fusion operations: 12 inches high, 6 inches wide, and made of strong iron plates. It was lined with two layers of lute. The grate was made of the same material as the crucibles used within it. Charcoal served as fuel.

Schmeisser could adjust the temperature at which it operated by means of bellows; but he also noted that he used a regulator for air admission that originated in Edinburgh (Schmeisser, 1795, v. 2, p. 373). He could insert a muffle if desired. In his *A system of mineralogy formed chiefly on the plan of Cronstedt,* Schmeisser included plates that illustrated all the furnace parts.

Cadet began his discussion of furnaces in his *Dictionnaire de chimie* with the following definition:

FOURNEAUX. Instrumens destinés à appliquer le feu aux substances que l'on veut traiter par cet agent. Les fourneaux sont de terre cuite, de brique ou de fonte: ils ont différentes formes, suivant les usages auxquels on les destine. (Cadet, 1803, v. 2, p. 544–545)[12]

He listed the types as reverberatory, fusion, assaying or cupelling, lamp furnaces, and the athanor. A few years later, Henry illustrated and discussed several wind furnaces as well as blast, reverberatory, and cupelling furnaces. He particularly liked a type of fixed furnace that could be used as either a wind furnace or for distillation with a sand bath (Henry, 1814, v. 1). Shortly after Sir James Hall died in 1832, his son, Captain Basil Hall, who had accompanied his father on a number of field excursions, presented a plan to the Geological Society of London for a furnace regulator that the elder Hall had devised:

The principle of the machine is such, that when any change of temperature takes place in that part of the furnace in which the material under experiment is placed [the muffle], a corresponding change is made in the current of air which maintains the heat. (Hall, 1834, p. 478)

The description of the device was complex, but it exploited the different coefficients of expansion of two metals, so that a spring would either coil or uncoil depending on the temperature. These were attached to parts that would increase or decrease the amount of air going through the furnace, depending on the heat emanating from the muffle in the furnace. There were great changes in chemistry, matter theory, and geological ideas, over the second half of the eighteenth century, but the fixed furnace and its operation changed only slightly. Even so, scientists continued to write detailed chapters about furnace construction and operation.

Portable Furnaces

There was a great advantage in having furnaces that could be moved from place to place to try ores or identify minerals. If available, local stones were just piled over a depression in the ground to make a crude furnace. The simple assemblage of several bricks with an iron stand could also serve. In 1689, Becher published his *Tripus hermeticus fatidicus,* which included the *Laboratorium portatile.* Becher's life included much travel, which entailed setting up his laboratory in various places. Made of iron,

---

[12]"Furnaces: Implements designed to apply fire to substances that one can treat with this agent. Furnaces are of terra cotta, bricks, or cast iron: they are different shapes, according to the uses for which they are destined" (my translation).

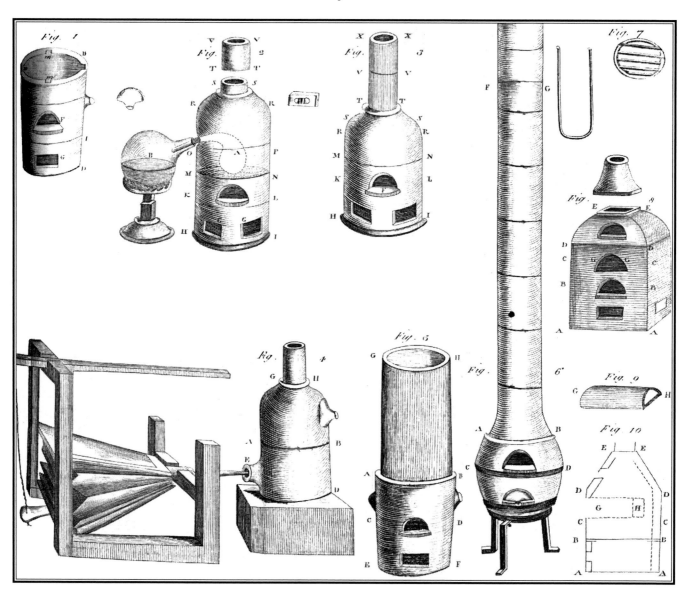

Figure 3.3. This Plate XIII from Lavoisier (1790) shows an athanor and some of its fittings.

the eight pieces of the furnace could be assembled in six different ways to be used for fusion, assaying, calcination, distillation of several kinds, and heating a water bath (Ferchl and Süssenguth, 1939, p. 164; Anderson, 2000, p. 17). His furnace had significance beyond its actual operations as denoted by the alchemical inscriptions on drawings of it and its parts (P.H. Smith, 1994). However, its use was not widespread.

A portable furnace supposedly modeled on Becher's, which gained considerable renown, was that of the Edinburgh chemist and friend of Hutton's, Joseph Black. His model was produced commercially, and a number of chemists spoke about using it. Also of iron, the design was popular for 150 yr (Anderson, 1978, p. 24). It was a relatively small furnace that could be regulated by means of a number of holes in a panel in the side that could be opened or closed for air flow. Henry said about Black's furnace:

It has been recently so far modified in London, as to deserve the name under which it is sold, by Mr. Accum, of the universal furnace, for, with it, every chemical operation (in the small way) may be performed. (Henry, 1814, v. 1, p. 344)

As mentioned, Black was recognized as an expert in the use of a wide variety of furnaces.

An ingenious furnace identified as "Mr. Aikin's portable blast furnace" was made of portions of three black lead (graphite) crucibles. One served as a base through which a blast of air could be introduced, and the second was the fire chamber. If desired, a third crucible could be inverted to cover the second. No lute was used so that the assemblage could be moved easily (Henry, 1814, v. 1, p. 335). Graphite was found to be very refractory, and this property was enhanced if a little clay was added when the crucibles were made. Such a furnace could withstand great heat, though most furnaces could begin to fuse if the heat became too great. By the end of the eighteenth century, furnaces used in laboratories were reliable, predictable (although the reactions in them might not be), relatively convenient, and available. They continued to be indispensable in mineralogical and geological work.

Lamp Furnaces

It was not a big leap to go from the use of candles and oil lamps for illumination to their use for heating samples. I also think immediately of the ubiquitous alcohol, or "spirit" lamp of my own early laboratory days, a simple reservoir with a wick coming through a cap. Experiments with both oil and alcohol as fuels, particularly alcohol, were considerably enlivened by lack of understanding of combustion, and want of the safety match. It is hardly possible to assign a time of first use of such a heating source in the laboratory. Lamps using both fuels were well known and being improved by the end of the seventeenth century (Ferchl and Süssenguth, 1939, p. 164).

In the appendices to both the 1770 and 1788 English translations of Axel Cronstedt's *An Essay towards a system of mineralogy*, Gustav von Engeström discussed the lamp furnace as part of the portable laboratory. Like the rest of that laboratory, the lamp furnace traveled in a fitted case. When in use, it was clamped to a rod so the operator could have his hands free. It was formed of a metal cylinder which served as a reservoir, with three holes to accommodate different sizes of wicks. The fuel was spirit of wine. Von Engeström included a number of miniature glass vessels in the traveling laboratory, and observed that with small samples the lamp furnace was sufficient for any solutions, digestions, or distillations that might be needed, and at much less expense than in other laboratories. The small sample size was also an advantage if the substance were rare or expensive (Cronstedt, 1788, v. 2, p. 992). The small furnace could also be used in conjunction with a blowpipe, which was fastened to the rod, with a piece of charcoal affixed to hold the sample.

Constant work was done to improve the small oil lamp furnaces. The lamp of the inventive Swiss, François-Pierre Ami Argand (1750–1803), with two concentric wicks that allowed air to flow through them, was more efficient than earlier models. The oil was held in a reservoir to the side. It was first used for lighting ~1780, but its utility for the laboratory was obvious. Several people in France and England claimed to have known the operating principles of this lamp before it was described by Argand, but this is not surprising, and variations for the laboratory were soon made. It was mentioned in a 1788 letter to Kirwan from the French chemist Louis Bernard Guyton de Morveau (1737–1816), who spoke approvingly of how quickly liquids could be boiled with few fumes (Grison et al., 1994, p. 184). The ideal was an efficient lamp that gave a concentrated heat and little smoke or soot (Anderson, 1978, p. 101–102).

A further development of the "Argand lamp," as it came to be known, was the lamp furnace. The body of the lamp with the

wick in it was suspended by a spring inside an enclosing cylindrical body. Placed on a stand at the bottom, it was fitted at the top with a sort of collar on which the laboratory vessels could be placed. The spring allowed the operator to raise and lower the lamp depending on the amount of heat desired. Lavoisier used the lamp furnace with oil or alcohol as part of his combustion studies with those substances. In the absence of matches, the ignition of volatile substances was an interesting problem in itself. The use of tinder, a piece of white phosphorus, or a red hot iron as used for solids and the heavier oils was too dangerous, as noted by Lavoisier ([1790] 1965), p. 420–421).[13]

Cadet gave a succinct description of the lamp furnace and stated that it was very useful whenever a small amount of heat was needed in the laboratory. It could also be used to heat water or sand baths, and it was most useful for the process called "digestion" (Cadet, 1803). It was by far the safest and most reliable limited heat source in the laboratory until the Bunsen burner was devised later in the nineteenth century.

Sand and Water Baths

Sand and water baths (Fig. 3.4) were used in conjunction with many kinds of heating devices in order to supply modest amounts of heat, often for a sustained time period. The concept had already reached industrial proportions in the sixteenth century when Biringuccio described the use of both types of bath in distillations. He related how useful distillation could be in metalworking. Low heat was also used to dry various materials including those used in women's cosmetics. There are illustrations of both sand and water baths built into furnaces. One water bath that accommodated a number of flasks consisted of a tank of water built on top of a furnace (Biringuccio, [1540] 1990, p. 127).

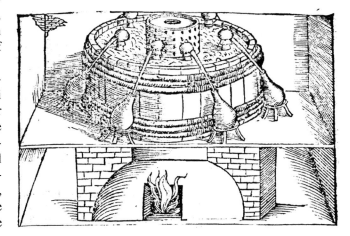

Figure 3.4. Biringuccio's ([1540] 1990) example of a multiple water bath for distillation.

---

[13]The description of the phosphorus used by Lavoisier makes it clear it was the white variety (Poirier, 1998, p. 64). Red phosphorus was most probably not made until 1848 (Anon., 1990, *A treatise on chemistry and chemical analysis*, v. 1).

Both of these mild heat sources were recognized as being safer than open flames, although some of the combinations that were heated would give present-day chemists pause. The sand bath was also used to ameliorate the effects of strong heat on glass vessels. As mentioned above, the baths were convenient to use in the laboratory in conjunction with a lamp furnace. An interesting object in the Playfair collection is a box of wood, open on one side, with a shelf partway up it. The sides of the bottom portion are scorched. It is conjectured that this was a heating bath (Anderson, 1978, p. 107). There must have been many schemes and pieces of apparatus to facilitate moderate and/or prolonged heating.

Burning Lenses and Mirrors

Burning glasses and mirrors (Fig. 3.5) were part of the arsenal of experimentalists, having been used in the seventeenth century for their high temperatures and lack of contaminants such as a furnace might introduce (Smeaton, 1987, p. 265). Polished metal was sometimes used for the mirrors. It was necessary that supports for the sample, and the sample containers, would be infusible in the heat developed. Because of the skilled work required to produce the correct curvature of lenses or mirrors, as well as the difficulty of finding flawless glass for the lenses, some of which were three or four feet in diameter, they were very expensive.

One of the first lenses used extensively for investigations that were, or became, scientific, was that of Ehrenfried Walther Tschirnhaus (1651–1708), the Saxon scholar and mathematician. Tschirnhaus was employed by Augustus the Strong, Elector of

Saxony, to survey minerals in the kingdom, and to investigate the feasibility of establishing a glass manufactory. Several glassworks were begun, and it was probably there that he got the glass for his lenses (Polak, 1975, p. 89). Hofmann has conjectured that Tschirnhaus produced his famous parabolic mirrors, made of polished copper, as well as lenses, as part of his mathematical investigations (Hofmann, 1976, p. 480). Georges-Louis Leclerc de Buffon called the use of burning mirrors, which focused heat from the sun, the third way of applying heat—the first two involving furnaces that delivered either concentrated or more diffuse heat. The burning mirror could deliver the strongest and most concentrated heat of all (Buffon, 1774, p. 61–62). Samuel Frederick Gray (1766–1828), the English botanist and chemist, dated one of Tschirnhaus's burning mirrors to 1687 (Gray, 1828, p. 150). The focusing lens now in the Deutsches Museum in Munich is dated at ~1700. These instruments had a part in producing the very high temperatures required to produce true porcelain. Tschirnhaus has been credited with initiating the investigations that enabled the production of porcelain at Meissen, accomplished a year after his death (Polak, 1975, p. 89; Atterbury, 1982, p. 82).

Darcet noted that Henckel had made use of a *miroir ardent* (Fig. 3.6), and reported the results in his *Flora Saturnizans* of 1722 (Darcet, 1766, v. 1, p. 87). Lavoisier cited the use of several of these instruments, including one of Tschirnhaus's. He pointed out the difficulties of using them, such as a cloudy day, despite the high temperatures that could be developed on a

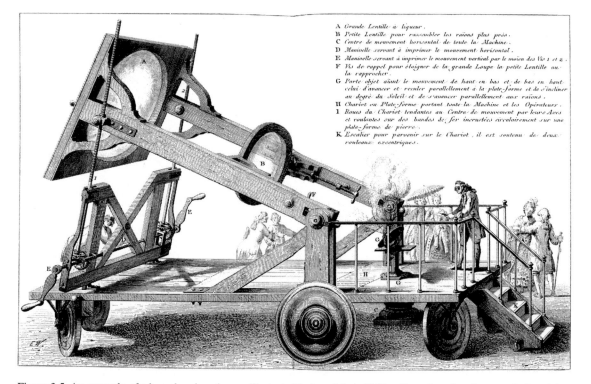

Figure 3.5. An example of a large burning glass as illustrated in Lavoisier's 1864 collected works, *Oeuvres de Lavoisier*, v. 3: Paris, Imprimerie imperiale, Plate IX. Courtesy of Othmer Library.

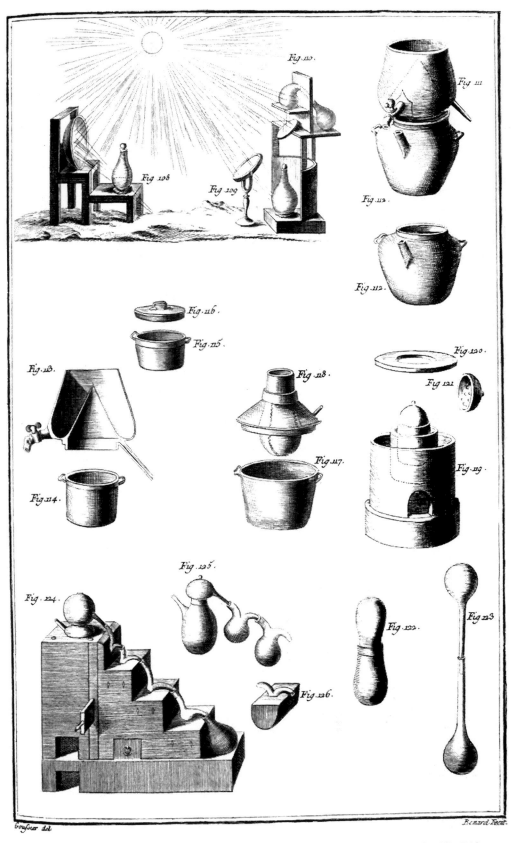

Figure 3.6. Use of burning mirrors and other equipment used in heating in the *Encyclopédie Diderot et d'Alembert* (1763): Paris, Planche 2.

day when the sun shone. He felt that the concave mirrors produced greater heat than the lenses, but noted the difficulty of suspending a sample near the focal point of the mirror where the reflected rays converged (Lavoisier, [1790] 1965, p. 474). Lavoisier used what must have been a smaller burning glass to ignite phosphorus to begin an enclosed combustion experiment (Lavoisier, [1790] 1965, p. 422), and used the lenses of Tschirnhaus and Trudaine d'Montigny at l'Académie Royale des Sciences at Paris in studies of the combustion of diamond in 1772 (Grison et al., 1994, p. 44 fn). Lavoisier and other scientists also used the lens to determine melting or combustion points of a number of metals and minerals (Poirier, 1998, p. 59). However, Lavoisier felt that although burning glasses and mirrors, especially the latter, could produce temperatures greater than porcelain furnaces, their expense and difficulty of use made it more desirable to use oxygen as a combustion agent in order to reach higher temperatures (Lavoisier, [1790] 1965, p. 474–475).

In his letters to Guyton de Morveau, Kirwan referred to the use of a *verre ardent*. Kirwan's instrument, made by John Parker of London, was "only three feet in diameter." Kirwan felt it was superior to that of Tschirnhaus for the production of heat. He claimed it melted a globule of platinum in 30 seconds, and it also fused mica and asbestos. Molybdenum resisted the heat, but wood could be easily carbonized (Grison et al., 1994, p. 43–45). Sometimes a smaller lens was used to further focus the light from a large lens. The mirrors and lenses were often used by scientists to attempt fusion of minerals and mineral mixtures. Considering the difficulty of measuring the heat developed by a burning glass or mirror, it is not surprising that there was disagreement about it. It was said that platinum could be melted very quickly, but it often contained other metals that lowered the melting temperature below the ~1770 °C of pure platinum (Smeaton, 1987, p. 270). Despite reservations about the utility of the mirrors and lenses, Gray said:

The formation of instruments for burning at short distances is, indeed, a subject of the first importance to science, and there is no doubt but they will, some time or other, be employed as the most powerful agents in chemistry and the arts. (Gray, 1828, p. 151)

CONTAINERS

Crucibles

All of the industrial and laboratory processes that required unusual high temperatures also required containers that did not melt, slump, or react with the substances contained in them. Not surprisingly, most containers were made of the same refractory materials used for the furnaces themselves, namely, iron, graphite, or clay mixtures. Kirwan mentioned that when Gerhard exposed siliceous earth to a violent heat in a crucible of highly refractive chalk, the mass vitrified where it touched the crucible (Kirwan, 1784, p. 13). The early metallurgists needed containers for their products, both molten metals and alloys such as brass and bronze,

the latter of which have been known since antiquity. Cupels, used to purify precious metals, were containers made of ash of various kinds, from bones or plants.[14] Glassmakers also required containers that would withstand heat and not fuse with alkaline additions. Biringuccio included instructions for making crucibles of resistant clay, possibly mixed with talc, some ram's horn ashes, and, surprisingly, an eighth part of ground and sifted iron scale. A later footnote rightly observed that this last would make the clay far less refractory (Biringuccio, [1540] 1990, p. 391).

The property of being unreactive as well as infusible became more important as the number of reactions, reagents, and reactants increased, and processes became more quantitative. Iron crucibles were not often used with very high temperature fusions or with many reagents because of their reactivity. In Pott's extensive investigation of the properties of various mineral mixtures, he commented on the damage to clay crucibles that resulted from some of them. Sometimes he could use small quantities of materials such as chalk or fusible spar with his mixtures so that perforation and leakage of his crucibles were prevented (Pott, [1746] 1753, p. 39–40). He also noted that in fusion the crucibles could supply material that altered the behavior of the mixtures. Lavoisier emphasized the importance of good crucibles for fusion, and stated that France could supply good and very refractory clay (Lavoisier, [1790] 1965, p. 461).

Decorative porcelain came into use in the eighteenth century and eventually similar high-temperature mixtures were used in laboratory equipment. Of course it would fuse if exposed to porcelain furnace heat. By 1772, Josiah Wedgwood advertised ceramic crucibles and other equipment for scientists, followed by some articles made of stoneware (Schofield, 1959, p. 182–183). Graphite continued in use for crucibles with admixtures used in furnace construction. Hall used these black lead crucibles for some of his experiments in melting whinstone and lava.

A new refractory and unreactive substance that became available by the late eighteenth century was platinum. Discovered and characterized in the middle of the century, it was difficult to work with until Wollaston, who also worked in chemistry and optics, added to earlier knowledge and discovered a process, not prohibitively expensive, for making it malleable (Ihde, [1964] 1984, p. 93). Platinum could be purchased in sheets and formed into the desired shape, including that of a crucible, with a lid made from another sheet. Kirwan mentioned the use of a platinum crucible in a letter written to Guyton de Morveau in 1785. He had used the crucible to evaporate phosphoric acid to a pasty consistency (Grison et al., 1994, p. 114). Sir James Hall used platinum crucibles because they didn't react with calcium carbonate as did clay.

---

[14]Cupels were used to assay an ore in a process to separate gold and silver from other metals. The ore was sometimes mixed with a flux, or just with the purest lead possible, and it was heated strongly in an oxidizing atmosphere. The lead and other metallic oxides would be absorbed by the ash cupel, leaving a button of precious metal. If gold was present, it could be separated by dissolving the silver in nitric acid. The process could reach an industrial level with use of much larger containers or trays made of ash (Biringuccio, [1540] 1990, p. 136–141, 161–169).

In the first decades of the nineteenth century, Henry summarized crucible materials and their use. "Hessian" crucibles (of clay) were best for sustaining great heat without melting, but they cracked if suddenly heated or cooled. Wedgwood porcelain crucibles were very pure, but also liable to cracking when the temperature changed abruptly, so Henry protected them by surrounding them with sand in a larger container. Black lead crucibles didn't react to rapid temperature changes, but could be destroyed if heated with some substances such as niter. They were consumed if heated in a current of air. Silver or platinum crucibles could be used if placed on a stand above the furnace grate (Henry, 1814, v. 1, p. 30).

Other Containers

Various boxes, dishes, spoons, and other containers were made of these refractory materials. They might be used to contain or support substances during fusion or evaporation. Glass containers covered with lute might also be employed. Scientists no longer had to fabricate all their own equipment by the end of the eighteenth century, as instrument suppliers began to provide it.

THE BLOWPIPE

The use of the blowpipe was common by the second half of the eighteenth century, and virtually all of the many books about mineralogy contained a description of the authors' favorite setups, a discussion of techniques, and reports on the results of its application. It was used more to characterize than to analyze mineral substances and their behavior, although similar reactions, colors, or products could indicate identity with a known substance. There were a few references to more quantitative work. An air stream from bellows had been the standard method of increasing furnace temperatures. The technique of blowing through a pipe to direct a flame and increase temperature evolved over hundreds of years in order to help artisans fuse smaller quantities of metals, enamels, and glass, but by general agreement it was first widely employed to investigate mineral substances in an orderly way before the middle of the eighteenth century by Swedish metallurgists (Szabadváry, 1966, p. 52–53).[15] About the blowpipe Cadet claimed:

Cet instrument, très-utile dans les essais des substances minérales, fut employé en 1738, pour la première fois, par André Swab, perfectionné ensuite par plusieurs minéralogistes. (Cadet, 1803, v. 2, p. 111–112)[16]

Cadet mentioned the section on the blowpipe published by von Engeström in English, but felt that Torbern Bergman had given the most rigorous account of its use.

While the blowpipe was sometimes just a single, tapered tube, Bergman described his blowpipe as being three parts that fitted together without the aid of threading. These were the tube which the practitioner blew into, a reservoir to trap moisture from the breath, and a delivery tube. Some practitioners felt the reservoir was not necessary. Over the years, there were many designs and several different materials were used for the mouthpiece, which was sometimes fitted onto the first tube so that lips would not get fatigued. The same was true for the third part of the tube, which required a narrow aperture. Bergman remarked that all three parts could be made of silver, but to spare expense the first could be made of iron and the others of copper (Bergman, 1784a, p. lxxxiij). Cadet said that if silver were used, some platinum could be added for increased hardness. Glass might also be employed for the tube, which could have a spherical portion for the reservoir (Cadet, v. 2, 1803, p. 112). Griffin quoted Berzelius as saying:

[B]lowpipes are best made of silver, or tinned iron plate, the beaks only being of brass. If the instrument be wholly of brass, it in time acquires the taste and odour of verdigris,—an inconvenience not entirely removed by making the mouth-piece of ivory. (Griffin, 1827, p. 12)

Griffin described the blowpipes favored by a number of well-known mineralogists. All were made of some combination of the materials mentioned, and all users were concerned with durability and continuous use. Besides breakage, the two points of major concern were elimination of water droplets from the breath that could condense and run out onto the sample, and the need to clear the very small aperture at the end of the tube when it clogged.

The first of those problems was eliminated when air was supplied via a bellows rather than by mouth. Supplying the airstream by mouth was a skill that required practice. The ideal was a constant stream of air, which necessitated breathing in through the nose, storing air in puffed cheeks, and letting it out through the mouth into the blowpipe without pause. Von Engeström noted that beginners blew too hard, resulting in frequent breaths, which could draw the flame into the blowpipe, thus allowing the sample to cool. He said:

The whole art consists in this, that while the air is inspired through the nostrils, that which is contained in the mouth, be forced out through the tube, by the muscular compression of the cheeks; so that the action of the nose, lungs, and mouth, resemble the action of bellows with double partitions. (Cronstedt, 1788, v. 2, p. 933)

Many practitioners acquired these skills. But not surprisingly, bellows were a welcome alternative. When intended to deliver a stream of air, the bellows might be controlled by pressure from an arm or foot. Arrangements for gas delivery ranged from those simple setups to Lavoisier's elaborate gazometer, illustrated and described in his *Traité elémentaire de chimie* (Lavoisier, 1789, v. 2). Gas delivery is also illustrated on Plate VIII in the

---

[15]Szabadváry stated that T. Bergman credited Anton Swab (1703–1768) with being the first to use the blowpipe to characterize antimony, but noted that Sven Rinman (1720–1792) had previously used it to examine tin ores. Cadet (1803, v. 2, p. 112) also said André Swab had used it in that way the first time.

[16]"This instrument, very useful in trials of mineral substances, was employed in 1738 for the first time, by André Swab, afterward perfected by many mineralogists" (my translation).

English translation of 1790 by Robert Kerr, *Elements of chemistry*, and described on pages 308–319. Both blowpipe and heat source might be mounted on supports so that the hands could remain free to manipulate the sample.

The flame source most often used was a candle. When blowing the air or gas stream through the flame a double cone resulted, as we see now in a well-adjusted burner which was not available until the 1850s. The hottest part of the flame is just at the end of the inner, blue flame. As mentioned earlier, there were also various lamps available that might burn alcohol or oil or other lipids. The sample was placed on a small hollow in charcoal, or sometimes held in a small platinum spoon.

Whether a mineral could be fused or not often served as a criterion in mineral classification. The effect of lowered fusion point by additions was common knowledge, having been employed in metallurgy, glass and porcelain making, and in mineral and rock testing. For identification in mineralogy, after trying with the mineral alone (per se), various fluxes might be employed, followed by solution and further reactions. The fluxes used by Bergman—microcosmic salt, soda, and borax—continued to be used.[17] The blowpipe worked best with very small samples, placed in a hollow in a piece of charcoal. Von Engeström noted that soda was not too efficient in this case, because it melted under the flame and was absorbed into the charcoal, which necessitated immediate addition of the sample (Cronstedt, 1788, v. 2, p. 943). Less finesse was required for borax and microcosmic salt. Von Engeström said they:

are very well adapted to these experiments, because they may by the flame be brought to a clear uncoloured and transparent glass; and as they have no attraction to the charcoal, they keep themselves always upon it in a round globular form. (Cronstedt, 1788, v. 2, p. 945)

The best proportion was to keep the sample to about a third the size of the amount of flux. Von Engeström gave information about making or buying each of the fluxes. Nearly thirty years later, Robert Jameson suggested making what he called nitrous borax "by dissolving common borax in hot water, neutralizing its excess of alkali by nitrous acid, then evaporating the whole to dryness, and, lastly, hastily melting it in a platina crucible" (Jameson, 1817, p. 290).[18] This somewhat problematical compound was used to produce a flux used with ore samples. He wanted the reaction to keep the metals in a high state of oxidation so that the characteristic colors of their compounds would be emphasized.

It did not take long after the discovery of oxygen, and recognition of how it enhanced combustion, to enlist its aid with the blowpipe. Oxygen was used by itself for a relatively short time, and the advantages of mixing it with hydrogen were soon seen. Until methods of controlling unwanted combustion

were in place, accounts of trials with one or both were often enlivened by accounts of explosions. Lavoisier recorded the comparative ease of fusion of many mineral substances with the oxygen blowpipe (Lavoisier, [1790] 1965, p. 477), and his hope of raising the temperature in fusion investigations was realized. English mineralogist Edward Daniel Clarke (1769–1822) wrote a short history of the use of hydrogen and oxygen together in "The gas blowpipe," which was included in an appendix in his extensive biography by William Otter (1825, v. 2, p. 413–439). Hydrogen is highly inflammable in the presence of oxygen, and reacts to combustion in air with a characteristic "pop."[19] The mixture of the two undiluted gases proved devastating to a good bit of equipment. If the stream of gases, mixed either at the blowpipe or in a tube leading to it, was allowed to diminish, the flame was drawn back into the reservoirs with disastrous results. Clarke stated:

Dr. Thomas Thomson, now Professor of Chemistry at *Glasgow*, made experiments with the mixed gases at *Edinburgh*, seventeen years ago; but was induced to abandon the undertaking, owing to the accidents which happened to his apparatus. (*in* Otter, 1825, v. 2, p. 416)

Clarke remarked that delivery of the gases from different reservoirs through different tubes was "as old as the time of Lavoisier" (*in* Otter, 1825, v. 2, p. 416), but that Robert Hare (1781–1858) of Philadelphia also claimed it. However, it has been pointed out that Lavoisier did not actually succeed in burning hydrogen in oxygen, but that Hare's apparatus allowed it to be done with some safety (Oldroyd, [1972] 1998, p. 226).

Detailed descriptions were given of proper blowpipe procedures for each kind of mineral, earth, or ore. Some required drawing out the fused mass while still warm to ascertain the color, particularly for metallic ores and/or compounds. A magnet was used if iron was thought to be present. The principle of oxidizing and reducing portions of the flame was understood in simple terms, and was better described as the role of oxygen was clarified. Haüy mentioned that a particular color of the flame could add information to observations (Haüy, 1801, v. 1). Other procedures depended upon solution of the fused mass in water or acid, with further reactions for identification. Jameson listed 22 results from blowpipe use—some with the mineral alone and some with additions—that could give information about the material (Jameson, 1817, p. 292–297). Despite the time required to learn proper technique, both equipment and reagents were relatively inexpensive, and these same methods, with additions, continued well into the twentieth century. With other heat sources (the burner), and more knowledge of materials, the elaborate and risky apparatus to deliver oxygen and hydrogen was no longer used. That stalwart of our laboratories, the Bunsen burner, was invented in 1853 by the versatile chemist Robert Wilhelm Eberhard Bunsen (1811–1899), who was searching for a colorless flame that would allow him to

---

[17]In modern terminology, microcosmic salt is sodium ammonium phosphate, $NaNH_4HPO_4 \cdot 4H_2O$; soda is sodium carbonate, Na_2CO_3; and borax is sodium tetraborate, $Na_2B_4O_7 \cdot 10H_2O$.

[18]The nitrous acid of that time is the nitric acid of today.

[19]If a large amount of hydrogen is present, the pop becomes a bang.

see clearly the colors of salts in mineral waters when they were flamed (Ihde, [1964] 1984, p. 233).[20] This was only possible after coal gas was available.

TEMPERATURE INDICATING DEVICES

Thermometers

In the last part of the eighteenth century, there were several ways of quantifying temperature, including the pyrometer, dilation thermometers, and the liquid-in-glass thermometer. We will return to the pyrometer later. For our present purposes, thermometers are interesting in two different ways: the first is simply how and of what materials they were made, and the parameters of those materials; the second is the kind of scale used, its divisions, and origins. At the time in question thermometers were useful for a number of reasons. One was simply to ascertain the temperature of a reaction or process. Temperatures could also be related to other properties, such as the expansion of gases. The known temperature of an accessible process could be used to infer that of another that couldn't be measured directly. And finally, those temperatures allowed comparison with or analogies to processes in nature.

After their history, which began with the thermoscope of Galileo and his contemporaries, thermometers came into their own in the eighteenth century. Eklund (1975, p. 15) has commented that chemists were slow to adopt their use in the mid-eighteenth century, even after the advent of more reliable instruments credited to Daniel Gabriel Fahrenheit (1686–1736), who worked in Amsterdam and was of course familiar with the work of his predecessors (Middleton, 1966, p. 66). It was necessary to find: a liquid that would expand uniformly over a range of temperatures; tubes of, it was hoped, uniform bore that did not expand or contract very much with changes in temperature; and a means of calibrating an appropriate scale. Alcohol or spirits of wine, a common thermometric liquid, has a constant coefficient of expansion over only a narrow range, and then expands far more per degree than does mercury. The usefulness of both alcohol and water for higher temperature measurements is limited by their boiling points. Fahrenheit adopted mercury, which, while harder to work with, had a more uniform expansion over a much broader temperature range (Turner, 1998a, p. 111). The origin of Fahrenheit's scale has often been discussed.[21] It evolved over some years as he worked on it, but finally, he used three fixed points: the first was that of a mixture of ice, water, and sea salt; the second was that of a mixture of ice and water; and the third was the point attained when the thermometer was placed in the mouth or under the armpit of a healthy man (Middleton, 1966, p. 75). On this scale the temperature of boiling water was the familiar 212°. For accuracy it would require that the bore of the tube containing the mercury was uniform, which of course was very difficult, if not impossible, to achieve. Most English thermometers used this scale in the eighteenth century, and it is still in use today in the United States.

The scale most used in France and the rest of Europe was that of René-Antoine Ferchault de Réaumur (1683–1757), well known for his work with steel and Réaumur's porcelain.[22] Réaumur's scale had one fixed point, the freezing point of water. He used "a degree defined volumetrically (instead of linearly) in terms of some fraction of the total volume of liquid in the thermometer bulb" (Gough, 1975, p. 330). Middleton remarked on how complex and verbose Réaumur's methods and report were, and said, "I am inclined to doubt whether Réaumur was really clear in his own mind about what he was doing. In any event, the public came to believe that 80 °R is the boiling point of water" (Middleton, 1966, p. 83).

For the thermometers to agree with each other, his use of alcohol necessitated standardization of its purity, as different purities have different dilation characteristics. The 80 degrees on Réaumur's scale were based on the fact that the alcohol he used would expand 80 degrees on his scale between zero and the temperature where the alcohol would boil in an open tube. His procedures were unbelievably complicated (Middleton, 1966, p. 79–82). It is not surprising that his instructions were misunderstood, and others made Réaumur thermometers with an 80 degree scale between the freezing and boiling points of water (Middleton, 1966, p. 116), disregarding his use of volumes. Thus, if "degrees Réaumur" are given in a report of an investigation, it is difficult to know what procedure was followed in making the thermometer or what the stated temperature actually signified.

Swedish astronomer Anders Celsius (1701–1744) followed others who had used the difference between the freezing and boiling points of water to determine the thermometer scale. His scale first used 0 for the boiling point, and 100 for freezing, but it was shortly inverted to the familiar 0–100. He preferred melting snow to indicate the freezing point, and noted, as Fahrenheit had, that the boiling point of water depended on barometric pressure. There were many other scientists and instrument makers who advanced the cause of accurate thermometers. A process to produce a tube of uniform bore was not developed until the mid-twentieth century (Burnett, 1998, p. 617). Prior to that time, each thermometer had to be individually calibrated. Jean André Deluc (1727–1817), the peripatetic Swiss naturalist, is credited with rigorous application of an adapted method to determine the bore of the glass capillary tube by insertion of a specific small amount of mercury and measuring its length at different positions along the tube (Wolf, [1952] 1961, v. 1, p. 295). The tube was sealed virtually without air.

---

[20]To us it will always be the Bunsen burner, but Ihde noted that after hearing what Bunsen wanted, the university mechanic, Perter Desaga, actually produced it (Ihde, [1964] 1984, p. 234).

[21]Just a few of the relevant references are Daumas ([1953] 1972); Burnett (1998); Wolf ([1952] 1961); Middleton (1966); Griffiths (1918); and Turner (1987).

[22]About the Réaumur scale, Middleton has written:

In contrast to the very meager documentation vouchsafed us by the instrument maker Fahrenheit, the well-born *savant* René-Antoine Ferchault de Réaumur has told us everything we could possibly wish to know about his thermometer scale in two memoirs of incredible verbosity. (Middleton, 1966, p. 79)

Thermometers to measure the temperature of Earth at depth were devised as early as the 1760s. There were many practical difficulties with these instruments, among them large stem corrections if an attempt was made to read the thermometer while immersed in the ground, doubts whether the true temperature was being shown due to soil disturbance, and lack of enough test depth to show real differences. It was recognized that the work done by an auger to drill a hole in which to insert the thermometer produced heat, thus it would be necessary to wait until it cooled for a reading. Middleton (1966, p. 139–141) has provided an interesting summary of these efforts through the nineteenth century.[23]

By the end of the eighteenth century, the use and value of thermometers were well understood. There had been many attempts to craft an instrument that was reliable under varying conditions. Liquids other than alcohol and mercury had been used, including oils and water. Calibration marks might be placed on the outside of the glass tube or behind it. There were many scales, with attendant need to correlate them (see Table 3.1 for examples).

Thermometers with several scales, usually Celsius, Fahrenheit, and/or Réaumur, were common. A backing board for a thermometer, made ~1758, had four scales: Newton, –6 to 33; De Lisle, 192 to 0; Fahrenheit, –4 to 212; and Réaumur, –16 to 84 (Morton and Wess, 1993, p. 470). Middleton noted a thermometer in the Museo Copernicano in Rome, made in 1841, that had eighteen scales (Middleton, 1966, p. 66). The increasing use of thermometers aided the advance of science as a whole. By the end of the eighteenth century, thermometers were in common use in the laboratories of the men who tried to answer geological questions when the temperatures were within the appropriate range. Few of those men were concerned

only with questions of rock origin, but as natural philosophers were likely to ask questions about other parts of the natural world. However, when they were concerned with fusion properties of rocks and minerals, they required an instrument that would register vastly higher temperatures. That instrument was the pyrometer, the thermometer having little application in mineralogy.

Pyrometers

The word "pyrometer" was first used to refer to any instrument associated with a measure of heat. Subsequently it came to refer to a means of measuring in the range where thermometers could not be used. The range of an alcohol thermometer depends upon the kind of alcohol used and its percentage of water. Currently, the normal student thermometer of this sort only operates to a little over 100 °C. A mercury column separates at temperatures above ~300 °C (Turner, 1983, p. 117). At first, higher temperatures were noted by comparison, such as the requirement that a substance be near the temperature of iron at red heat, or at the fusion point of a particular material. The property of expansion with temperature increase was well known even in solids. Physicist Petrus van Musschenbroek of Holland (1692–1761) is credited with introducing the term "pyrometer" in 1731 when he described experiments in which he heated rods of different metals to determine their thermal expansion (Mortimer, 1747, p. 676–677; Anderson, 1978, p. 82). Some instruments included a dial that advanced as the temperature increased, and were designed to illustrate expansion as might be demonstrated in a public lecture (Morton and Wess, 1993, p. 94).[24] Clock makers were interested in metallic

---

[23]This early work was done in Zurich by Ott, who worked on thermometers from ~1762–1767, in Scotland by Robert Ferguson and James David Forbes, and in Belgium by Lambert Adolphe Jacques Quetelet. Saussure employed such a thermometer with a small door in the shaft into which the thermometer was inserted so that it could be read without exposing the bulb. For reference to the earth thermometer, see the footnotes on pages 139–149 in Middleton (1966).

[24]One instrument that was called a pyrometer consisted of an iron frame into which a test bar could be placed near the base. One end of the test bar was fixed, but the other was attached to a watch spring wrapped around the barrel on a shaft that had a long vertical lever. A thread was fastened to the top of the lever and turned around the shaft that carried the indicating hand over the dial. When the bar was heated along its length, the elongation showed on the dial (Morton and Wess, 1993, p. 159).

TABLE 3.1. THERMOMETER SCALES (in °)

Water	Newton	Hooke	Rømer	Fahrenheit	Réaumur	Celsius	De Lisle
BP	34		60	212	80	0 later inverted	0
		–5 hottest weather					
	12 blood heat			96			
		45 temperate					
FP	0		65	7.5	32	0	100	150
				0			
							205 FP of mercury

Note: Comparative temperatures from Middleton (1966).
BP—boiling point. FP—freezing point.

expansion with change in temperature, as it could alter pendulum length and thus the period. There were also early bimetallic thermometers, but at that time they were of little use in indicating temperatures (Daumas, [1953] 1972, p. 214). It has been argued that these pyrometers should be called "dilatometers," because that is the property they measured.

The first instrument worthy of the name "pyrometer" used in relation to metallurgy, mineralogy, and glass and ceramics making, was invented by Josiah Wedgwood of Etruria, friend, correspondent, and laboratory ware supplier to well-known scientists of his day, including Joseph Black, Joseph Priestley, Sir James Hall, and Lavoisier, and many others. Wedgwood was both a fine potter and an excellent businessman, who wished to maintain the quality of his known wares and initiate new forms, slips, colors, and glazes. He read the exhaustive experiments of Pott, but found that he could not replicate the colors or textures of Pott's experiments or even his own, because he could not standardize furnace temperatures or know what sequence of heating/cooling should be followed. As Wedgwood ([1782] 1809a, p. 279) rightly said: "a red, bright red, and white heat, are indeterminate expressions." He recognized that the different parts of a furnace might have different temperatures. In his first effort at a better measure, he took a mixture of calces of iron mixed with clay, formed circular disks an inch in diameter and a quarter of an inch thick, and placed them in a kiln

in which the fire was gradually augmented, with as much uniformity and regularity as possible, for near 60 hours; the pieces taken out at equal intervals of time during the successive increase of heat, and [were] piled in their order on each other in a glass tube. (Wedgwood, [1782] 1809a, p. 279–280)

The pieces demonstrated a gradation of colors that at first Wedgwood felt could be used to indicate the heat to which a mixture of the same kind had been exposed. Practically, however, Wedgwood found that both color perception and description were not exact enough for his purposes. He then said:

In considering this subject attentively, another property of argillaceous bodies occurred to me; a property which obtains, in a greater or less degree, in every kind of them that has come under my examination, so that it may be deemed a distinguishing character of this order of earths: I mean, the diminution of their bulk by fire; I have the satisfaction to find, in a course of experiments lately made with this view, that it is a more accurate and extensive measure of heat than the different shades of color. (Wedgwood, [1782] 1809a, p. 280)

Wedgwood gave precise directions as to where the clay used in his pyrometer pieces should be mined, and how it should be prepared. Eventually, because natural samples were not uniform enough, he used a mixture of clay and alumina (Morton and Wess, 1993, p. 591). At first the pieces were made in individual molds, but later they were cut to a precise length from a rod of specific diameter extruded from a box with several holes of the correct diameter in it. Wedgwood constructed a tapered gauge of two 24-inch-long pieces of brass on a brass plate, the long pieces being an inch apart at the top, and 3/10ths of an inch apart at the bottom, graduated from 0 at the top to 240 "degrees of Wedgwood" (Wedgwood, [1782] 1809a, p. 281). However, the brass expanded when hot pieces were used with it, so Wedgwood began making the gauges of stoneware by 1786 (Reilly, 1989, v. 1, 133). Figure 3.7 demonstrates that pyrometers were valued far past the eighteenth century. The pyrometer piece,

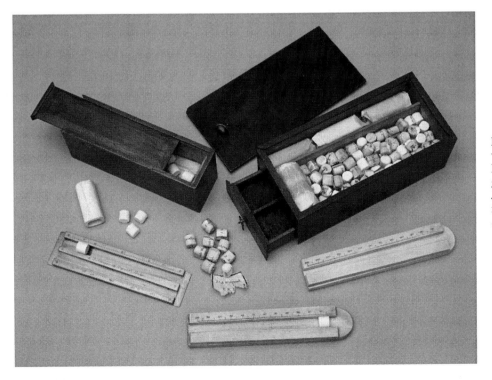

Figure 3.7. Two examples of Wedgwood pyrometers held by the Science Museum, London. The larger one was given to King George III in 1786 by Josiah Wedgwood. The smaller one is inscribed "J. Newman, 122 Regent Street, London," and was produced from 1828–1856.

dried and contracted in the heat, was taken from the furnace and placed at the top of the gauge. The farther it went down toward the narrow end, the higher the temperature to which it had been exposed. Wedgwood published his article, "An attempt to compare and connect the thermometer for strong fire, described in Vol. 72 of the *Philosophical Transactions* with the common mercurial ones" in 1784 in which he reported that his highest temperature, 240 °W corresponded to 32277 °F, and on down to the freezing point of water at –8 and 42/1000 °W or 32 °F. Zero on Wedgwood's scale was "red-heat fully visible in day-light" (Wedgwood, [1784] 1809b, p. 578).

Despite Wedgwood's hopes, and the optimistic reports of contemporaries, his Fahrenheit values were incorrect; the relations between scales could not be reduced to a simple function, and the amount of contraction of the clay on heating was not linear. In fact, his scale has been described as "horribly nonlinear" (Smith, 1969, p. 334). Chaldecott (1975) pointed out that Wedgwood was perhaps unjustly judged by later workers because he used values for metal melting points and for the boiling point of mercury that were generally accepted during his time, but that were nearly impossible to ascertain. Other factors added to the problems of calibration as well.[25]

Even so, Wedgwood's pyrometer pieces were the first even remotely reliable way to quantify the heat of furnaces. Sets of them, supplied in a box with a gauge, were generally internally fairly consistent. Each piece, placed next to the sample as it was heated or protected from it by a porcelain tube, was supposed to be used just once, although sometimes use was repeated. A ledger of Wedgwood's mentioned 33 sets sold in eight months in 1786 and 1787. Some of the men who used the sets were recognized as contributing to geology and mineralogy, for instance, Bergman, Black, and Wollaston (Chaldecott, 1975, p. 11). The Italian scientist Lazzaro Spallanzani (1729–1799) used a set in his investigations of basalt (Dolman, 1975, p. 562). Sir James Hall made his own pyrometer pieces after Wedgwood's death in 1795, but in 1805 sent a set to Etruria to be compared to those made there. Josiah Byerley—the son of Wedgwood's nephew Thomas Byerley—who had become a partner in the pottery when Wedgwood's first partner died, wrote his assessment to Hall, saying that Hall's pieces contracted more than they should, but that they were consistent with one another (Hall, 1805a). Wedgwood's pyrometer was still in use in the second half of the nineteenth century, when degrees Wedgwood were given for temperatures at which various grades of pottery and their glazes should be fired (Jacquemart, 1873).

It is difficult to say when the property of fusion temperatures of different ceramic mixtures was first used. The pyrometer had a sort of successor in Seger cones, invented toward the end of the nineteenth century, by the German ceramicist Hermann August Seger (1839–1893). Seger cones were made of clay of somewhat different compositions, and were designed to melt in a sequence of temperatures in a kiln (Sorrenson and Burnett, 1998, p. 498). But in its day, virtually all of the scientists who experimented with rock fusion used Wedgwood's pyrometer pieces, and Wedgwood's scale was frequently reported.

For his research, Loysel made use of a different property of clay, namely, tenacity, to measure temperature. He recognized that making glass required both furnaces and containers that resisted the heat and the flux action of the materials used to make glass, and stated the testing steps to determine the quality of the clay and mixtures to fabricate the parts that required a certain degree of tenacity. This effort involved another case of multiple variables. He recognized that the heat of the furnace was most important, but he was also concerned with the interaction of composition, the weight that was required to be supported, and the temperature to which it was subjected. Needing to be sure the parts could sustain furnace heat, he originated a method to determine the temperature of his glass furnace that exploited the tenacity of a clay mixture, noting that all parts of the furnace didn't require the same tenacity. Loysel made small pieces (*bâtons*) of the same composition and size, using at least four or five in each test, and dried them first at 7 or 8 degrees Réaumur, raising it to 25° over several days. The pieces consisted of a rectangular rod projecting from a cube. The size of the pieces was carefully adjusted with a template. The pieces were then clamped to the edge of a table from which a basin hung, with sand added to the basin until the rod broke. He repeated the experiment with several bâtons of the same composition dried for the same amount of time; then repeated them with different compositions. He derived an expression for the tenacity that, for his variables, became t(tenacity) = $P/6$, where P was the weight required for the rupture (Loysel, [1800] 2003, part 2, p. 56).[26] He calculated the tenacities for various mixtures of argil (clay) with dried and powdered argil of different alumina content, in order to determine which had the greatest tenacity, and thus would be most useful for fabrication of containers and furnaces.

For the same mixture, the tenacity differed according to the degrees of heat to which the bâtons had been subjected, and Loysel found a regular relation between the tenacity and the heat to which they had been exposed. He began by calculating tenacities, the result of many trials, at temperatures measurable by the Réaumur thermometer in a sand bath. He then constructed a curve where the abscissa represented the degrees, and the ordinate the weights required for rupture, and determined that the curve was a parabola. By extrapolation, he extended the temperature to Wedgwood ranges, and compared his figures to those determined by Wedgwood pieces expressed in Réaumur degrees (Loysel, [1800] 2003, part 2, p. 57–58). There were differences, which he thought were the result of the variable times to which the Wedgwood pieces were subjected to heat. His method was simpler:

---

[25]The reception and use of pyrometer pieces are discussed in Chapter 4.

[26]The derivation is given in note I of the second part of the essay.

Il suffit d'éprouver la ténacité de ce mélange à trois degrés différents de température pour obtenir l'équation d'interpolation, et de trouver la ténacité d'un morceau des vases cuits ou des parois du fourneau. On en conclura le degré de chaleur qu'ils ont éprouvé. (Loysel, [1800] 2003, part 2, p. 58)[27]

Perhaps unfortunately, Loysel's method was not widely applied. It was an excellent example of the use of the interrelations of well-understood variables married to mathematics, and it sought a link between the thermometer and the high temperatures that might be registered by pyrometers.

Later, there were pyrometers that depended on the expansion of metals in heat, some to test the accuracy of Wedgwood's pyrometer, among them those of Guyton de Morveau and the mineralogist Alexandre Brongniart (1770–1847) (Chaldecott, 1972b, p. 350). Guyton de Morveau's was explained in 1803 and employed platinum, a rod of which rested in a groove formed in a piece of well-baked refractory clay:

The rod was in contact at one end with the terminal wall of the groove, whilst its free end pressed against the short arm of a bent lever; the long arm of the lever acted as a pointer, the tip of which, moving over a graduated arc, gave an indication of the angular displacement of the lever caused by any thermal expansion of the platinum rod. (Chaldecott, 1972b, p. 350)

Guyton de Morveau investigated the properties of his pyrometer and others in the following years, and in 1808 read the first of a series of reports on pyrometry to the Class of Physical and Mathematical Science of the Institute of France (Chaldecott, 1972b, p. 353). Those reports reviewed the history of pyrometry from its beginning, and continued to comparison of Wedgwood's and Fahrenheit's scales.[28] As will be discussed in the next chapter, his meticulous and detailed work, although at some points subject to calculation errors and still incomplete, was terminated by his death in 1816. The wish to quantify high temperatures continued unabated through the nineteenth and twentieth centuries. Some of those early technologies are still in use in ceramics or glass operations.

RESULTS

Temperatures Attained

Despite the lack of accurate instruments in the eighteenth century to determine what were called "the higher degrees of heat," there were still quite educated guesses about fusion conditions. And if we know currently measured temperatures, we can extrapolate from earlier reports of fusion and fusion products to get a fairly accurate estimate of the temperature conditions under which investigations were carried out. Burning glasses and mirrors sometimes fused platinum, a point now measured at 1769 °C. While there may be some variability due to impurities, in furnaces iron is fluid at ~1530 °C, while copper melts at 1083 °C. A porcelain furnace needs to attain ~1430 °C. Wedgwood had "red heat fully visible in daylight" as the zero on his scale (Wedgwood, [1784] 1809b, p. 578). He assigned values of 32, 28, 27, and 21 °WW, the approximate temperatures at which gold, fine silver, Swedish copper, and brass melt, and explained that furnaces must attain those values to melt them (Wedgwood, [1784] 1809b, p. 578).[29] Despite difficulties with knowing the absolute values for Wedgwood degrees, after the publication of such values, other workers used those melting points to determine the temperatures to which their furnaces could be brought. Hall had no problem with maintaining temperatures of 15–55 °WW in his reverberatory furnace. For various stages in firing a glass furnace, Loysel used his *épreuve des bâtons* to list 900–1000 degrees on Réaumur's scale for the initial temperature, the lowest temperature of a large fire was around 8000 degrees, and was 15,000 for a temperature that couldn't be exceeded or the pots containing the glass would melt (Darcet et al., 1791, p. 137).

The results of applying the blowpipe to many minerals were often given in descriptive terms. Saussure's method to correlate the size of a globule of mineral fused by the blowpipe to Wedgwood temperatures is discussed in Chapter 10 along with the evaluation of Wedgwood pieces. The temperature developed by a blowpipe was conjectured to be anywhere from 125 to 150 °WW (Cleaveland, [1816] 1978, p. 65; Bakewell, [1819] 1978, p. 59). Additionally, Gray recounted a number of efforts to determine the temperature of lamp furnaces and reported a range of possibilities in degrees Réaumur (Gray, 1828).

CONCLUSION

While heat production was not much altered, the "quantifying spirit of the eighteenth century" (Fränsgsmyr et al., 1990) was exemplified in a change in the way temperatures were reported during the eighteenth century and into the nineteenth. As with other variables, reporting became more quantitative. Even before numerical scales came into wide use, there was never a sense of randomness, but of order in the comparative sense. Before Fahrenheit's thermometers were available, Lémery listed four comparative degrees of fire in 1696 (Eklund, 1975, p. 7). Nearly 100 years later, before Wedgwood pyrometer was devised, Gellert (1776, p. 75–77) had six divisions of heat:

1. From the bottom of Fahrenheit's scale to 80°, or temperatures that supported plant life;

---

[27]"It is sufficient to test the tenacity of this mixture at three different temperatures to obtain the interpolation equation, and to find it from the tenacity of a piece of the dried receptacle or from the lining of the furnace. One will conclude from it the degree of heat that one has tested" (my translation).

[28]The papers, and summaries of some of them, were published in a variety of publications from 1803 until 1814, for which see Chaldecott (1972b, p. 350–356 fns). Chapter 4 will discuss this further.

[29]It is difficult to translate Wedgwood degrees into either centigrade or Fahrenheit measures. An attempt, using Wedgwood's correlative Fahrenheit degrees, yields values for centigrade that range from 90 to nearly 100 degrees centigrade for each Wedgwood degree.

2. the region of animal life, from 40° to 94°;
3. from 94° to 212°, where reactions with vegetable and animal substances could be carried out;
4. from 211° to 600°, where oils and other substances boil, lead and tin fuse, charcoal forms, and some substances sublime;
5. between 600° and the temperature where iron could be fused, and various listed reactions occurred; and
6. the heat of burning mirrors or lenses, where all things could be fused, and to which he felt there was no upper limit.

Fourcroy linked four degrees of heat less than that of boiling water to portions of Réaumur's scale. Above the temperature of boiling water he listed five degrees that began with burning wood and melting sulfur, through fusion of some metals, and finally to the heat of a burning glass. He then commented that Wedgwood had devised a means to determine those temperatures (Fourcroy, 1789, v. 1, p. 160–164). On the other hand, Kirwan listed five degrees of fusibility: emollescence; porcelain state; slags, scoria, enamels; semi-transparent; completely transparent (Kirwan, 1794a, v. 1, p. 43). He then linked the five degrees to Wedgwood's scale:

very easily fusible	30–40 °WW
easily fusible	100–125 °WW
moderately fusible	125–135 °WW
difficultly fusible	135–150 °WW
very difficultly fusible	150–160 °WW.

(Kirwan, 1794a, v. 1, p. 44)

Mineralogy textbooks continued to list comparative scales of fusibility well into the twentieth century.

At our distance from the eighteenth century, especially when our concern is with the reasoning used to conjecture about rock origin, it is perhaps easy to assume that most of the experimentation was a kind of hit-or-miss matter of putting minerals, mineral mixtures, or rocks under a blowpipe or in a furnace to see what happened. A more detailed investigation shows that application of heat involved furnaces or other heat sources whose characteristics and fuels were well understood. Containers and supports were carefully chosen, and the reality of interaction between container and contents provided for. The thermometer gained acceptance in the search for replicable results, and there was a great deal of concern with calibration of thermometric scales, and the relation of different scales to each other. Records were meticulously kept, and information exchanged across national borders. Of course, not everyone who fused or smelted materials was a scientist, or was searching for truths about nature. However, there was a great deal of knowledge about how to produce, supply, and control heat, all necessary to determine how the engine of Earth worked.

It can be argued that the very detailed nature of that knowledge and its parameters actually held back the cause of the Plutonists, if not the Vulcanists, who could find little in what they knew of subsurface earth to explain the results they could produce in the laboratory, other than the conjecture of the existence of so-far-undiscovered coal seams under volcanoes, which at least could explain the fact that there were inactive volcanoes. When the coal was all burned, activity ceased. Pieces of the puzzle were tantalizing, but they didn't fit together. And, as will be seen in Chapter 5, the other undeniable but uncontrolled and uncontrollable variable was pressure. Next, we will see how these methods of describing heat were evaluated.

CHAPTER 4

Evaluation: Degrees of heat

INTRODUCTION

This chapter, perhaps a sort of "inter-chapter," will discuss the efforts made to ascertain how much reliance could be placed on the Wedgwood pyrometer pieces, both alone and in conjunction with the little-discussed blowpipe. The evaluation of other instruments has been frequently reported on in the history of science. The story of thermometers is well known, and there are numerous books and articles about how the different sorts were calibrated, and what the correspondences between them were. The same is true for balances and microscopes. Recent efforts have been made to duplicate some of the instruments to ascertain how well they might have functioned. However, explorations of the reliability of the blowpipe and the pyrometer of Wedgwood—the first very old, the second a then-new invention—have not been much reported on, although they are particularly relevant to thoughts about rock origin.

It has also been noted that laboratory and field investigations, many of the latter with instruments, have been little discussed with respect to geological questions, and generally have been dismissed as inconclusive. A very few laboratory inquiries, such as those of Hall with slow cooling of basalt and his purported fusion of carbonates under pressure (1805a, 1812), have subsequently been considered more influential than they were at the time. The people discussed in these chapters are those who were themselves well acquainted with the laboratory and/or field tests. They were numerate and thoughtful. Moreover, they had the eighteenth-century desire to question the natural world and to spend time devising instruments and procedures to do so. It is noteworthy that even thermometers had not been trustworthy much before the mid-eighteenth century, and just the act of reliably taking temperatures at various depths in rivers, lakes, the sea, and at various depths in the earth, by means of insulated thermometers, could yield new knowledge that fed into an overall theory of how Earth might function.

THE PYROMETER

Wedgwood's pyrometer was by no means the first instrument to measure higher temperatures. Glassmakers, iron masters, and other metalworkers, as well as natural philosophers, all recognized the need for consistency in furnace conditions. Thermometers that depended on the expansion of a metal bar had been proposed since the early part of eighteenth century. One of the best known was that of the English physician Crom-

well Mortimer (c. 1700–1752) who continued Musschenbroek's efforts with a dilatometer. After a discussion of the thermometers then available, Mortimer published a method of relating the expansion of heated metal rods to a metal circle calibrated to freezing and boiling temperatures of water and/or mercury, on which the markings could be extended to record higher temperatures (Mortimer, 1747).[1] Practical application of the device would have been difficult as Mortimer recognized. Zero on the circular indicator wheel was to be set at "normal" temperatures. The steel rod designed to expand or contract would need to be set near the source of heat or cold. He suggested that for use in a smelting furnace, it might be placed next to the fire chamber and a small aperture would allow the melted metal to contact the bottom of the rod. It is clear that the entire complex mechanism would be subjected to differing temperatures. The design of the lever system to magnify expansion would later be echoed by David Mushet (1772–1847) in a somewhat more practicable system.

Interestingly, after Mortimer proposed the use of various metals for the expanding rod, depending on the range desired, he suggested making a rod of what he called tobacco pipe clay for some of the highest temperature degrees desired, to be calibrated with boiling mercury instead of boiling water. That followed his assertion that all materials expanded when exposed to heat. He didn't mention clay shrinkage. Of course the potters of the time were well aware of shrinkage. A metal-expansion pyrometer also was an idea whose time had come, as evidenced by early references to Musschenbroek, Desaguliers, and Ellicot (Mortimer, 1747, p. 676–677) and to Ellicot, Whitehurst, and Smeaton (Meteyard, 1865, v. 1, p. 161). Chaldecott (1972a) discussed German expansion thermometers of that time, and described instruments that are still exhibited in various museums.

Josiah Wedgwood's observation of the color change of heated clay that contained some iron, which property comprised his first efforts at a pyrometer, was also common knowledge, as would be expected after thousands of years of firing pottery pieces. In fact, the laudatory comments about Wedgwood's first efforts rarely include mention of precursors in that effort, including those from the pottery dynasty of Wedgwoods, his cousins John and Thomas. Around 1740, Thomas and John (Long John) Wedgwood used "pyrometrical beads" to give some notion of the

---

[1] Apparently Mortimer presented his invention to the Royal Society of London in 1735, but it was not published until the later date (Chaldecott, 1972a, p. 87).

heat of furnaces. The small clay beads acquired a coloration on firing, and were then used to test subsequent firings. Of course the color could differ because of different conditions, but skilled potters could improve the accuracy of firing conditions by using them (Meteyard, 1865, v. 1, p. 159–160).

Wedgwood read his first paper about his pyrometer to the Royal Society of London in 1782. Extensions and corrections followed in 1784 and 1786, as discussed in Chapter 3. Despite its problems, the device was greeted with enthusiasm by both scientists and ceramicists. As might be expected, it was scrutinized by both groups to verify its properties, and his methods were also examined in the twentieth century to see how reliable they may have been. We will return to the latter, but now will look at some of the reactions of his contemporaries. As can be seen from the dates of their reactions, its utility was quickly recognized over much of Europe. Wedgwood supplied more than 60 people with pyrometer sets in 1786 and 1787 (Chaldecott, 1975, p. 11).

Fourcroy, a contemporary, listed the different degrees of heat needed for the fusion of various materials, and noted that it would be highly desirable to be able to indicate the exact degrees of heat that were required. He then described Wedgwood's pyrometer and indicated how it worked. For this he cited an article in *Journal de Physique* in 1787 (Fourcroy, 1789, v. 1, p. 164). In 1791, Darcet, Fourcroy, and Berthollet commented on Loysel's work on methods of managing a glassworks, and noted that his work on the tenacity of clay pieces contrasted with that of Wedgwood, who used the property of the shrinkage of clay when exposed to heat.[2] These men, who reported on the efforts of the glass manufacturer Loysel to make a furnace temperature indicator by means of the tenacity of clay pieces, were familiar with the Wedgwood device. They noted that Wedgwood's method using the contraction of heated clay pieces required less equipment than did Loysel's methods, which required mathematics as well. Loysel's method only tested between 17 and 234 degrees Réaumur and they thought it must be extended to 14,000 (Darcet et al., 1791, p. 137).

In 1794, Kirwan used Wedgwood degrees for all the fusibility trials that he reported on, and recorded them in his mineral characterizations. When discussing lava he said:

The heat of most volcanoes (I exclude those that for the most part produce only vitrified substances) seldom reaches 100°, of Wedgewood [*sic*]; the proof of which is, that almost all real lavas, whether cellular or compact, are vitrifiable at that degree. (Kirwan, 1794a, v. 1, p. 407)

He apparently had a rather high degree of confidence in both the ability of the pyrometer pieces to show measured temperatures, and also in the reliability of Wedgwood degrees as converted to Fahrenheit, as he used both. In his comments about rock origin

in 1799, he quoted Saussure's use of Wedgwood degrees without questioning them (Kirwan, [1799a] 1978, p. 454).

La Métherie noted that what he called "Weedgewood's" clay for the pyrometer could be heated quite strongly without vitrifying. Vitrification would interfere with the hoped-for regular contraction. However, he found it could be vitrified in a current of what he called "pure air" (La Métherie, 1795, v. 1, p. 304). The science establishment of the time was certainly aware of Wedgwood's contribution. For example, Spallanzani had the specific purpose in his *Voyages*, published in 1792,[3] of determining the temperature required to fuse volcanic rocks, and stated that the Wedgwood pyrometer was ideal for this. Some of his difficult and dangerous investigations with basalt will be discussed in Chapter 10.

David Mushet (1772–1847) of the Clyde iron works designed an assay furnace into which he could insert a metallic-rod pyrometer complete with calibrated disk on a platform adjacent to the chimney, as shown in the bottom left illustration of Figure 4.1. The pyrometer is at the left side of the drawing of the entire furnace, which has the chimney (H) truncated in order to fit into the picture. The crucible that was to hold the sample is at C. The metallic rod (LL) that expanded with heat and actuated the lever reached from the bottom to the top of the firebox. The rod was enclosed in a wrought iron box, the small space between the rod and box being filled with powdered charcoal to prevent oxidation of the steel. Steel proved to be the most satisfactory for the rod, as copper fused too easily, and malleable iron remained expanded after heating and converted to form steel itself. The rod, with a sharp tip, touched the lever (CC) one inch past the fulcrum. That end of the lever was weighted to counteract the weight of the longer end. The left side of the lever was 10 inches long. If the rod expanded to push the short end up half an inch, the long end was depressed 5 inches. The rod pointed to the arc DD that he divided into 50 equal degrees. Mushet suggested using that pyrometer in conjunction with Wedgwood's scale so that

any given substance may be melted, and the degree marked upon the [Mushet's] scale: let this be compared with the degree of heat at which the same substance melts by Wedgwood's scale; then let the greatest degree of heat of the furnace be measured by one of Wedgewood's rolls, and compared with that pointed out by the index, the one difference divided by the other will give a scale of comparison. (Mushet, 1799, p. 258)

Mushet also noted that although there were many variables, with attention the assayer could become well acquainted with his own furnace. For greater accuracy, three or more of the metallic pyrometers might be installed around the furnace.

In a letter to Guyton de Morveau, the German chemist Alexander Nicolas (Aleksandr Ivanovich) Scherer (1771–1824) noted the inclusion of information about Wedgwood's pyrometer and

---

[2] I have not yet discovered the original date of Loysel's work. This comment on it is dated 1791, so clearly Loysel was published beforehand, but all dates I find are 1800.

[3] I've used the 1798 translation.

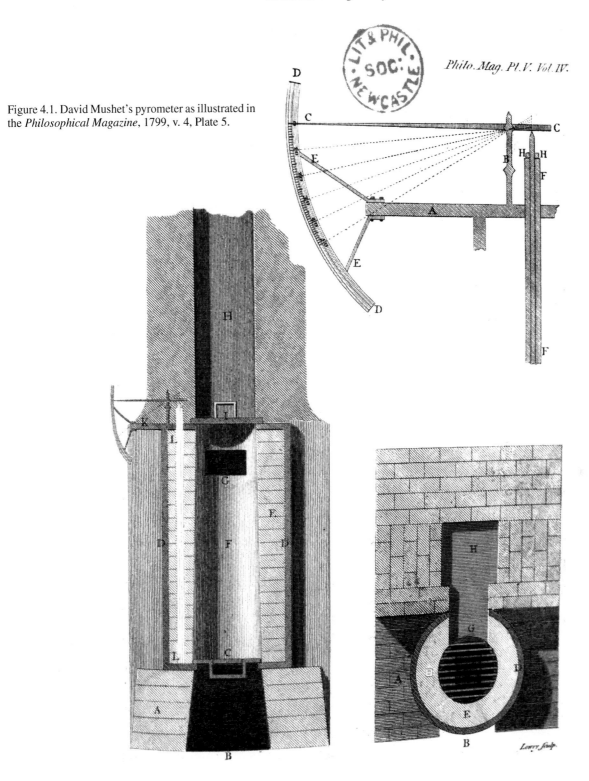

Figure 4.1. David Mushet's pyrometer as illustrated in the *Philosophical Magazine*, 1799, v. 4, Plate 5.

the improvements made to it by Tiberius Cavallo (1749–1809), who had included it among the subjects of his Bakerian lectures for the Royal Society of London (Bertucci, 2004, p. 591). In a footnote in the published letter, Guyton recorded the results of chemical analysis of Wedgwood's pyrometer pieces, which he found to be:

Alumine	54.705
Silice	43.764
Magnésie	trace inappreciable
Oxide de fer	*idem.*
Parte (loss)	1.531

(Scherer, 1799, p. 172)

In the attempt to get reproducible results from Wedgwood's pyrometers, a good bit of attention was paid to their composition. References to Nicolas Louis Vauquelin's (1763–1829) analyses of pottery products were quite frequent. Different requirements for different uses of various ceramic products were listed in the *Philosophical Magazine* report of his work, which had originally been published in the *Bulletin des Sciences* (Vauquelin, 1799). Vauquelin noted both the chemical composition and size of particles advantageous for glazed and unglazed ware were determined by the use, such as firing for ceramic pieces, melting iron or copper, etc., the conditions they would be subjected to, and the need for strength. He did analyses for Hessian crucibles, washed and unwashed natural clay, porcelain, and Wedgwood pyrometer pieces. His values for Wedgwood pieces were:

Silex:	64.2
Argil	25
Lime	6
Oxyd of iron	0.2
Water	6.2

(Vauquelin, 1799, p. 290)

Because Guyton's results differed considerably from those of Vauquelin, Guyton asked him to repeat his analysis. Vauquelin's figures had been widely disseminated, and his later, better values came much closer to Guyton's values, but were not published until 1810 (Chaldecott, 1975, p. 12; Guyton, 1810c, p. 131).

In 1800, Isnard Gazeran said that Wedgwood's pyrometer was still the only instrument that could measure the highest temperatures. He thought it was so good that some experiments should not be done without using it to indicate the temperature. However, since he said that Wedgwood had not given the composition of his pieces, Gazeran thought it useful to the scientific and technological public to determine it. Gazeran consulted Vauquelin's analysis of Wedgwood's pieces, and tried to replicate it. He found the best clay for the purpose of making the pieces was one found in France. He passed 150 parts of the clay through the finest screen, and added 63 parts of washed and ground sand from Fontainebleau. These he mixed thoroughly, then ground the mixture for 2 hours. It was next dried in the air until it lost 170 parts of the original 200 parts of water he had added. He then molded it into small cylinders, using tin cylinders that were 15 mm in both diameter and height. After he removed the pieces from the molds, he put them in an oven to dry for 24 h at a temperature of 40 °R. He trimmed the pieces to adjust (as Wedgwood had done) to 0° on the pyrometer scale (Gazeran, 1800, p. 101–103).

Two of the pieces Gazeran made, which weighed within a centigram of those of Wedgwood, were placed in a crucible along with two pieces from England, and all were subjected to a very high temperature for an hour and a half. The two Wedgwood pieces registered 158° and 160 °WW, or an average of 159°. The pieces Gazeran had made registered 159° and 160 °WW, or an average of 159½°. Gazeran noted that Wedgwood pieces could vary as much as 4–9 degrees, while his showed only a 1 degree difference. His pieces also showed no sign of vitrification after being exposed to

a temperature that would melt iron (Gazeran, 1800, p. 103). Thus, he found if he arranged the composition of the pieces carefully, he would get the same results as from Wedgwood's pieces.

Murray called Wedgwood's pyrometer "the only one yet invented which can convey to the mind any accurate ideas respecting high degrees of heat, and enable us to compare them with each other and with lower degrees" (Murray, 1801, p. 153). After a precise description of the pieces and gauge, he remarked on efforts to calibrate it:

By means of a similar gauge, but by employing the expansion of a piece of silver by heat, Mr. Wedgwood succeeded in obtaining a knowledge of the intermediate degrees of heat between that of boiling mercury and the zero of his own scale; from which he ascertained, that one degree of his was equal to 130° of Fahrenheit's scale, and that the zero of his corresponded to 1077½ F. Consequently, to accommodate the results obtained by the pyrometer to Fahrenheit's scale, all that is necessary is to multiply the pyrometric degrees by 130, and to the product to add 1077½. (Murray, 1801, p. 154)

Would that it had been that simple! As comments from the eighteenth century to the twentieth make clear, the scale was not linear, nor did others find the same ratio between thermometer scales and the pyrometer. Murray noted that pieces made more recently did not give the same results as Wedgwood's, and thus attention had to be paid to their composition.

Murray recorded the composition of the clay that Gazeran had used, and cited the result of his trials. However, he questioned why natural clays should be used at all, since they were sure to vary, even in different parts of the same bed. He saw no reason why pure ingredients could not be combined; and even if the argil portion was still subject to a little variation, if it was prepared in the same way every time the variability would be controlled. Thus, he hoped that pieces would be made that agreed with Wedgwood's, as there was already a great deal of information available about those. For instances when that couldn't be done, he suggested two ways in which equivalent pieces could be produced, which basically involved adjusting the gauge so that the readings corresponded to Wedgwood's.

In 1803, there was a report published in the *Journal des Mines* about an investigation of Wedgwood's pyrometer by Alexander Miché (1755–1820), chief engineer of French mines. He and three colleagues had been charged by the Conférence des Mines to investigate various combustibles, especially peat, for which they had made use of Wedgwood's pyrometer. They found many anomalies with use of the pieces, and felt that the results were not comparable. In a footnote in the 1803 report, they described the pyrometer, and said they were confident that those they employed were truly representative of Wedgwood's because they had been loaned to them by Dolomieu. However, they felt that they had too blindly rushed into their earlier work. Miché (1803) described the conditions under which they had again tested a number of the pieces, and produced what they considered a better table. It is easier to understand their comments if the relations in the table, reproduced here, can be seen:

TABLEAU des expériences faites pour éprouver le Pyrométre de Wedwood [sic]

Épreuves	Numéros	LETTRES	DEGRÉS			
			Indiqués par les cylindres			
	D'ordre des expériences	De dénomination des cylinders	Avant l'épreuve	Après l'épreuve	Acquis par le retrait des cylindres	Du thermomètre de Réaumur
Première épreuve	5	A	1	23	22	73
	6	B	1	12	11	72
	8	C	6	13	7	73
	9	D	5	12	7	67
	10	E	2	12	10	
	11	F	4.5	12	7.5	60
	12	G	6.5	32.5	26	76
	13	H	7	13.5	6.5	
	14	I	1	25	24	76
	15	K	5	25.5	20.5	76
	16	L	7	8.5	1.5	
	17	M	10	14.5	4.5	74
	18	N	7	12	5	76
	19	O	8.5	25.5	17	78
		P	1	28	27	78
	20	Q	8	22.5	14.5	80
		R	10	23	13 ⎫	
		S	7	43	36 ⎪	
	21	T	6	44	38 ⎬	80
		U	3	21.5	18.5 ⎪	
		V	4	24.5	20.5 ⎪	
		X	6	26	20 ⎭	
	22	Y	8	13.5	5.5 ⎫	68
		Z	8	14	6 ⎬	
	23	W	7.5	10	2.5 ⎫	76
		AA	7.5	20	12.5 ⎬	
Seconde épreuve		A	23	31	8	78
	19	G	32.5	36.5	4	id.
		I	25	28	3	id.
		L	8.5	26	17.5	id.
		R	23	23	0	63
		S	43	44	1	id.
	22	T	44	48	4	id.
		U	21.5	26	4.5	id.
		V	24.5	25	0.5	id.
		X	26	26	0	id.
	23	Y	13.5	17.5	4	−6
		Z	14	22	8	id.
		R	23	23	0	76
Troisième épreuve		S	44	44	0	id.
	23	T	48	48	0	id.
		U	26	26	0	id.
		V	25	25	0	id.
		X	26	26	0	id.

(Miché, 1803, p. 49)

In their extensive investigation, Miché and his coworkers had made great efforts to achieve similar conditions for the pieces they were comparing. They compared placement in several places in the furnace, and found it made little difference. Wedgwood temperatures, which would correspond to size, were recorded before and after the trials. The authors pointed out that pieces, such as Y and Z, W and AA, and T and X, which began with the same values, ended with different ones. And pieces such as O and P that began with a difference of 7.5, after identical treatment ended with a difference of 10, one

having been changed by 17 and the other by 27. The differences between the pieces as recorded in the table was sufficient for them to come to the conclusion that little confidence could be placed in an instrument that was so variable in its effects (Miché, 1803, p. 48).[4]

---

[4]At first, the use of Réaumur temperatures is confusing, as they could not indicate the usual temperatures involved when using pyrometer pieces. But it appears they heated the pieces in a "hearth" to which heat was conveyed by a heated liquid. Plunging the thermometer into that liquid resulted in the recorded Réaumur temperatures (Miché, 1803, p. 46).

Citizen Fourmy (also called Fourmi) recognized the need to know the temperatures of *opérations pyrotechniques* (Fourmy, 1803, p. 423) for which he would use thermometers. But the usual mercury or alcohol thermometers that depend on liquid dilation could not record the relevant temperatures. The thermometers that depended on the dilation of metals did not serve his purpose.[5] But clay's contraction with heat was well known, and Fourmy noted that the "illustrious Wedgwood" had attempted to use that property. He said that Wedgwood had assumed that the diminution was proportional to the intensity of caloric, and quoted from his papers. However, Fourmy saw that vitrification of the clay pieces would be a problem, and proposed to investigate the point at which the pieces would melt. For his investigation, he sought a cause that produced a specific effect uniquely and invariably from that cause, and that the effect must be proportional to the cause (Fourmy, 1803, p. 426). However, the effect of heat on the clay pieces depended not only on the application of heat, but also how long it had been applied, and how rapidly. Also, fabrication of the pieces was subject to variability, which he discussed in some detail.

Fourmy then tried to clarify whether the effect was proportional to the cause. For this, he used arguments about the variable relation of different amounts of silex and how they behaved with respect to the clay in different intensities of heat and lengths of exposure to it. The result was that the diminution of the pieces appeared to be due to several causes, and that the pyrometer did not uniquely and invariably reflect the intensity of the caloric. Some reasons for error included the fact that the composition of the pieces was much more complicated than had been assumed. Fourmy could not believe that the shrinkage of all pieces was the same. He questioned the composition that Wedgwood spoke of for his pieces, not knowing if it was only theoretical, or if it varied by accident (Fourmy, 1803, p. 433). It might vary in successive trials, and pieces made at different times could differ according to the natural variation in the clay: "Et ce qui prouve sur-tout que ce système n'est pas fondé sur une propriété invariable des mixtes alumineux" (Fourmy, 1803, p. 433). And, "Il est donc évident qu'un pyromètre fondé sur la retraite des mixtes alumineux, ne peut être considéré comme un instrument doué d'une certaine exactitude" (Fourmy, 1803, p. 433).[6] However, Fourmy felt that despite its imperfections, the pyrometer was still a useful tool, and there was nothing better. He suggested several ways of improving the process of making the pieces to diminish the possibilities of error, but pointed out a series of things that were not yet sufficiently known. Meanwhile, use of the pieces could serve to give a rough estimate. He referenced the Miché article discussed above.

Fourmy published his second *Mémoire* on pyrometers (or thermometers *en terre cuite*) in 1810. In it, he tested further the relations between duration and intensity of heat applied to pyrometer pieces. He had been unable to find enough pieces, so he arranged to employ some already used by Brongniart and supplied by him (Fourmy, 1810, p. 429). Fourmy stated that there were two ways to proceed: the first was to subject the pieces to a shorter or longer time of sustained heat; the second was to repeatedly subject the pieces to high temperatures for a short time. He recorded what Loysel had found out about the pieces by using the first method. Fourmy used several furnaces in his eight experiments using different pieces, some of which were used several times. His conclusion after careful investigation was that the pieces could not be used as a universal thermometer, but could be reliable if the same compositions were used strictly under the same conditions. They were not reliable if used under variable conditions (Fourmy, 1810, p. 442).

Recognizing the value of an instrument for measuring high temperatures, Guyton had invented a platinum pyrometer, partly in order to check the reliability of Wedgwood's pyrometer. A platinum rod was supported in a groove in a block of refractory clay, with a pointer pressed to a bent lever at the moveable end. An attached graduated arc gave the angular displacement caused by expansion (Chaldecott, 1972b, p. 350). This small pyrometer was intended to be placed within a furnace, protected by a muffle, or under an inverted crucible. As would be expected, Guyton anticipated and tried to counteract possible errors, as well as presenting the work of a number of other scientists who had attempted to make expansion thermometers designed for estimating very high temperatures. One large problem was the question of whether expansion was uniform over the required temperature range, which a number of scientists had checked with various metals. Guyton presented a table giving the expansions of 14 metals, one of which was platinum, between the freezing and boiling points of water, done by ten other workers (Guyton, 1810a, p. 263). There was also the question of whether expansion of both glass (in the thermometer) and the mercury were the same over their range. Wedgwood had determined that each of his degrees was equal to 130 °F (Wedgwood, [1784] 1809b, p. 577). But Guyton found that Wedgwood, who had used the expansion of silver for that comparison or calibration, had not checked the work of others, and had not seen the variations reported (Chaldecott, 1972b, p. 355). Again, in the light of what could be called the "known unknowns," it is not surprising that the origin of rocks in their natural habitat was a matter of mystery, not yet understood by experimental procedures.

Guyton (Fig. 4.2) read two papers about Wedgwood's pyrometer to "la Classe des sciences physiques et mathématiques de l'Institut" in April and May of 1808, which were published later in 1810 (Guyton, 1810b, p. 18–46, and 1810c, p. 129–152). He first corrected the impression that Mortimer might have applied the idea of clay shrinkage before Wedgwood did, saying it was Wedgwood's idea, and the result of much experimentation. Guyton recounted the history of Wedgwood's invention,

---

[5]The pyrometer based on metal expansion is briefly discussed in this chapter. Bimetallic spirals to indicate temperature had been described in 1779, but they were not suitable for this purpose (Middleton, 1966; Anderson, 1978, p. 84).

[6]"And this proves above all that this system is not based on an unchangeable property of the aluminous (clay) mixtures" (my translation).

"It is thus evident that a pyrometer based on the contraction of aluminous mixtures cannot be considered as an instrument endowed with a certain accuracy" (my translation).

Figure 4.2. Louis Bernard Guyton de Morveau.

and thought it astonishing that there was no consensus about its value when scientists had been using the method for 20 years. He remarked that it was a gift from the "arts," that is, technology, to the sciences that used fire as a means of research, attested to by the research of 11 prominent scientists that he named (Guyton, 1810b, p. 20). Some of them thought it useful, others felt they could not use it with confidence. Guyton saw three reasons for the ambivalence:

1. Wedgwood's publications were not well known, and the French versions were incomplete.
2. The true causes of some variations were not evident.
3. The difficulty of obtaining clay that would shrink uniformly.
(Guyton, 1810b, p. 21)

Guyton investigated these difficulties, particularly the third concerned with getting a suitable clay, and listed the degrees Wedgwood that various workers had found for their furnaces along with anomalies in the results. He noted that Loysel had apparently not been aware of some of that work, and that the commissioners of the Academy remarked that he had attempted to extend the scale after only noting correspondences from much lower Réaumur temperatures (Guyton, 1810b, p. 26). It was difficult to compare results produced by different people using differ-

ent materials and different furnaces. Guyton also mentioned the effect produced by putting a hot piece of clay in a metal gauge, which resulted in expansion of the metal, and noted that Wedgwood turned to making the gauges of "pipe earth," or clay used for tobacco pipes combined with ground charcoal in order to ameliorate that problem.

In the first part of the last paper (1810c), furnished with much experimental information, Guyton continued to interweave statements of the various problems and anomalies, Wedgwood's efforts to surmount them, and data that stretched from Mortimer's era until the time that this paper was written. The second part of the paper was again concerned with the considerable problem of attempting to determine the value of Wedgwood degrees compared to other scales. It was here that the discrepancy in analysis figures from Vauquelin was pointed out. An error had been made in the previous report, and Vauquelin's re-analysis resulted in values that were much closer to Guyton's own values:

Silice	47.55
Alumine	44.29
Eau	8.56.

(Guyton reporting, 1810c, p. 131)

Guyton continued his efforts to determine the reliability of Wedgwood's pieces in a paper, "De l'effet d'une chaleur ègale, longtems continuée sur les pièces pyrométriques d'argile," read in 1808 and published in 1811. Guyton recorded his reactions to Fourmy's work, considering whether the intensity or duration of applied heat influenced the pyrometer pieces. He noted that the pieces only achieved their maximum shrinkage if the heat had been applied for a sufficiently long time. Looking closely at the data in Fourmy's table, which he reproduced, Guyton pointed out the weaknesses and inconsistencies in both the data and Fourmy's conclusions about it and remarked that Fourmy didn't distinguish sufficiently between the effects of intensity and duration. Guyton noted that Fourmy had not stated the temperature of the metal gauge either before or after the trial, so that, as he acknowledged, his results were not strictly comparable (Guyton, 1811, p. 77). Some of the pieces used for the same test, which Fourmy thought had been subjected to the same degree of heat, had been split or deformed, showing that the effects on each had not been the same. According to Guyton, Fourmy hadn't observed a decrease of weight along with the decrease in volume, as Saussure had mentioned should happen, and he hadn't paid sufficient attention to the different temperatures in different parts of a porcelain furnace.

As might be expected, after Wedgwood's death in 1795, sources of pyrometer pieces were problematic. The Wedgwood factory continued to make them for a time, but the political upheavals of the time limited their distribution (Chaldecott, 1975, p. 11). However, as we have seen, Sir James Hall used the pyrometer pieces extensively in his experiments. He mentioned the pieces he had made himself in a footnote to his fusion efforts with carbonates: "Necessity compelled me to undertake this laborious and difficult work" (Hall, read in 1805, published in 1812, p. 85). His pieces differed from those sold by the Wedgwood factory, but they were consistent with one another. He therefore made a table for himself so that he could compare his pieces to those of Wedgwood. His sets apparently indicated the fusion point of silver as 22 °WW, while Hall said that Wedgwood's table of fusibilities listed it as 28 °WW. The explanation for Hall was that Wedgwood's observations must have been made with a different set of pieces (Hall, 1812, p. 85).

In 1814, Guyton published a two-part paper concerning corrections to Wedgwood's table relating his scale to other thermometer scales, the fruit of his own detailed work with the platinum pyrometer for higher temperatures. Among many changes was the notation that zero on Wedgwood's scale should be 517 °F, and not 1077 °F as suggested by Wedgwood. He also said that the "valeur de 130 degrés de Fahrenheit, qu'il a assignée à chacune des divisions de sa jauge pyrométrique, y soit réduite à 62.5" (Guyton, 1814a, p. 115).[7]

Guyton corrected the melting temperatures for a number of materials, as well as the boiling points of some. In the tables in the second part of the paper, he listed the expansion dilation values and related them to centigrade degrees, compared the results of different authors, and listed Wedgwood temperatures as compared to the centigrade and Fahrenheit scales. This extensive sequence showed that not only was Wedgwood's seemingly crude invention useful, but also that it was tested and retested over a period of many years by innovative and painstaking experimentalists.

We are now so accustomed to looking up any bit of information we need that it is difficult to conceive of a time when many parameters such as melting point were simply not known. Wedgwood was of course particularly interested in the behaviors of clay, silica, and glazes. His work was part of the much larger effort in his time to establish standards of all kinds for weights, measures, temperatures, pressures, and magnetic force, not only for practical purposes, but also so that generalizations about theoretical possibilities would be more reliable. Despite the best efforts at standardization of the clay, silica, and water mixtures, and of the size and length of the pyrometer pieces, the inherent variability of preparations and the peculiarities of specific furnaces, the pieces were never as accurate as was hoped for. Comparisons might be valid within a particular set of pieces, but the pieces could not be reused, and they often disagreed not only with a later set made by the same person, but also, inevitably, with those from other sources.

RELATION TO THE BLOWPIPE

There were attempts to expand the usefulness of the Wedgwood pyrometer to blowpipe and burning lens/mirror fusions. In 1794, Saussure published an amendment to his 1785 paper on the use of the blowpipe, an instrument with which he was quite skilled. There were difficulties with finding a proper support for the mineral or rock fragment that was to be subjected to the blowpipe flame. Glass tubes had been used, or fragments of "verre à vitre" (window glass), but difficultly fusible substances sank into the fused glass. Saussure initiated the use of fibers of kyanite on which to mount the samples. He mentioned that it only required a touch of saliva or a dilute gum solution to affix the fragment. If the alkali in the saliva or the phosphorus in the gum could upset a delicate analysis, pure water would suffice. Some fragments might fall off, but one could plunge the moistened point of the kyanite thread into the powdered mineral. The kyanite was fused to a glass tube to hold the assemblage in the flame. True, the sample might be so small that the result would need strong magnification to be seen, but that was no obstacle, especially if the microscope was fitted with a micrometer (Saussure, 1794, p. 4–5). In addition, Saussure discussed the difficulty of pulverizing some very hard minerals, as well as the difficulty of fusing some very refractive samples. He had to look for somewhat different signs of fusion in minerals that were the most difficult to fuse, or which changed their characteristics during fusion, such as partially melting and then resisting fusion. He could also very closely observe the effects of combinations of minerals in influencing fusion temperatures, and listed six ways in which simple or more complex fusion could occur.

At this point, Saussure had observed enough fusions for him to propose a relation between the diameter of the globule formed and the difficulty of fusion—the most easily fused minerals forming the largest globules. The most fusible substances required the least amount of fire or caloric, so it followed that for a standard size (a cube one line on each side),[8] a standard amount of caloric would be required for each sample type. And:

l'inverse de cette proposition doit être également vraie, c'est que, si une quantité donnée de calorique peut tenir en fusion le double plus d'un corps que d'un autre, ce premier corps, peut être regardé comme le double plus fusible que le second. (Saussure, 1794, p. 9)[9]

Then, if one considered that the flame of the blowpipe delivered a constant amount of caloric in a given time, one could equate the mass of the sample with its fusibility via the diameter of the globule (Saussure, 1794, p. 9). He also looked for a means of relating diameter to Wedgwood's scale, and chose the lowest heat that would result in fusion of a cube of window glass with sides of 1 inch. This resulted in a globule of a certain diameter. Saussure then compared it to the size of a fused cube of feldspar

---

[7]The "value of 130 degrees Fahrenheit, which he had assigned to each of the divisions of his pyrometer gauge, could be reduced by 62.5" (my translation).

[8]We recall that a French "line" is 3.175 mm. The French "pouce," which translates as "inch" had the value of 1.065977 English inches (Lavoisier [1790] 1965, p. 487). We remember here that significant figures were not treated with the rigor that they are now.

[9]"the inverse of this proposition must be equally true, so that, if a given quantity of caloric can hold in fusion more than double of one body than of the other, this first body can be regarded as doubly as fusible as the second" (my translation).

of the same size. After much cogitation and many experiments where placing the sample at the end of the kyanite fiber didn't work well, Saussure began to fuse his samples on a charcoal block as was frequently done with the blowpipe. He had decided to concentrate on silver, because he could obtain the metal in a good state of purity, and Wedgwood had determined the fusion point to be at 28 degrees on his scale. Saussure found the globule of silver to be 2.7 lines in diameter, so that 2.7 was equivalent to 28 °WW. He checked his relation by choosing to use those values to calculate the WW temperature of cast iron, which figure Wedgwood had found experimentally. The cast iron formed a globule of 0.6 lines diameter. Saussure used the ratio:

$$0.6 : 2.7 = 28 : x$$

from which x, the Wedgwood temperature of the fused iron, was 126°, which Saussure felt vindicated his choice of the ratio of globule diameters. Wedgwood's temperature for the fusion of cast iron was found to be 130 °WW.

Using that formulation, Saussure published a table of experimentally determined diameters and the corresponding Wedgwood fusion temperatures for 113 rocks and minerals. He listed another 10 for substances that fused with great difficulty, and 11 more for metals and metallic substances. All depended on Wedgwood's ratio, so all would later be found to be too high. Saussure linked what he found from those studies with statements about volcanic rocks, and tried to correlate laboratory and field conditions, as we will see in Chapter 11. When the last two volumes of Saussure's *Voyages* were published in 1796, he added the Wedgwood temperature to his characterizations of the rocks that he had collected. For a porous lava, he noted that under the blowpipe it formed a black enamel with a diameter of 0.8 inches, which corresponded to 71 °WW. However, the most refractory of those lavas didn't fuse until 105 °WW (Saussure, 1796, v. 3, p. 323). A brown rock from Avignon fused with difficulty, forming a globule of diameter of 0.3 lines corresponding to 189 °WW. After treating that rock with acid during which it effervesced, the remainder fused with greater difficulty to a diameter of only 0.13, representing 581 °WW (Saussure, 1796, v. 3, p. 341).

How accurate this relation of Saussure's was is open to question, as was Guyton's correlation of Wedgwood degrees with the Fahrenheit and centigrade scales. The variability of Wedgwood's pyrometer pieces has been noted. Despite efforts to standardize thermometers and recognition of the factors that led to variability in their readings, there was no overall standard for them. With respect to the materials tested, we now have data showing that pure substances have sharp melting points, a characteristic that is used in identification. However, one has but to look at the materials that were tested to understand that a sample of, say, calcite from one place might differ considerably from a sample from another place, while retaining the external and chemical features that enabled its identification. The problem increased for rocks with multiple minerals that retained the same name.

CONCLUSION

The interweaving of science and technology had continued from well before Josiah Wedgwood's efforts. Furnaces, glassware, and ceramics, metal-working techniques, and practical experience all contributed. Repeatability and accuracy were clearly regarded as important. Communication between workers in different countries, although sometimes slow, was generally quite good. About the time of the eighteenth century, more and more detailed, repeated, and tested laboratory observations were made that supplied parameters for theoretical flights of fancy about rock origin.

It is clear that determination of the measure of heat and temperatures was important in many facets of science and technology through the eighteenth century and into the nineteenth. But there are of course better-known controversies and problems in the history of the science of that time. The nature of heat itself, and its propagation, was a conundrum. The devices for measuring it came more from the practical world of technology than from science, and iron masters and glass- and porcelain makers contributed. There might be some surprise that people such as Vauquelin, Fourmy, and Guyton, known best as chemists, were so concerned with questions that were approached from the viewpoint of technology. Aqueous chemical reactions were carried out at far lower temperatures, although the importance of heat in those reactions was recognized. But dry analysis, or analysis "by fire," was carried out at elevated temperatures. Much of the chemical analysis for constituents was not only of chemical compounds, but of those and other compounds occurring naturally, namely, minerals.

This is where these quantitative, or near quantitative, methods became important in geological theory. It might seem a simplistic answer to the basalt controversy, or even to part of it. The temperature of basalt fusion had been a matter of much conjecture and values were disputed. Attempts to use pyrometer pieces in the field were inconclusive. The temperatures attained in furnaces in which basalt was fused were uncertain. The growing understanding of heat and heat conduction was almost an impediment, because of the number of uncontrollable conditions and the unknowns in the great furnaces of nature that defied rigorous investigation by the investigators. And even if fusion temperatures could be determined precisely, there was still the better-controlled evidence of aqueous solutions and reactions, the other suggestion for rock origin. As the above sequence of articles showed, interest was maintained among Wedgwood's contemporaries and followers for more than 30 years. Work to determine extreme degrees of heat continued apace through the nineteenth and early twentieth centuries. Quite possibly because of the difficulty of determining firm parameters for fusion, the lack of reliable instrumentation that gave repeatable results, recognition of the lack of knowledge about heat itself, as well as the sheer enormity of the amounts of material involved for large earth processes, aqueous analysis was under far better control and seemed to be a more plausible explanation for rock origin.

Those who worked on the problem of temperature determination identified virtually every variable that could be tested. Composition of the pyrometer pieces was approached in several ways. Wedgwood's pieces were analyzed, and both natural clay and mixtures of purer compounds were used to form the pieces. The difficulty of both vitrification and occasional swelling of the pieces from included gases when heated was addressed, as was the effect of small amounts of impurities that might cause fusion changes. The temperature of the metallic or clay gauge itself was considered. The expansion of metal when heated led Wedgwood himself to change to ceramic gauge pieces, as metal gauges could fuse or could otherwise be altered. The impact of heating pyrometer pieces more than once was addressed. Fourmy and others investigated the effect of the length of time a piece was held at a high temperature, as well as of the intensity of the fire. The role of pressure began to be considered. In a relatively few years, arguments about rock origin went from large and speculative statements that might include biblical events and chronology to precise statements about tested and controlled parameters. After more of "the basics," Chapter 11 will discuss the comments of the experimenters themselves about rock origin. Another of the variables, pressure, will be discussed next.

CHAPTER 5

Pressure and rock origin

INTRODUCTION

By the middle of the eighteenth century, although there was a wealth of practical knowledge about what drove chemical reactions, the variable of pressure was not well understood. However, experimentalists recognized that it did have an effect. After mention of two arguments against his theory of rock origin from fusion, namely, the inclusion of water in some crystals and of volatiles that remained in limestone, Hutton famously remarked: "The effect of compression upon compound substances, submitted to increased degrees of heat, is not a matter of supposition, it is an established principle in natural philosophy" (Hutton, [1795] 1972, v. 1, p. 94).

If an experiment were to be carried out under light pressure, volatiles would separate, but they would not under the conditions obtained in the regions deep within Earth where minerals were formed. Hutton claimed that this circumstance had not yet been considered by other chemists and naturalists (Hutton, [1795] 1972, v. 1, p. 140). The means of producing the higher pressures in Hall's experiments was heat applied in a closed system, so that escaping volatiles would supply the pressure. As we will see, he did use some of the enclosing equipment used in experiments with gunpowder. Manual pressure was used for the folding experiments.

Methods of measuring pressure were still being evolved. The phenomenon of atmospheric pressure had been recognized and demonstrated from the time of the invention of the barometer, initiated in 1643, by the Italian mathematician Evangelista Torricelli (1608–1647). His instrument consisted of a glass tube first filled with mercury and then inverted so that the open end was under the surface of a bowl of mercury. Torricelli had been an assistant to Galileo, and had mentored several of the men who later formed the *Accademia del Cimento* in Florence (in existence from 1657 until 1667), known for its experimental approach. He thought that atmospheric pressure might be responsible for the so-called abhorrence that nature had for a vacuum, apparently present at the top of the tube of mercury in a barometer. There were alternative hypotheses about what could be holding up the column of mercury or the water sometimes used in the instrument. Those questions about the nature of the vacuum led Otto von Guericke (1602–1686), mayor of Magdeburg, who was a Copernican and student of Descartes' works, to devise a better vacuum pump (Krafft, 1972, v. 5, p. 574). His very public demonstrations of atmospheric pressure, especially those conducted in Magdeburg in 1657, inspired many others to investigate the properties of pressure, and to improve the vacuum pump.

The skilled experimenter Denis (or Denys) Papin (1647–~1712) also worked with the pump, sometimes in conjunction with Robert Boyle and the Dutch physicist and astronomer Christiaan Huygens (1629–1695). He is credited with making an advance in the steam engine. Our interest in this chapter centers on his steam digester, a vessel designed to operate with a sample under elevated pressure. Papin stated the reason for his publication, *A new digester or engine for softening bones*:

Some Experiments of the screwed Balneum Mariae have already been printed in the second continuation of the Physico-mechanical Experiments of the Honourable R. Boyle put forth this year 1680, but that Book being writ in Latine, and not giving the Description of the Engine, nor the ways how to use it safely for want of sufficient Tryals, I thought it would not be improper now to make upon that Subject a separate Treatise in the vulgar Tongue for the use of such Housekeepers and Tradesmen as may have occasion for it. (Papin, 1681, Preface)

This instrument still exists in the form of our pressure cooker. (See Figs. 5.1A and 5.1B for the digester and its explanation.) Papin originated the first safety valve, which was a rod at the top of the digester that would be displaced if the pressure got too great. Like our kitchen utensil, Papin applied the digester mainly to cook food, among other things using it to make edible jelly from bones that would otherwise go to waste. He suggested that the digester could help chemists and conducted some experiments, but they chiefly involved putting solids into solution with the aid of the higher temperatures possible under pressure. At first, temperatures were not directly observed with a thermometer, but Papin timed how long it took a drop of water on the lid to evaporate, and observed the dependence of the boiling point on pressure (Papin, [1681] 1966, p. 39–40).

Unlike experiments on fusion, there had been relatively little experimental work with pressure. With the former, there was the convenient laboratory or industrial furnace, understood and controllable, and also the centuries of experience with earth materials. Instrumentation to measure pressure in the laboratory was much less familiar and more difficult to devise. The majority of early investigations were concerned with the demonstration of atmospheric pressure and what happened when it was decreased. Little was done with increased pressure. Papin used a value of Boyle's to estimate atmospheric pressure at about 12 lbs./in.^{2} (Papin, [1681] 1966, p. 3). He then put a weight on a lever over the rod that extended into the digester and served as a safety valve. If the valve just lifted, he could calculate the internal pressure using the diameter of his rod, the length and weight of the lever, and the weight on it. He wrote:

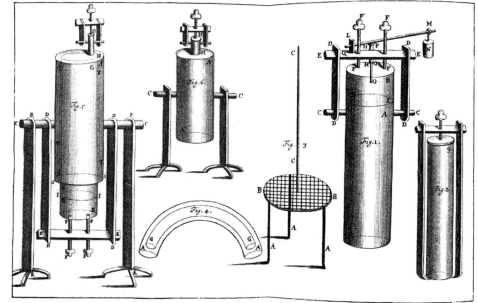

Figure 5.1A. The plate showing Papin's (1681) digester.

Therefore, when there is one pound weight hanging in M [on his lever], and yet the water gets out under the valve, one may conclude that the *inward* [internal] *pressure is about eight times stronger than the ordinary pressure of the Air.* (Papin, [1681] 1966, p. 3)

Papin varied pressure by controlling the temperature to which he subjected his digester, and his report indicated that he could adjust it with relative ease between 5 and 12 atms.

Papin's digester remained part of laboratory equipment for a long while. Joseph Black's lectures included mention of a digester (Simpson, 1982, p. 57). William Henry noted that not only did the boiling point of liquids decrease when the pressure on them decreased, but it was also easy to demonstrate the converse—the increase of boiling temperature with increase in pressure—by means of a Papin's digester (Henry, 1814, v. 1, p. 104). By early in the nineteenth century, instruments to show the increase of pressure with temperature had been devised, and were in scientific collections (Ambler, 1969, p. 54–55).

Pressure and its effects had thus been investigated quantitatively for about 150 years before pressure was considered as an experimental geological variable, but it had not been so well regulated or documented as temperature. At the beginning of the eighteenth century, barometers were not corrected for temperature. By the final decades of the century, the best of them could be read to a few thousands of an inch, and were corrected for temperature, capillarity, and the curve of the meniscus, therefore giving far better values (Heilbron, 1990a, p. 7). Thus, by the time Hutton's theory was tested, pressure instruments had evolved, quantitative reasoning was increasingly accepted, and calculations were carried out more carefully than ever before. However, the methods used to estimate greater than atmospheric pressures by the American-born inventor Lord Rumford (Benjamin

Thompson, 1753–1814) in his investigation of the power of gunpowder, which were subsequently adapted by Hall for geological purposes, were very similar to those of Papin (Rumford, 1797).

For anyone who looked closely at rocks, geological puzzles could be seen everywhere. Limestone layers were contained within lava or basalt layers, and yet retained carbon dioxide (fixed air). And no matter whether the preferred rock consolidation method was precipitation of minerals or cements from solution, or fusion, an explanation for rock layers that were not horizontal was needed. Wernerians explained them as the result of precipitates and sediments deposited on the originally uneven core of Earth. The unevenness of layers in stalagmites, stalactites, and chalcedony was noted. Huttonians or Vulcanists applied the documented expansive power of heat, which in the laboratory could be illustrated in many ways including early metal pyrometers. In nature, volcanic activity was invoked as a force that would raise and disrupt rock layers, especially when aided by water vapor and other volatiles that would provide more pressure. Hutton's heat source presumed a deeper and more pervasive source of heat than did the Vulcanists. As recognition of the forces grew, the forces needed also increased, from those that operated locally to those that might affect an entire mountain range.

LIMESTONE, HEAT, AND PRESSURE

From the first public expositions of Hutton's theory in 1785 and 1788, there were serious disagreements about the nature, source, and strength of the forces he invoked. But he saw no need to investigate them experimentally: He saw their results, evident to him in hand specimens, and the results proved their reality. Known and/or more closely observed causes could also be applied to earth disruptions. Richard Kirwan attributed the

CHAP. I.

The Description of the Engine, and how to use it safely.

AA. Is a Brass Cylinder hollow within, shut up at the bottom, and open at the top.

BB. Is another hollow Cylinder of the same bigness as the other, but much shorter, being to cover and shut the same by applying both their apertures to one another, as you may see in the Scheme.

CC. Are two Appendices or Ears cast to the Cylinder, AA. as the Tronions of a piece of Ordnance.

DDDD. Are two pieces of Iron put upon the Appendices CC. at one end, and the Iron bar EE. at the other.

EE. Is an Iron bar put through the ends of the Iron pieces DD. and so may easily be taken off or put on, when we have a mind either to open or to shut up the Engine.

FF. Are two Screws, which being fitted to the holes in the bar EE. serve to press both the Cylinders AA. BB. against one another.

GG. Is another hollow Cylinder made of Glass, Pewter, or some other Material, fit to receive those things that are to be boiled : this being filled and stopt with a cover exactly ground to it, and pressed upon it with a Screw, as you see in the Figure, is to be included in the Cylinders AA. BB. with water all round about it.

TO use this Engine with conveniency and ease, it ought to be fitted in a Furnace built on purpose for it, and should go in as far as the Appendices CC : so the fire being underneath, and the

Figure 5.1B. The page in Papin (1681) with an explanation of the parts of the digester.

enormous force of volcanic eruptions to water that was converted by heat to vapor "whose elastic force is known to be several thousand times greater than that of gunpowder" (Kirwan, 1784, p. 391). But this was not sufficient for the experimentally minded Sir James Hall.

Hall's investigations about the effect of heat combined with pressure are surely one of the great experimental sequences in the history of geology. The belief that pressure was a sufficient factor to enable water or vapor retention within minerals or rocks, as stated by Hutton in the quote at the beginning of this chapter, was not shared by many. Hall himself felt that only recent advances in chemistry allowed some geological facts to be tested. For him,

explanations that used water as the agent of rock formation had to fail, as the substances were insoluble and Earth didn't have enough water to have put all the rock material into solution. Fire had seemed equally inadequate, since many earth materials are destroyed, rather than formed, by fire. Neither water nor fire could be scientifically shown to be the prime cause of the origin of earth materials and their changes. In Hall's view, this situation changed when Black determined the nature of limestone. Hall said: "Of all mineral substances, the *Carbonate of Lime* is unquestionably the most important" (Hall, 1812, p. 73). Not only were there large amounts of limestone or marble in every country he knew of, it was also present, he said, in all kinds of other stones. But the loss of the substance that formed what Hall called carbonic acid upon heating had been proved. Therefore, any theory had to account for its presence in rocks that had either been fused, or that existed near a volcano or within its volcanic strata.

Hall reduced the portion of Hutton's theory that he proposed testing to three short statements:

I. THAT Heat has acted, at some remote period, on all rocks.
II. THAT during the action of heat, all these rocks (even such as now appear at the surface) lay covered by a superincumbent mass, of great weight and strength.
III. THAT in consequence of the combined action of Heat and Pressure, effects were produced different from those of heat on common occasions; in particular, that the carbonate of lime was reduced to a state of fusion, more or less complete, without any calcination.[1]

The essential and characteristic principle of his [Hutton's] theory is thus comprised in the word *Compression* (Hall, 1812, p. 74)

Hall admitted that it took him a long while to understand the implications of Hutton's theory, and in order to understand it, he "conceived that the chemical effects ascribed by him to compression, ought, in the first place, to be investigated" (Hall, 1812, p. 75). It was here that Hall differed from Hutton in feeling that investigations in a controlled manner on a small scale could yield useful information. This extensive series of experiments was carried out from 1798 to 1805. Hall's "compression" paper was read to the Royal Society of Edinburgh in 1805, but he communicated his initial results in 1804. The Royal Society paper was published in 1812.

Hall's aim was not simply to fuse limestone or marble without loss of carbon dioxide. He also wanted to know "what compressing force is requisite to enable it to resist any given elevation of temperature" (Hall, 1812, p. 77). Experiment was the only way he could find out. He noted that some compounds of what he called "lime with acids" (Hall, 1812, p. 77–78) are fusible, and others are not. His hope was that the carbonate of lime would be fusible at no more than the heat required to fuse whinstone, which he had learned from his previous work was 28 or 30 °WW (Hall, 1812, p. 78).[2] This seemed reasonable because nodules of limestone are

---

[1]Limestone assumed its importance for two reasons: There are very large masses of it that need to be accounted for in their normal state; and Hutton's presumption of heat would apparently negate the presence of the carbonate.

[2]This would be *very* roughly ~2750 °C.

sometimes found in whinstone. He conjectured that the two substances had been in a liquid state at the same time, but had not mixed, just as oil and water do not mix. He predicted that he could fuse the carbonate if he could heat it to the fusion temperature of basalt under such a pressure that the volatile gas was retained.

The Experiments

The initial premise may seem simple enough. Hall noted that what he called carbonic acid remained combined in the solid carbonate, but became volatile under the influence of heat. He stated his case:

It is evident, that were the attractive force of the lime increased, or the volatility of the acid diminished by any means, the compound would be enabled to bear a higher heat without decomposition, than it can in the present state of things. Now, pressure must produce an effect of this kind; for when a mechanical force opposes the expansion of the acid, its volatility must, to a certain degree, be diminished. (Hall, 1812, p. 77)

Therefore, his plan was to heat the sample in a confined space so as to raise the pressure and then observe the result. But in practice this was not simple.

Samples

Hall listed the carbonate materials he used as chalk, common limestone, marble, spar, and shells. Spar was a name used for several minerals, calcite being one of them. Hall's description of what he called "pure calcareous spar …, remarkably transparent, and having a strong double refraction" (Hall, 1812, p. 133) leaves little doubt that it was calcite. He found it more refractory than chalk. After many closely observed experiments, Hall noted discrepancies in the temperatures that somewhat consolidated samples had achieved after being heated under pressure. He realized that mixtures might fuse because of contact with parts of the apparatus, particularly tubes made of Cornish clay; and also that samples of natural carbonates might have impurities that could lower their fusion temperatures. Since he was concerned with natural rocks, he supposed that using those natural rock samples would give relevant results. Still, he asked several friends who were practiced in chemical analysis, which he claimed to have not mastered himself, for pure carbonates of lime for his investigations (Hall, 1812, p. 132). One friend was mineralogist and geologist Sir George Steuart Mackenzie (1780–1848). The others were British mineralogist Thomas Allan (1777–1833), and analytical chemist Charles Hatchett (1765–1847). Not surprisingly, Hall found their pure samples to be more refractory than his samples of natural ground chalk.

Hall measured the length of a piece of the carbonate before and after heating under pressure by inserting it in the gauge of his pyrometer. He found that it shrank three times as much as the pyrometer pieces did when heated to the same temperature, that it no longer absorbed water, and its specific gravity increased

(Hall, 1812, p. 96). Partly because of those properties, Hall generally pounded the chalk, marble, and spar samples until they were fine (he referred to them as powdered), passed them through the finest sieves also used for fabrication of pyrometer pieces, sometimes (depending on the experiment) weighed this substance carefully, and rammed it into the sample tubes. He followed a similar procedure with both the chemists' carbonates and oyster shells, and in a few instances, put large pieces of periwinkle shell into the inner tube.

Containers

After the first investigations, Hall put the carbonate sample into a small tube that in turn was enclosed in a larger container which was placed in the furnace. Hall did some of his most innovative thinking about these vessels. Both inner and outer containers had to be strong enough to withstand the pressures developed as well as the heat necessary to reach fusion conditions. The inner container needed to be sealed so that any volatile material formed would be retained. The initial concept was simply to enclose the small sample container in a tube that could be heated in a furnace. Hall wrote:

When I first undertook to make experiments with heat acting under compression, I employed myself in contriving various devices of screws, of bolts, and of lids, so adjusted, I hoped, as to confine all elastic substances. (Hall, 1812, p. 79)

But in January 1798, Hall thought of a far simpler device to answer his purposes of strength and tightness, namely, a gun barrel, closed at one end with an iron plug. Gun barrels had been used previously as containing vessels where a somewhat high pressure was desired, for example, in the work of Priestley in his investigation of the expansion of water at high temperature (Priestley, [1790] 1970, v. 3, p. 519). Hall placed his sample in the closed end of the barrel, filled the middle with a refractory substance, and left an empty space at the other end. (See Fig. 5.2.) The empty end was heated and then could be plugged with fused iron or sealing wax, which have quite different melting points. The different substances could be used according to how the barrels were placed in the furnace. At first the barrels were put into the furnace horizontally, with the end containing the sample in the furnace. Hall's comment was, "In these trials, though many barrels yielded to the expansive force, others resisted it" (Hall, 1812, p. 81). Yielding to expansive force included having the barrel burst open and the furnace blown to pieces. Some barrels remained intact, but some were split and some swelled. In many cases, the desired pressure was not maintained.

Because of these difficulties, Hall determined to use porcelain tubes he acquired from Wedgwood, made of the same material as white mortars (Hall, 1812, p. 89). He varied his procedure somewhat with them. Not surprisingly, there was difficulty in getting these tubes to sustain a high pressure, and Hall dealt with the problem by attempting to surround the sample holder or to

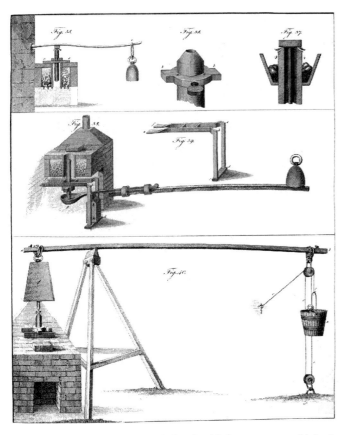

Figure 5.2. Hall's (1812) Plate IV, showing his furnace setup with horizontal and vertical placements and counterpressure methods.

was too small to insert both his samples and a pyrometer piece, he had to place the piece outside when the bar was in the furnace. Hall had many experiments he considered successful with it, but the internal size decreased due to scale formed by oxidation (Hall, 1812, p. 110). Wishing to have more bars like it, he discovered that the bar he was using was made of a Siberian iron called "old sable." By talking to a workman, Hall learned more about the properties of different kinds of iron bars, this "old sable" being the most resistant to heat. He was successful in acquiring more of them and, despite the fact that one cracked along its length, he generally was able to confine his samples and achieve the desired temperatures.

The containers that held the samples were also the subject of extensive investigation. At first he enclosed the carbonate sample in what Hall called "a cartridge of paper or pasteboard" (Hall, 1812, p. 80) to keep it from contact with the iron barrel. He had surrounded the sample with baked clay that was found to have been stained black throughout the barrel by material from the paper. Hall thought this demonstrated that the pressure had not confined the volatiles. In an effort to keep the carbonate from contamination, he decided to enclose the sample in a small tube made of glass or of Réaumur's porcelain. He mentioned in a footnote that glass was too fusible, and he had begun using tubes of "common porcelain" (Hall, 1812, p. 84). However, he did report on a number of trials made with both bottle glass and Réaumur's porcelain as sample holders.

The small porcelain sample tubes reacted at times with the carbonate. Hall tried to contain the sample in charcoal, but it absorbed both air and water irregularly (Hall, 1812, p. 132). He then tried sheets of platinum. (See Fig. 5.3.) He folded one sheet into a cup, as we still do today with filter paper. He covered it with another sheet of platinum, bent at the edges. Sometimes he wrapped the platinum around a cylinder, the ends of which could be covered in the same way. The first arrangement could retain a thin liquid, while the second could not. Hall thought that was not too important, as the results of carbonate fusion would remain (Hall, 1812, p. 132). The results Hall obtained with platinum pleased him because the sample was not contaminated by contact with iron or porcelain, and because he had only to unwrap the platinum sheet from around the sample after fusion instead of breaking the tube to remove it. At times the platinum was contaminated with the fusible metal used for a seal, and it partially melted (Hall, 1812, p. 138).

In all of these trials, Hall's objective was for the sample container to remain sealed during fusion so that the contents would remain under pressure and carbon dioxide could not escape. When he began, with the carbonate sample in pasteboard, he rammed the outer tube full of baked clay. That sample was somewhat consolidated after heating, and still effervesced in nitric acid, but staining indicated that volatiles had been lost. He next tried surrounding the sample with a fusible metal, consisting of a mixture of eight parts of bismuth, five of lead, and three of tin, which melted at the temperature of boiling water (Hall, 1812, p. 82). Since only one end of the barrel was in the heat, the other end remained cool, and he hoped the metal would prevent the escape

coat the tube with various substances that would not allow the gas to escape. After several years, he determined to get more reliable porcelain tubes, and so experimented with mixtures of porcelain clay to produce his own, which he found were better. But in the constant effort to keep the tubes tight, he sometimes succeeded so well that the porcelain tubes formed what Hall called a "minute longitudinal fissure" (Hall, 1812, p. 94), which allowed escape of some products. The porcelain tubes had the advantage of not adding iron as an impurity with unpredictable results as the gun barrels had, but it proved impossible to expose them to the strongest heats he desired, and he could not retain all of the carbonic acid (carbon dioxide). So, despite a number of what Hall considered reasonably successful outcomes, he returned to using iron for the containing barrels again.

In the new series, Hall tried placing gun barrels in the furnace vertically instead of horizontally. He was pleased with some of the results, and felt that the iron barrels could indeed bear more pressure than could porcelain. However, barrels were still being destroyed. He had new ones made by wrapping thick iron plate around a mandrel.[3] These did not perform well, so he then had a solid iron bar bored. This was strong, but since the internal cavity

---

[3] A mandrel is a metal rod that serves as a support for metal cast around it.

of carbonic acid. It could then be easily removed to release the sample. However, when he ran a trial with just the fusible metal in the barrel, and heated just the one end, the metal was extruded in drops, and eventually in streams, from minute pores in the barrel. The barrel was destroyed.

In his next attempt, Hall allowed a little air to be in the barrel with the fusible metal "which, by yielding a little to the expansion of the liquid, would save the barrel" (Hall, 1812, p. 85). Sometimes he got good results with this method, but the appropriate amount of air to include was not easy to determine. He also filled the sample tube halfway with the carbonate, and the other half with what he termed "pounded silex or with whatever occurred as most likely to prevent the intrusion of the fusible metal in its

liquid and penetrating state" (Hall, 1812, p. 85). He placed that tube, another like it with only air in it, and a pyrometer piece in a cradle of iron attached to a rod. The rod allowed the assemblage to be withdrawn from the outer barrel after the fusible metal was melted either by insertion into a horizontal furnace, or a tub of hot brine. He described several experiments where he was very pleased with the results of that procedure.

When Hall used porcelain tubes for his outer container, he knew the porcelain was porous. He thought he could put the carbonate into the tube, and then fill it nearly to the end with pounded flint. At the end he put borate of soda. He fused the borax into glass, and then reversed the tube to heat the carbonate. He found that the carbonic acid from the carbonate passed through the silex,

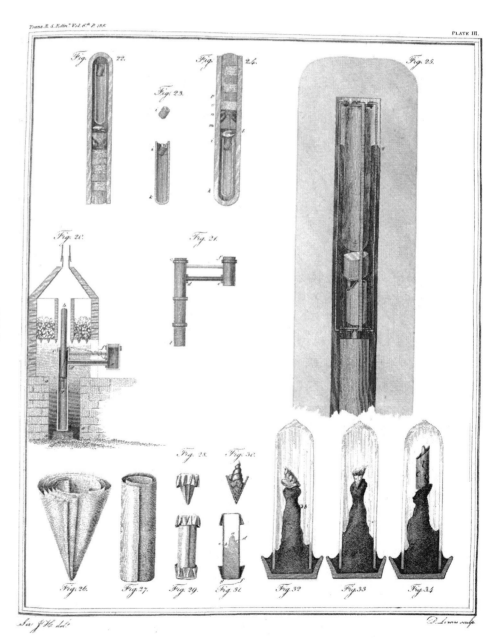

Figure 5.3. Hall's (1812) Plate III which shows several sample enclosures and sample placement, as well as one furnace arrangement and a platinum cup shape.

and that the borax glass cracked when it cooled, thus letting the volatile material escape. He did get some good results, but often products were lost, and he tried multiple combinations to prevent this. He was able to coat the inside of the outer porcelain tube with melted borax, and this was satisfactory for some trials.

Later, Hall began using water instead of air to increase the pressure in the barrel. The expansive power of water on changing to vapor was well known. He added the water to a small piece of chalk in a small tube that did not hold the experimental sample. That tube and the sample tube were placed in the cradle and the whole inserted in the barrel that had been filled with the fusible metal just above melting temperature. The metal cooled

and solidified immediately, thus trapping the water and vapor. Not surprisingly, he had great difficulty in regulating this innovation, and recounted many trials of several arrangements. Some gave good results. Hall later tried other volatiles to increase compression, among them the carbonate and the explosive nitrate of ammonia, gunpowder, and paper impregnated with niter (Hall, 1812, p. 124). It is quite amazing that Sir James and the people he employed survived all this!

Furnaces and Position

The furnaces Hall used, and the position of the barrel within them, were modified as he saw need for it during his long series of experiments (see Fig. 5.4). He began with "a square brick

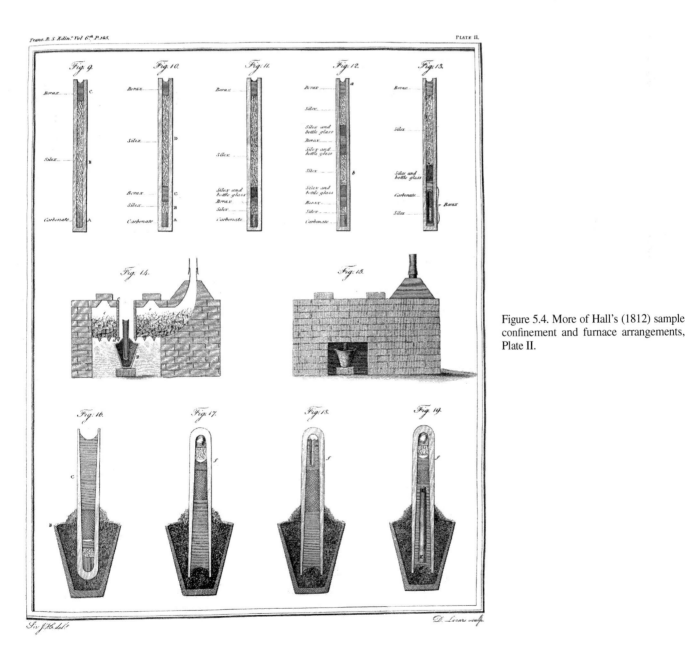

Figure 5.4. More of Hall's (1812) sample confinement and furnace arrangements, Plate II.

furnace, having a muffle traversing it horizontally and open at both ends" (Hall, 1812, p. 86). The muffle was surrounded by fire. The barrel was inserted with the sample end in the furnace and the other end extending outside, cooled by wet cloths. The fusible metal was gently melted at the end of the experiment. Later, with the porcelain tubes, he used a furnace in which the muffle was vertical, and surrounded by fire. The barrel was placed vertically, and in this position the surrounding borax would not all fall to the lower side of the tube. Hall later had a pit dug under the vertical muffle to hold a receptacle of water in which the cool end of the experimental barrel could be contained.

Measurement and Calculations

Use of pyrometer pieces. Hall used Wedgwood pyrometer pieces to estimate the temperatures for most of his trials, although he sometimes did not include them in the initial phases of a series. He was, of course, most concerned with getting an indication of the temperature, but it was not always possible to place the pieces in the optimum position. When he used gun barrels, he sometimes placed the piece in the cradle that held the sample tube in the barrel, next to the sample tube, or sometimes in the muffle next to the barrel. In one trial, he put a piece in each of those places and found that they registered the same temperature. At a much later trial with two pyrometer pieces, they differed by 4 °WW. If the pyrometer piece had to be placed outside the barrel because of lack of space inside, he placed it as close as possible to the location of the sample tube inside. In some trials without a pyrometer piece, he could give estimates of the temperature that were comparatively accurate, considering his experience with the properties of his furnaces, apparatus, and fusion products. We saw what the contemporary and current evaluations of the pyrometer were in Chapter 4.

There has been some difference of opinion about whether Hall succeeded in actually fusing calcium carbonate (Eyles, 1972, p. 54). He is credited with being the first to experiment with minerals at elevated pressures (Yoder, 1980, p. 6). But the first unquestioned fusion of calcium carbonate was done in 1912 by H.E. Boeke at 1289 °C and 110 atms of pressure (Eyles, 1961, p. 216). What is perhaps more important is that Hall did subject various forms of calcium carbonate to high temperatures under pressure, and that some definitely retained carbon dioxide, as shown by effervescence in acid after cooling.

In some of Hall's first experiments, he found that powdered limestone was "agglutinated into a stony mass, which required a sharp blow of a hammer to break it" (Hall, 1812, p. 81). That same sample effervesced violently in nitric acid. Another, still in the gun barrel series, produced a hard, crystalline substance, semitransparent at the edges. The pyrometer piece in the muffle with the barrel indicated 33 °WW (Hall, 1812, p. 87–88). Hall also had success with some of the porcelain containing–barrels. One product was a marble, which according to him, the workmen he employed to polish it couldn't tell from natural marble (Hall, 1812, p. 97). However, there were many failures when Hall tried to increase the heat applied to the porcelain. He determined that going beyond 27 or 28 °WW was not possible. The old sable iron proved able to sustain higher temperatures, even to 78 and 79 °WW. It took a number of trials for Hall to find out how to prevent the escape of carbon dioxide at those temperatures, in the course of which he had samples reduced to what he called froth, which to him indicated fusion (Hall, 1812, p. 118). However, when cooled, some products did not effervesce in acid.

It is difficult to judge how successful these experiments were (see Table 5.1). The variables were many and interacting, and included the temperature and pressure, the time they were applied, the composition of the sample material, shielding materials and the tube, and the method and substance used to increase the pressure. Some products appeared to be hard and well consolidated, and some had shiny, faceted surfaces. Some, despite having been subjected to rather high temperatures, still effervesced in acid. In the opinion of Victor Eyles, who did extensive biographical work on Hall, evidence shows that he did succeed in some fusion of the carbonate (Eyles, 1972, p. 55). But Hall did not draw his conclusions based solely on appearances. Along with the wealth of observations, he calculated the specific gravity of reactants and products, the percent loss in weight, and the pressure that was applied, as well as measuring shrinkage of the samples.

The whole sequence of experiments has generally been underreported, both during Hall's time and since. In his day, reac-

TABLE 5.1. HALL'S PARAMETERS FOR CALCIUM CARBONATE UNDER PRESSURE/HEAT

Sample tube	Enclosed by	Barrel	T in W	Success?
Paper	Baked clay	Gun barrel	25	Yes
Glass	Fusible metal	Gun barrel	33	Yes
Réaumur's porcelain	Fusible metal	Gun barrel	23	Yes
None	Borax and silex	Porcelain	not >27	??
Porcelain	Baked clay	Gun barrel	32	No
Porcelain	Chalk	Bored iron	64	Melted
Porcelain	Chalk	Bored iron	63	No
Platinum	Chalk	Bored iron	30	Yes
Porcelain	Borax and sand	Bored iron	25	Yes

Note: Success might be indicated by marked effervescense in acid or in a resultant compact surface showing the angles of calcite.
T—Temperature; W—Wedgwood °.

tions to the work were of two sorts. Huttonians declared that rock origin by fusion had been proved, by removing the objection that carbon dioxide could not have been retained in heated carbonates. Wernerians found the whole premise doubtful. They failed to see how the delicate details of fossil shells could be preserved under such a regimen. Even though repetitive experimentation with control of variables was the ideal of late eighteenth-century science, few people made any attempt to emulate Hall's work. One exception was Étienne-Marie-Gilbert de Drée (1760–1848), a French count who was the brother-in-law of the celebrated French scientist Deodat de Dolomieu (1750–1801). Drée's work, which did not duplicate Hall's, will be discussed in Chapter 10. Later authors tended to assume that Hall's work was proof of igneous origin for basalt, and of the effectiveness of pressure in retaining carbon dioxide in limestone. There was little discussion about the experiments themselves.[4]

Specific gravity results. As discussed in Chapter 2, specific gravity was an important means of identifying minerals. Hall determined the specific gravity of his samples before and after they were heated under pressure, so that comparison of the numbers would allow conclusions to be drawn about, say, whether the sample was compacted or lost carbon dioxide during the procedure. But questions remain as to whose investigations allow a look at calculation practice in Hall's time. How good were the numbers used? The answer to this question has two components, as everyone knows who has done such rudimentary work as determining a density in a laboratory setting. First, to what numbers of significant figures may the instruments be read? And, second, do the number of places in the result reflect both the answer to the first question *and* the effects of the mathematical processes that have been used? We can look at both the numbers upon which Hall based his conclusions as well as the processes by which he arrived at them.

Hall himself was conscious of the need to use figures with care. In his 1812 paper, he included two appendices, the first of which explained his procedures for determining specific gravity, and the second, how he arrived at the table that gave the pressures to which his samples had been subjected and the depth of sea that those corresponded to. (See Fig. 5.5.) He also stated that he and an associate, Mr. Jardine, had separately calculated the tables for both specific gravity and pressure (Hall, 1812, p. 185). The calculations themselves are in his MS. 5020 (Hall, 1805a) in the Scottish National Library. He carefully explained his procedure for weighing a sample tube, then the tube plus the sample. Next he said:

After the experiment, the tube, with its contents, was again weighed; and the variation of weight obtained, independently of any mutual action that had taken place between the tube and the carbonate. The balance which I used, turned, in a constant and steady manner, with one hundredth of a grain. (Hall, 1812, p. 112)

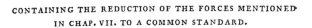

| 184 | EFFECTS of HEAT | | | | [APPENDIX.] |

No. II.

TABLE,

CONTAINING THE REDUCTION OF THE FORCES MENTIONED IN CHAP. VII. TO A COMMON STANDARD.

I. Number of experiment referred to in Chap. VII.	II. Bore, in decimals of an inch.	III. Pressure in hundred weights.	IV. Temperature by Wedgwood's pyrometer.	V. Depth of sea in feet.	VI. Ditto in miles.	VII. Pressure, expressed in atmospheres.
1	0.75	3	22	1708 05	0.3235	51.87
2	0.75	3	25	1708.05	0.3235	51.87
3	0.75	10	20	5693.52	1.0783	172.92
4	0.75	10	31	5693 52	1.0783	172.92
5	0.75	10	41	5693.52	1.0783	172.92
6	0.75	10	51	5693.52	1.0783	172.92
7	0.75	10	—	5693.52	1.0783	172.92
8	0.54	2	—	2196.57	0.4160	66.71
9	0.54	{ 4 / 8.1	—	4393.14 / 8896.12	0.8320 / 1.6848	133.43 / 270.19
10	0.75	3	21	1708.05	0.3235	51.87
11	0.75	4	25	2277.41	0.4313	69.70
12	0.75	5	—	2846.76	0.5396	86.46

EXPLANATION.

Figure 5.5. Hall's (1812) table showing depth and pressure calculations.

The weights Hall used were recorded to the second decimal place, or hundredths. However, in his calculations he reported more significant or, in Heilbron's words "(in)significant," figures than we would regard as warranted or relevant. In the table, Hall did not resort to the "crowds of irrelevant decimals" (Heilbron, 1993, p. 31) prevalent at about this time as physics, and science in general, embraced quantitative reasoning.[5] In his other calculations, he often ended with more decimal places than would now be acceptable, but MS. 5020 reveals his meticulous attention to both the precision indicated by the numbers and to the calculations.

Many mineralogy books include figures for the specific gravity of marble and limestone. Hall explained his procedures to determine them thus:

As many of the artificial limestones and marbles produced in these experiments, were possessed of great hardness and compactness, and as they had visibly undergone a great diminution of bulk, and felt heavy in the hand, it seemed to me an object of some consequence to ascertain their specific gravity, compared with each other, and with the original substances from which they were formed. (Hall, 1812, p. 177)

---

[4]Two early examples are in Lyell ([1830] 1990, v. 1, p. 62, 348, 472), and Bakewell ([1833] 1978, p. 146, 210).

[5]There is a good discussion of the rise of consciousness about the use of significant figures in Jan Golinski's (1995) chapter titled "The nicety of experiment."

Specific gravity was determined as described in Chapter 2, by weighing the object in air and again when immersed in water. The difference in the two weights represented the weight of the water that the sample displaced so that specific gravity of the carbonate samples could be calculated as the dry weight divided by the weight of the same volume of water. Hall recognized that the carbonates might have had voids in them, so he weighed some samples again after they had been immersed in water. He added the difference between dry and wet weight, or the absorption, to the difference in immersed and dry weights, to get a more reliable specific gravity. He also made sure that all bubbles had ceased rising from the chalk. In MS. 5020 there are lists of specific gravities of both chalk and natural limestone. After calculating data for many samples of both, Hall noted the variability in similar samples, and suggested that natural limestones might be more or less calcined. There is a full page of long division that shows each of his specific gravity calculations (Hall, 1805a, p. 175).

The results, however, were not clear. The evidence of successful trials was that samples became more compact and dense after heating. However, in some cases, the specific gravity decreased. Hall suggested that this might be because partial fusion of the sample could close off pores within it, so that water could not penetrate. He also used the gauge for the Wedgwood pyrometer pieces to measure how some of the samples decreased in size during heating, which of course would tend to increase specific gravity. Historians of science in the twentieth century often express confidence that Hall was successful in some of the trials as a result of his critical assessment of every possible parameter.[6]

Force. Hall wished to relate the pressure supplied by his apparatus to conditions in the natural world that might supply the pressure. He had first to design an apparatus that would provide measurable pressure, then to calculate that pressure, and finally, to translate the pressure to natural conditions. Deluc made it sound easy:

In order to determine the degree of resistance, which, in the course of these experiments, the fixed air and the aqueous vapour underwent at different degrees of heat, and their effects upon the calcareous matter, Sir James Hall produced the resistance, by means of moveable weights, in the manner of the valves applied to the boiler of the steam engine. (Deluc, 1809, p. 361)

Actually, for the first five years of Hall's experiments, 1798 until 1803, there were no pressure measurements. But in June of 1803, Hall began to try to determine the limits of the force developed in his tubes "in a mode nearly allied to that followed by Count Rumford, in measuring the explosive force of

gunpowder" (Hall, 1812, p. 139).[7] As mentioned above, these measurements were done mechanically in a way quite similar to that of Papin.

Thus, in order to calculate the restraining pressure, Hall needed to know the bore of his sample tube and the weight that just prevented the escape of gases from within. He tried three different arrangements of lever arms and weights applied to the barrel whose bore area he knew, as was illustrated by his Figures 35, 38, and 40. (See Fig. 5.2.) From the pressure reading he could calculate the pounds applied to the square inch. His methods of calculation are interesting to modern eyes. When Hall calculated specific gravity he had just two numbers, albeit of four to six places, to be concerned with. For force, he recorded the restraining weight in hundredweights, and next converted to pounds avoirdupois by multiplying by 112 pounds per hundredweight. He calculated the area of the cross section of his tubes, the height of a column of seawater that represented that force in feet and miles, and then converted to atmospheres.

Each calculation was done in a manner that would probably not occur to us. Since Hall's force investigations were done with circular bores in the test barrel, we would probably apply Area (circle) = πr^2, where r is the radius. Instead, Hall recorded the area of a circle with a diameter of unity, or one, which value is 0.785398. He then squared the diameters of his two different bores (0.75 and 0.54 in.^{2}) and multiplied the results by the area of the circle of unity to get the areas of those circles. His interest in having all his calculations on record in his laboratory book extended to the inclusion of the calculation of bore area at the beginning of each of his calculation sequences, instead of calculating it once and using the resultant number for the rest. Unlike Rumford, at least as stated in the published paper, Hall didn't use what we would call a "ballpark" figure for atmospheric pressure. Instead he went back to first principles, and in his list of data used to calculate the table, included Laplace's figure for the mean height of the barometer at sea level, which was 29.91196 in. of mercury, and Brisson's for the specific gravity of mercury, 13.568 (Hall, 1805a, p. 179). With those, he calculated the weight in pounds of 29.91196 in.^{3} of mercury and, by means of a long numerical series, ended with the number of atmospheres of pressure on his apparatus. By back calculating, I found that the

---

[7]Count Rumford was the American Benjamin Thompson (1753–1814), only part of whose highly varied career included subjects of military interest. Thompson arranged his pressure apparatus so that a weight closed the bore of a gun barrel. He set off measured weights of gunpowder, and found the confining weight that would just contain the "elastic force" generated. He calculated the area of the bore of the gun barrel he used (which was 0.049088 in.^{2}) and said:

[A]ssuming the mean pressure of the atmosphere upon 1 superficial inch equal to 15 lbs. avoirdupois, this will give 0.73631 of a pound avoirdupois, for that pressure upon 0.049088 of a superficial inch, ... consequently the weight expressed in *pounds avoirdupois* which measured the force of the generated elastic fluid in any given experiment, being divided by 0.73631, will show how many times the pressure exerted by the fluid was greater than the mean pressure of the atmosphere. (Rumford, [1797] 1870, v. 1, p. 140)

Thompson used ever-increasing weights of gunpowder and calculated atmospheric pressure well over 10,000 atm.

result of using his numbers gave a value of 14.66 lbs./in^2, a better value than Rumford's use of 15.

Using the value of 1.0272 for the specific gravity of seawater, Hall calculated the weight of a cubic foot of seawater in pounds. Again, without duplicating his calculations, I first converted that to the weight on a square inch of a column of seawater 1 foot tall. Then, using Hall's weight in hundredweights for one trial and the diameter of that bore, I converted the number of feet of seawater that the force represented, as in the fourth column of his table. It was gratifying to see that the same number that Hall recorded emerged, although not without rectifying at least one error that I had made. A comment here: As noted above, use of significant figures had not been standardized at the time. Some of the results in Hall's table are given with six digits, while some numbers he has used in his calculations have only one.[8]

There are at least three interesting things about Hall's calculation sequences. The first was the use of base values for each variable and, at least in the recorded sequences in his laboratory book, the initiation of each calculation from those values. The second was the actual form of the calculations, which were all expressed as ratios. This is an effective method that has been nearly abandoned in at least the United States at present. The third revelation elicited an "aha!" from an experienced engineer who is familiar with the history of mathematics. After Hall set up his density calculations, there was a neat list of logarithms, recorded in eight digits, for each of the major numbers used in the calculations (Hall, 1805a, p. 179 and facing page). The aforesaid engineer understood immediately. The numbers Hall used generally had three to six digits. When he recorded the result of a multiplication or division the answer, to be used in another calculation, was carried to even more digits. Without a calculator, it was immeasurably simpler to add or subtract logarithms than to do interminable multiplication or division. Hall was most probably following a normal practice of his time, but in twentieth-century works on the history of geological theory and the fieldwork of that time, there are virtually no comments on calculation methods, or serious studies of the treatment of significant figures, as is beginning to be done in the history of chemistry. One could argue that there was very little quantitative work done, and thus there is little need. I have previously cited others who noted that there have been few studies of experimental geology (Newcomb, 1990a, p. 161).[9] I felt then that even the published sources of experiments were little examined, and that it was perhaps too soon to look at laboratory notebooks and manuscript material. This short investigation of the calculations that lay behind particular conclusions has

proved that our understanding will not be complete until that is done. Meanwhile, the study of the origins of Hall's two tables in 1812 certainly inform that understanding.

Results

The results of Hall's experiments on carbonates were met with general acceptance that carbon dioxide *could* be retained under the conditions he employed. That the powdered rock was compacted was fairly well accepted, but actual fusion remained in question.[10] What it meant to geological theory was far from clear (Newcomb, 1990a, p. 194–195). Deluc thought the retention of carbon dioxide in limestone under pressure of a solid could be supported. He summarized Hall's work by saying that a calcareous powder was changed by heat to limestone under a pressure of 52 atms; under 86 atms to marble; and under 173 atms it could be fused (Deluc, 1809, p. 361). However, if heat were applied to limestone under water, the combination of the very low specific gravity of the carbon dioxide and its solubility in water would mean it couldn't be retained (Deluc, 1809, p. 365). To answer that objection, Hall pointed out that he had used water vapor itself to provide compression, and that the pressure increased before the temperature reached the point of calcination (Hall, 1815a, p. 90). While questions remained, Hall had clearly shown that pressure could act to retain the gas, and had translated experimental data to real-world conditions. Having proved to his satisfaction that pressure could have produced the anticipated effect on limestone, Hall's next experimental investigation was much simpler. Figure 5.5 summarized some of the conditions that obtained in Hall's work with limestone. And while heat was then the source of pressure, in the next inquiry he applied only lateral pressure, and no heat, to folding.

FOLDING, LATERAL PRESSURE

The folds in solid rock are hard to ignore, whether they are the relatively gentle or overturned folds in sedimentary rocks or the folded, refolded, and sometimes refolded-again-and-again layers of metamorphic rocks. From the earliest time, humans had looked at Earth and tried to explain what they saw. Over the sweep of millennia, the kinds of explanations varied, from the work of multiple gods to the divine attention to system and the power of one God. Whether postulated as caused by design of a Creator for the benefit of man, or a simple field observation of facts (volcanic uplift), the record was there to be read. And read it was.

It wasn't necessary to live in an area of geological interest to be aware of Earth. During much of the eighteenth century

---

[8]Results with my calculator and current awareness of significant figures generally yield answers very close to Hall's. I have also wondered what conventions of rounding up or down were, not knowing if Hall's lower answer than mine in the last place in one instance represented a different rounding convention, or simply that he didn't carry the calculation further.

[9]See also general comments and background by Rudler (1889), Bascom (1927), Mathews (1927a), Manten (1966), Yoder (1980), Yearley (1984), Fritscher (1991), Wyllie (1998), and the sequence by Yoder (1993).

[10]Both V.A. Eyles (1961) and B. Fritscher (1988) have examined some of Hall's samples in collections at the Geological Museum of London and the British Geological Survey headquarters in Keyworth near Nottingham. Both note that some samples are clearly solidified, and accept that carbon dioxide was retained. Eyles was sure that Hall's temperatures exceeded 1000 °C (Eyles, 1961, p. 215).

and before, published accounts of geological travels of the more empirically minded natural philosopher might center on minerals and/or mines, often tied to exploration for precious metals and stones, as well as useful substances. There was also an enormous literature by casual travelers, focused on the description of dramatic or pleasing vistas. Books of travels, often well illustrated, were extremely popular among the literate population, having somewhat the same role as our television documentaries. It wasn't necessary to live in an area of uplift or folding, or even to go and see it, to understand that it did exist.

James Hutton reasoned from instances of that kind. Having already determined to his satisfaction that heat was responsible for fusion of loose sediments into rock, and having seen rock strata "in every degree of fracture, flexure, and contortion" (Hutton, [1785] 1973, p. 17), it was obvious that heat would also be the cause for that. He expanded on the theme in 1788: "There is nothing so proper for the erection of land above the level of the ocean, as an expansive power of sufficient force, applied directly under materials in the bottom of the sea" (Hutton, [1788] 1973, p. 262). That power was his prime mover—heat—and he thought he needed only to see the effects to know it was there. In 1795, he noted that if water under pressure could "endure the heat of red hot iron without being converted to vapour in man's operations, how much more powerful nature must be" (Hutton, [1795] 1972, v. 1, p. 94). Even though Saussure suggested, as quoted by Hutton, that layers might have been distorted by forces of crystallization and pressure when they were still flexible, Hutton said Saussure simply had no idea of the possibility that strata could be formed under the sea and then softened by heat (Hutton, [1795] 1972, v. 2, p. 42). Actually, Saussure's ideas were not all that different from Hutton's, although he produced no overall theory of Earth (Carozzi, 1975, p. 122).[11] Saussure, referred to as "a celebrated Neptunist" (Murray, [1802] 1978, p. 113), and others who generally believed in Neptunist origins for rocks still understood that heat and expansion could be a factor in distorting rock layers.

In his [1802] 1964 *Illustrations of the Huttonian theory of the Earth*, John Playfair included Note XII titled "Elevation and inflexion of the strata." The examples of folding in the 30-page chapter were drawn from many contemporary writings as well as his own observations. He marshaled all of them to prove that Hutton's agency of heat could be the only explanation of the change from horizontality. Murray saw no difficulty in attributing strata that were other than horizontal, or indeed vertical, to successive crystallization "on the sides of masses already produced" (Murray, [1802] 1978, p. 108). He also referred to Saussure's observation of the layers in stalactites and alabaster that clearly formed by crystallization, but "instead of being in a straight line, are found with all these varieties of bending and waving which in the strata are conspicuous on a large scale" (Murray,

[1802] 1978, p. 108). The redoubtable Humphry Davy remained open-minded about the two possibilities of rock formation, with perhaps a slight preference for the Huttonian model, but could still say that in the case of some arched calcareous strata he had seen, deposition on an uneven surface was more likely than that heat had played a role (Siegfried and Dott, 1980, p. 93). In one more instance of the persistence of the debate over rock origin, Mackenzie, a pupil of Jameson's, in 1812 still found it necessary to convince his listeners that trap rock (basalt) had been formed by the agency of heat and was analogous to lava. With evidence that lava had flowed under seawater, he could also say that the volcanoes responsible for that could have thrust the layers above sea level (Mackenzie, 1815).

At just the time that Mackenzie was reporting his trips in conjunction with other members of the Royal Society of Edinburgh, Sir James Hall investigated a portion of the coast of Berwickshire, a county in southeast Scotland, where rock layers could be seen to be strongly folded. He noted the near-vertical position of some beds, and saw that the layers were successively concave and convex upward, with axes horizontal and parallel (Hall, 1815a, p. 82). (See Fig. 5.6.) In other places, the curves were more complex. He had been curious about how such things could occur, and thought "a mechanical force of a sufficient strength" (Hall, 1815a, p. 84) might do it. He wanted both to test that assumption, and also to show that geological forces, namely, volcanic, could provide the force. The first he would do with experiment, the second with reasoning. About the folds he said:

[I]t occurred to me, that this peculiar conformation might be accounted for, by supposing that these strata, originally lying flat, and in positions as nearly level as might be expected to result from the deposition of loose sand at the bottom of the sea, had been urged when in a soft, but tough and ductile state, by a powerful force acting horizontally; that this force had been opposed by an insurmountable resistance upon the opposite side of the beds,—or that the same effect had been produced by two forces acting in opposite directions; at the same time that the whole was held down by a superincumbent weight, which, however, was capable of being heaved up by a sufficiently powerful exertion. (Hall, 1815a, p. 84)

Others conjectured for many pages either on forces they could not see, or sought to find field evidence that demonstrated the phenomena, such as a volcanic intrusion contiguous to tilted beds. Sir James Hall had the advantage of extensive field experience, including his long tour among the volcanoes of Italy, and also had the ingenuity and the ability to translate from a very large scale to a small, controllable experimental investigation.

Hall's first effort to simulate the effects of a horizontal force on rock layers consisted of layers of cloth of different thicknesses piled rather thickly on a table. He put a nearby door, which was conveniently off its hinges, on top of the pile, and loaded it with weights. Next, he placed a board vertically at each end, the width of each board being slightly less than the height of the layers. To provide the force, he pounded on the boards with a mallet. As the boards were forced together, the cloth layers were folded to a

---

[11]Saussure's promised theory of Earth, although published widely, was only a list. His fuller exposition was apparently never written, although A.V. Carozzi has used Saussure's field notes and memos to further investigate his thinking (Carozzi, 1989).

PLATE II.

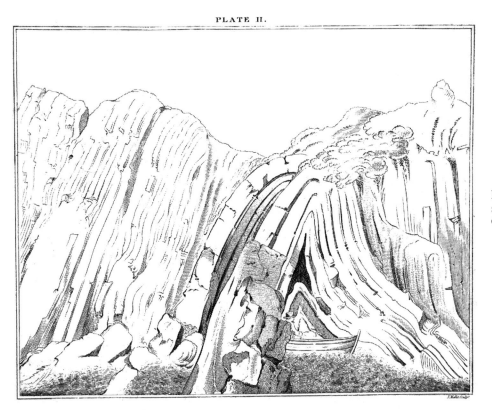

Figure 5.6. Hall's (1815a) Plate II, showing folded strata near Brander Cove (p. 81).

state which he thought not unlike the rock layers he had seen, the horizontal door being forced upward. For the sake of clarity, he used cloths of different colors so that the layers could be traced. When he showed this simulation to the Royal Society of Edinburgh, he used layers of clay. The paper was accompanied by illustrations of rock layers he had sketched (Hall, 1815a, p. 86). Hall noted that natural rock layers would be different in tenacity, firmness, brittleness, pliability, and ductility, depending on the materials that made them up and the conditions under which they were consolidated (Hall, 1815a, p. 89).

In the rest of the paper, Hall cited reports in current travel and scientific literature, his travels and observations, and those of other geologists to describe folded layers and how they might be the product of lateral thrust. In the field, he had noted the intrusion of granite. In another geological controversy, namely, whether veins were filled from above (Werner) or below (Hutton), Hall postulated that intrusion from below could supply the upward pressure necessary for folding in an area where there were no volcanic rocks. In what could be classified as a geological experiment (Yearley, 1984, p. 3), he cleared overlying layers away from a granite outcrop and employed workmen to dress the contact of granite with the country rock and water it to simulate a polish, thus making the structure clearer. He felt this clarified the sort of event that could have occurred to cause the folding.

Different kinds of products and degrees of folding could be expected due to variability of thrust force and the rock layers. Heat was constantly a part of those discussions. In a review of

structural geology published in 1927, Hall was given credit for the first experiments that reproduced folded strata by means of compression. The next experimenter in Hall's mode published in 1888 (Mathews, 1927, p. 147–148).[12]

CONCLUSION

The arguments about pressure are, necessarily, somewhat conflated with arguments about heat, since heat was often the agent of pressure, and was always a factor at the pressure location. Pressure might be due to an overlying column of seawater; an overlying mass of rocks; expansive forces generated by the heat of volcanism; or, finally, expansive forces generated by other intrusive rocks such as granite. Heat was present in the final three instances. In the first of those, the heat was presumed, as demonstrated by the normal increase of temperature with depth. That instance, and the final case of intrusive rocks, imply a deep, pervasive, and perpetual heat source. Explanations for volcanic heat ranged from plutonic heat to localized instances of flammable coal and pyrites.

As the subject matter of geology enlarged and diverged from more traditional areas such as mineralogy, there was relatively less laboratory testing of theory, especially in the new areas, and fewer scientists willing to attempt it, even as the work with

---

[12]The Scottish geologist Henry Cadell did experiments similar to those of Hall.

minerals continued and became more sophisticated with the advances in chemistry. In a sense, Sir James Hall's work, especially with fusion, was an extension of earlier inquiries and a broadening of contemporary investigation. However, to the best of my knowledge, his pressure investigations were nearly unique in his day. He was typical of the empiricist tendency of his time in that carefully designed experiments were considered the best way to gain understanding. Along with Hutton he agreed that reliable scientific arguments could only be originated from the appropriate evidence (Donovan, 1978, p. 187). Unlike Hutton, he believed that experiment, even that concerned with earth theory, was a valid method. Hall's purpose, in all four of his "experimental" papers (1805b, 1812, 1815a, and 1826), was to prove two tenets of Hutton's theory, namely, that rocks were both consolidated and elevated by heat. Pressure played a necessary role in each, although each was initiated by heat.

Murray had pointed out objections to the Huttonian theory of limestone origin by fusion, noting that Lavoisier and Saussure had been unable to fuse it, and he quoted Kirwan as saying there was not sufficient heat available to melt mountains (Murray, [1802] 1978, p. 29–30). He quoted Playfair's suggestion that even if quicklime (CaO) could not be fusible, the carbonate might well be fusible if heated under pressure. Even though Black had suggested that since the carbonate of barium was more fusible before the carbon dioxide (fixed air) was driven off, the same might apply to calcium carbonate, Murray rightly pointed out that calcium carbonate loses carbon dioxide well before it is fused. Additionally, Murray emphasized the necessity of great heat, but quoted the remarks of Playfair that combustion could not be supported under the conditions of pressure under Earth (Murray, [1802] 1978, p. 42–43). In this case, pressure to retain carbon dioxide was present, but there was not sufficient heat to fuse limestone. Later, Murray questioned the efficacy of heat as the sole agent of elevation. He said that even Neptunists might admit to an expansive force from below, the most likely being heat. But he thought even that did not prove the Huttonians correct, because the layers, while elevated, were not fused. Murray gave numerous instances from bed layering and composition that he felt showed the Huttonian insistence on fusion must be wrong. He suggested that subsidence after cooling might also result in folded or distorted strata. Also, since clearly stratified rocks could be seen folded, Murray didn't see how fusion under pressure could allow preservation of the "original division into beds" (Murray, [1802] 1978, p. 153).

It can be seen from these few instances of the arguments of the time that Hall's evidence that carbon dioxide in calcium carbonate most probably could be retained if heated under pressure did not result in immediate theory change. The role of pressure was admitted and debated before and after Hall's papers on confining and horizontal pressure. But the experiments themselves were elegant in concept, if not always in execution. That latter comment implies no fault, but is merely an observation that, when truly new ground is being investigated under novel conditions, with new designs and equipment whose parameters are not known, there will inevitably be unexpected results. Hall was admirable not only for his ingenuity at experimentation and his grasp of theory, but also for his ability to synthesize from a broad base of experience and reading. The design and execution of the series of more than 500 experiments with confining pressure brought together chemistry and geology, true, but he also organized and oversaw workers, ordered materials, sequenced, calculated, and thought deeply about his aims in the work. A great deal more was learned about pressure and the materials of Earth.

CHAPTER 6

Chemistry in the service of geology: Equipment

INTRODUCTION

By the latter part of the twentieth century, science had undergone well over a century of categorization into disciplines: physics, chemistry, biology, geology. Each had subdivisions of various sorts: optics, solid state physics, plasma physics; astronomy; organic, inorganic, bio-, analytical, physical chemistry; botany, zoology, and molecular biology; and structural geology, petrology, mineralogy, seismology, and sedimentology, etc. While boundaries between the sciences could be fluid, there often was a sense of working within a discipline until the last decades of the twentieth century when the advantages of more interdisciplinary work, as well as its necessity, were again perceived. In geology, this has culminated in what is called earth system science. Even the United States Geological Survey has a biological division. Thus, we currently see the advantages of knowledge and techniques from other sciences being applied to problems in earth science.

At the end of the eighteenth century, at the dawn of geology as a science, it could be argued that geology *only* had experimental methods from other sciences. Geology was just beginning to develop the tools to describe, quantify, and characterize the materials of Earth and their positions.[1] Because of this, and because of the eclectic nature of the studies and interests of those we recognize as workers in the field, it is not surprising that methods from both physics and chemistry should have been employed, as well as botany and zoology when fossils were considered. Interest in minerals had been strong for many years, because minerals were useful, valuable, or both. Minerals were the constituents of rocks, about whose origin there were many questions and much interest. A key to understanding minerals was chemistry, which was going through its own advances in the eighteenth century, a process that a number of men interested in earth materials either contributed to, or followed with great interest. Among them were those who commented on the promise of chemistry in geology, samples of which follow.

1753: Pott clearly felt that chemistry was necessary for his investigation of the reactions of the minerals and earths used making porcelain:

Sur ce principe, j'ose espérer que les nouvelles Expériences que je présente au Public, seront agréables à ceux qui versés dans la Chymie, cette Science si réelle, sçavent combien des recherches exactes sur la nature des corps, des Expériences vraies, des découvertes de rapports entièrement inconnus, & enfin des distributions convenables des productions de la nature, doivent être préférées à des idées chimériques, à des spéculations douteuses, & à des abstractions métaphysiques. (Pott, [1746] 1753, p. iv)[2]

1755: The French natural philosopher and art connoisseur, Antoine-Joseph Dezallier d'Argenville (1680–1765), wrote a natural history that included earths, stones, metals, and minerals, and was distinguished by its beautiful plates. About chemistry, he said:

Les Chymistes, dont l'usage est de joindre & de décomposer les corps pour en connoître les parties essentielles, disent que c'est le seul moyen d'entrevoir la façon dont la nature opère, & que pour expliquer cette nature, il la faut contrefaire par des représentations de choses connues, auxquelles on fait produire le même effet: nous avons l'exemple du volcan de Lémeri.

Les Naturalistes qui emploient ordinairement peu de Chymie, raisonnent d'une manière toute différente. (Argenville, 1755, p. ix–x)[3]

1770: After praising Pott's methods of studying mineralogy, namely, by chemical experiment, and to learn more of minerals, Cronstedt said: "To obtain this end, chemical experiments are without doubt necessary" (Cronstedt 1770, p. 278 [misnumbered as 178]).

1774: However, Werner said:

The internal [chemical] characters [of minerals] cannot be as accurately known and determined as the external characters, because this requires an accurate knowledge of chemistry (a science which itself has not yet been worked out completely). (Werner, [1774] 1962, p. 4)

Werner also objected to the amount of apparatus and materials that was needed for analysis, and the time it took. He remarked that quantities were often too small, or that one would not be

---

[1] It is in the experimentation itself that increased quantification was being introduced.

[2] "On this principle I dare to hope that these new experiments that I present to the public will be pleasing to those who, experienced in chemistry, this science so true, knowing how exact investigation into the nature of bodies, the proper experiments, the discovery of relations entirely unknown, and finally the proper distributions of nature's productions, must be preferred to fanciful ideas and doubtful speculations, and metaphysical abstractions" (my translation).

[3] "Chemists whose practice is to join and decompose bodies in order to know their essential parts, say that it is the only means to catch a glimpse of the way nature operates, and that in order to explain that nature it is necessary to imitate it by representations of known things which can produce the same effects, for example the volcanoes of Lémery.

"Naturalists who usually use little chemistry reason in a very different manner" (my translation).

allowed to subject a specimen to such procedures (Werner, [1774] 1962, p. 4 and 105). However, Werner's supporter, Jameson, constantly referred to chemical experiments and procedures in the various versions of his mineralogical treatises.

1776: In his work as a metallurgist, concerned with minerals, Gellert remarked:

Metallurgic chemistry is an art which teaches how subterranean or fossil bodies, by means of proper agents, may be changed, separated or compounded, that [*sic*] so we may discover the several particular parts of their composition, as [*sic*] also understand their effects. (Gellert, 1776, p. 1)

Chemistry was clearly an essential tool.

1795: The mineralogist Schmeisser said:

It must be confessed, that Mineralogy would not have been even in its present state, if fortunately a Vogel, Wallerius, Bergman, Cronstedt, &c. who cultivated the study of Chemistry, had not clearly shewn, how much to this last science the history of minerals was to be indebted.

To those men we have to look up for many discoveries and improvements, which have put us in the way of rendering the science more useful and more perfect. If mineralogists before or after them, had adopted the same plan of chemical investigation, certainly many operations of nature in the mineral kingdom, still unknown to us, many ores which are still uninvestigated, and which only gratify the eye of curiosity in cabinets, would have been ascertained and described, together with the useful purposes, to which, in arts and manufactures, they might be turned. (Schmeisser, 1795, v. 1, p. ii)

1802: But doubt persisted in the person of John Murray, for example:

We should not indeed err much, perhaps, if we considered the greater number of bodies which are at present the subjects of our knowledge as compounds. Chemistry is but in its infancy; within a few years only has it discovered the composition of a number of substances; and shall we believe that it has already attained the end of its researches, and that the varieties of matter which analysis has discovered are truly elementary? (Murray, [1802] 1978, p. 85)

1809: However, when speaking of the progress made in explaining geological productions (rocks), Deluc remarked: "It was chiefly after having attained this point in its progress, that geology stood in need of the aid of chemistry" (Deluc, 1809, p. 51). He then mentioned that some workers did, and some did not, think that chemistry would be an aid, among the latter being Hutton.

1811: Pinkerton had strong opinions, and his view of chemistry was what we'd expect:

But as it is now universally allowed by all mineralogists, however different their systems, that the whole science rests upon chemistry alone, and that no certainty can be found except by chemical analysis … (Pinkerton, 1811, v. 1, p. vii)

And, with respect to the mineral kingdom, "[T]here is no infallible guide but Chemistry; upon which alone a rational and durable system can be founded" (Pinkerton, 1811, v. 1, p. xliv).

1812: Hall broadened the necessity of chemistry to portions of geology other than mineralogy. He saw the promise of success:

One principal cause of this failure, seems to have lain in the very imperfect state of Chemistry, which has only of late years begun to deserve the name of a science. While Chemistry was in its infancy, it was impossible that Geology should make any progress; since several of the most important circumstances to be accounted for by this latter science, are admitted on all hands to depend upon principles of the former. The consolidation of loose sand into strata of solid rock; the crystalline arrangement of substances accompanying those strata, and blended with them in various modes, are circumstances of a chemical nature, which all those who have attempted to frame theories of the earth have endeavoured by chemical reasonings to reconcile their hypotheses. (Hall, 1812, p. 72)

1831: Finally, Thomson judged the effectiveness of chemistry in mineralogy: "Klaproth, during a very laborious life, wholly devoted to analytical chemistry, entirely altered the face of mineralogy" (Thomson, [1830–1831] 1975, v. 2, p. 206).

One might imagine that all who studied minerals, many of whom were concerned with mineral assemblages and rock origin, were skilled chemists. That would be wrong. It is interesting that neither Werner nor Hutton particularly supported the efficacy of chemistry as a means of approaching geological truth, though not necessarily for the same reasons. Hutton, although a chemist of ability and experience, thought that the operations of Earth were too large and complex to be mimicked in the laboratory. As mentioned previously, Werner simply didn't think it appropriate for his methods of observation. Nevertheless, most of the men who investigated rocks and minerals either used chemical methods themselves, or requested help in doing so from their colleagues.

As we have seen, Hall believed progress in geology depended on chemistry. Despite his obvious ability to design and carry out experiments, he felt himself to be a poor chemist, and asked Robert Kennedy to analyze his whinstones and lavas. When he wanted pure samples of carbonate of lime, he requested them "from such of my friends as have turned their attention to chemical analysis" (Hall, 1812, p. 132). The importance of mineralogy and its chemical investigative methods to the growth of geology as a science has been stressed by Laudan (1987). Klein (1996) has emphasized the contributions of metallurgy and other workshop traditions to the development of chemistry itself in the eighteenth century.

IN THE LABORATORY

In earlier chapters, the necessity for laboratory containers resistant to heat and reaction during fusion, and/or pressure was discussed. Some of the same, plus additional, attributes were required for equipment used for chemical reactions and precise weighing of geological materials or their products. While much of the equipment for geological testing was adapted from previous uses, some new equipment and techniques were devised. We have already mentioned thermometers and pyrometer pieces, barometers, and the blowpipe. Procedures were also modified to take into account the sometimes slow pace of reactions with earth materials, the difficulty of obtaining solution of some substances, or the need to deal with impurities.

This chapter will describe the materials of the chemistry laboratory as used for work, particularly chemical, which ultimately aimed to clarify questions in geology by means of composition or physical properties that might supply parameters for rock origin. Before we make too many twenty-first-century assumptions, it is useful to be reminded of some components of eighteenth-century thought. Laudan has pointed out that

Werner, like most other taxonomists, drew a sharp line between identification and classification. Identification is the development of a repertoire of techniques for recognizing further examples of an individual that has already been described and named. Classification is the process of assigning the entities in the world to a place in a conceptual network. They are distinct, if related, activities. (Laudan, 1987, p. 82)

Laudan noted that Werner *classified* minerals by composition, which requires analysis, but he *identified* them by external characteristics. In analysis, we now can readily associate the products of analysis as characteristic of a particular mineral, with the usual caveats about substitution in crystal lattices. For Bergman and mineralogists who followed him, the end products of analysis, often the "earths," were not the ultimate building blocks of matter, but they might be the elements of minerals, identified with a certain class of minerals, and empirically identifiable (Oldroyd, 1974a, p. 511–512).

The chemist Gellert, who worked with geological materials at the inception of both geology and the new "French" chemistry has provided these definitions:

The *laboratory* is a place where the chymical operations are performed. This must be spacious, light, airy, a stone-building, and provided with chymnies of a good draught.
By *chymical apparatus* are understood those instruments and vessels by which the chymical agents, *fire, air, water, earth,* and the *dissolvant-menstrua,* perform their intended effect upon the bodies. (Gellert, 1776, p. 143)

Few laboratories had the attributes cited by Gellert. Laboratories were often dark and unheated, sometimes located floors above or below easy access points. Books written about alchemy or chemistry were commonly illustrated with plates showing apparatus, often with explanations of its use (Ferchl and Süssenguth, 1939; Greenberg, 2000). Brock has compiled a list of apparatus available from the ninth century:

Among the apparatus described and used were beakers, flasks, phials, basins, crystallization dishes and glass vessels, jugs and casseroles, candle and naphtha lamps, braziers, furnaces (athanors), files, spatulas, hammers, ladles, shears, tongs, sand and water baths, hair and linen filters, alembics (stills), aludels, funnels, cucurbits (flasks), and pestles and mortars—indeed, the basic apparatus that was to be found in alchemical, pharmaceutical and metallurgical workshops until the end of the nineteenth century. (Brock, 1993, p. 23)

The definition of "apparatus" has been a fluid one, and, like much else in writing history of science, our use of the term may not coincide with earlier use. Anderson found an additional distinction helpful: *Apparatus* is necessary for an experiment, but isn't used for measurement; *instruments* are employed for measurement (Anderson, 1985, p. 217). Because this distinction was not always made, in the section that follows the terms may sometimes be used interchangeably.[4]

Containers and Reaction Vessels

Glass

The laboratory employed for assaying or metallurgical work generally had less glassware than that of an iatro- or pharmaceutical chemist. Sturdier vessels would necessarily have been used. Reactions concerned with a broad range of substances, including rocks and minerals, were carried out in the laboratories of the fifteenth and sixteenth centuries with the expected plethora of glassware (see Fig. 6.1). Glass itself had been produced since before 4000 B.C. In Chapter 3, the chapter on fusion, the means to prevent glass breakage in heat, namely, luting and slow heating, were mentioned. In a laboratory, breakage was, and is, a problem, but glass could be easily blown or molded into convenient shapes. It was relatively inexpensive, and its properties were known. Many kinds of glassware were in use for hundreds of years with little change in form. Vessels pictured in alchemical texts continued to be employed. However, Anderson has remarked that because an engraving was made of a particular form of apparatus, there was no guarantee that the apparatus had ever been constructed, and also that illustrations of alchemical laboratories were often stylized (Anderson, 1985, p. 219).

Macquer stated the ideal situation:

Les vaisseaux qui servent aux opérations chymiques devroient pour être parfaits pouvoir éprouver sans se casser, une grande chaleur & un grand froid appliqués subitement, être impénétrables à toute matières & n'être altérables par aucuns dissolvans, être invitrifiables, & pouvoir supporter la plus violente chaleur sans entrer en fusion; mais jusqu'à présent on ne connoît point de vaisseaux qui rassemblent toutes ces qualités. (Macquer, 1749, p. 275)[5]

He described the glassware illustrated in his plates, and listed appropriate uses. Unfortunately, a better, non-expanding glass would not be available for a long while. The Dutch physician and chemist Hermann Boerhaave (1668–1738) and Macquer both pointed out that glass vessels are unreactive (relatively—we will return to this), non-porous, and transparent, so reactions

---

[4]Anderson noted the additional distinction that instruments and apparatus might be used for research, for routine measurements, or for teaching.
[5]"In order to be perfect, the vessels used in chemical operations should be able to undergo great heat and cold applied suddenly without breaking, to be impenetrable to all matter and not to be changed by solvents, to not be vitrifiable, and to be able to support the most violent heat without entering into fusion; but until the present there are no vessels that bring together all these qualities" (my translation).

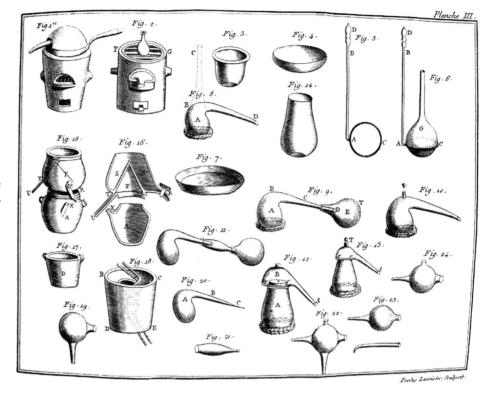

Figure 6.1. One of the examples of the varieties of glassware available (Lavoisier, 1789, Planche III).

could be observed (Eklund, 1975, p. 8–9, 46). During the eighteenth century, there was also a general progression from having glassware made in or near the laboratory itself to purchasing it from a supplier, although common items had been supplied from glassworks as early as the fifteenth century (Anderson, 1978, p. 142). Examples, such as those pictured in the book on the Playfair collection at the Royal Scottish Museum, frequently show features, which indicate that someone less skilled than a master glassworker made them (Anderson, 1978, p. 142). Special adaptations were still made in the laboratory, and until late in the twentieth century, chemistry students were commonly required to produce a few simple forms.

What would we see were we to visit a laboratory in the eighteenth century or at the beginning of the nineteenth? As Holmes and Levere (2000, p. xi) have pointed out, apparatus pictured in the later eighteenth century didn't differ a great deal from that shown in the sixteenth century, either changing in small ways or remaining nearly static. In 1766, Joseph Black, upon leaving the university at Glasgow to go to Edinburgh, left a list of laboratory glassware that included "27 different shapes and sizes of vessels, ranging from one four Quart Retort down to 128 Vials" (Swinbank, 1982, p. 34). Glass vessels were used both as containers and reaction vessels. Shapes were designed for solids, liquids, or gases, for numerous uses. Some containers, which might be of transparent or colored glass, were as mundane as glass bottles of different shapes with stoppers of various kinds. While visiting Kirwan's laboratory, Jameson remarked on the vials with "neatly ground heads with glass cov-

ers which are very useful preventing the entrance of dust &c" (Sweet, 1967, p. 110). Bottles designed to trap gases might be fitted with tubes to let gas in or out, might have a tap fitted on the stopper, or could have a manometer attached (Anderson, 1978, p. 132). Some were graduated so that volumes could be read. Glass tubes were used both as test tubes closed at one end, which might rest in a support, or as connective tubing. During the eighteenth century (and later), a wide-mouthed drinking glass was referred to as a beaker. However, the flat-bottomed, wide-mouthed vessel of that name, with a pouring lip, also made its appearance in the laboratory, sometimes graduated. Experimenters employed thickly walled bell jars to enclose evacuated spaces, just as we do today.

Flasks of various sorts might be used as containers or for reactions. These were not yet our familiar round-bottomed, Florence, or Erlenmeyer flasks, although those were not far in the future. Determination of which shape was given which name is not always easy, because there were variations. The third American edition, from the sixth English edition, of Henry's *Elements of chemistry* (1814) included nine plates illustrating chemical apparatus (see Fig. 6.2). This frequently cited book must have been popular, because numerous editions were published on both sides of the Atlantic. Perhaps we may take it as a summary of what was available and used. Henry's retort was a glass globe with a long tapered neck extending to the side and into the mouth of what he called a simple receiver, which was just a round, necked bottle. Gellert identified a retort as "a vessel with a round belly and a neck bent downwards, used for distilling such substances which

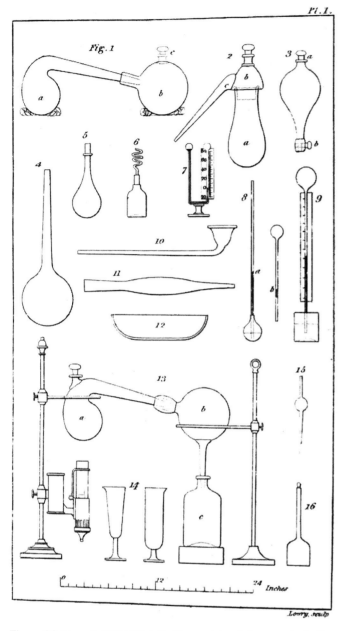

Figure 6.2. Henry's (1814) Plate I with a variety of glassware including some distillation setups.

collect the vapors that resulted from heating different substances, which then ran through a pipe to a receiver. The alembic seemed always to include a head that would trap and direct a distillate. Also illustrated was a matrass, shown as a round-bottomed flask with a long, vertical, slender neck, which Henry said was useful for dissolving bodies that required long heating or digestion. The long neck stayed cool so the flask could be shaken (Henry, 1814, v. 1, p. 331). A narrow glass bottle was fitted with a ring so that it could be suspended over a lamp.

There were many adaptations of glass for special purposes (see Fig. 6.3), and equipment might be joined for a series of reactions. Digestions were carried out in heated flasks topped by a column of connected bulbs. The bulbs, sometimes called "aludels," either of glass or earthenware, were also used to collect sublimates. Glass elements were joined together by lute. Work with gases could be done with a series of collecting bottles. Insoluble gases were collected over water, soluble gases over mercury. A eudiometer, conceived in the 1770s, was a graduated glass tube designed to measure the volume of an evolved gas, and was used in gas analysis. Glass siphons were used to remove the liquid above solids from an evaporating pan (Lavoisier, [1790] 1965, p. 362). Something as mundane and necessary as an adapter, designed to connect a retort to a receiver, was also made of glass. Funnels might also be of glass. All of this apparatus suffered from the tendency of glass to break when exposed to sudden heat, as well as to soften at somewhat elevated temperatures. Because it is a mixture, glass doesn't have a sharp melting point, but softens over a range of temperatures around 800 °C and above, depending on the composition of the glass. It is this same property, however, that allows glass to be manipulated into multiple shapes.[6] Much of the glassware we are familiar with was already in use, although perhaps in a slightly different form. There was an increasing tendency to have it calibrated to show volumes.

Smeaton pointed out other important advances in working with gases, although this new ability certainly was useful in a variety of procedures. One was the use of the ground-glass stopper. These could be ground to fit a particular bottle, and were an improvement over cork and leather for corrosive substances. However, connections to apparatus to convey gas or liquid to other vessels were problematic. For this, ground-glass taps were devised and improved during the 1770s (Smeaton, 2000, p. 219–221).

Glassworks were relatively common industries, located near, or near transport for, sources of raw material and fuel, and where there would be a market. Glass could not be shipped great distances unless it was unusually fine so care was taken to secure it. Anderson (1978, p. 145) listed 15 glassworks in Scotland on record before 1800. Some wares, although not made in the laboratory, could still be crude or flawed. By 1797, the Leith glassworks was producing both green (relatively crude) and flint (transparent,

rise over with difficulty. Some are of glass and others of clay" (Gellert, 1776, p. 173).

What Henry called an "alembic" was a rather narrow flask capped by a round head which was ground to fit well into the flask, and which had a tube extending downward. This was used for distillation—the distillate running out the tube and into a receiver. Alembics and distillation were "certainly known to Greek chemists working in Alexandria" (Anderson, 2000, p. 5). Sometimes the lower part was termed a "cucurbit," and the top part a "helmet." Gellert also called a helmet a "still-head," a "round hat of glass with a long pipe" (Gellert, 1776, p. 173). Its purpose was to

---

[6]Glass also reacted with some substances such as lead calx (oxide) and hydrofluoric acid, as reported by Carl Wilhelm Scheele in 1771 (Boklund, 1975, p. 146).

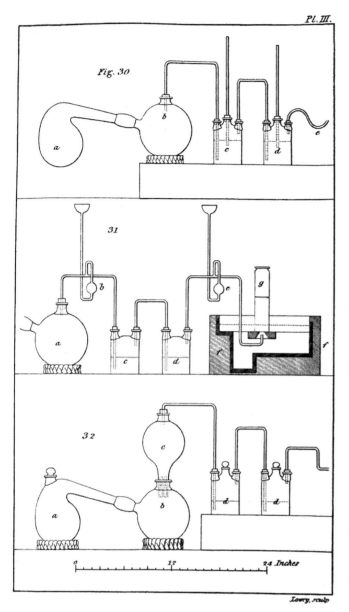

Pl. III.

Fig. 30

31

32

Lowry, sculp

Figure 6.3. More complex arrangements of glassware for several-step reactions, Henry (1814), Plate III.

He described an instrument that would allow damaged glassware to be altered for this purpose, namely, an iron ring affixed to a wooden handle. If the ring was heated to red heat, applied to the glass vessel, and a little cold water was splashed on the glass, the glass would usually break along the ring. These vessels cost far less than new ones ordered from glass manufacturers (Lavoisier, [1790] 1965, p. 377). Anderson (2000) contains a nice summary of the archaeological evidence for glass and other laboratory ware at periods several centuries earlier than the time at which chemistry is supposed to have emerged as a discipline. However, we are here principally concerned with equipment used in rock, mineral, mineral water, and gas investigations.

Earthenware or Pottery

Like glass, earthenware had advantages and disadvantages in the laboratory. It was heat resistant, didn't break as easily as glass, and also could be molded into virtually any shape desired. It had the drawback of being porous enough to absorb liquids, which led to incorrect volume readings, and the absorbed material could also react with flask contents in further chemical reactions as well as in fusion (Eklund, 1975, p. 7). However, glazed pottery, which was more expensive and more brittle, could also be used. Such alembics fitted with helmets were illustrated in a publication of the early sixteenth century (Ferchl and Süssenguth, 1936, p. 86). Pottery or earthenware might begin with a composition somewhat similar to porcelain, but it was not fired at nearly as high a temperature.

Beside his description of aludels, Gellert mentioned a cementing box with a cover, made of earthenware. In metallurgy, the process of cementing consisted of heating two substances in contact with each other so that they could react. He also made use of what he called an earthen prism, which was put into a muffle in the furnace to cool the reaction (Gellert, 1776, p. 175–176). When Black went to Edinburgh from Glasgow in 1766, among the inventory of laboratory equipment he left were "5 muffles, some Clay Doors for Ashpits of furnaces & some earthen Pots for a Vent" (Swinbank, 1982, p. 33).

Clay crucibles were used, but were breakable and induced much error through porosity or reactivity. Lavoisier covered a glass retort with what he called a "dome of baked earth" in order to protect the glass during evaporation. However, if any cooler liquid struck the glass, it would break (Lavoisier, [1790] 1965, p. 378). Lavoisier also utilized what he called "pans of earthenware" when he concentrated and/or purified substances by allowing slow evaporation of the solvent.

Porcelain and Stoneware

Despite being breakable, porcelain was (and is) used for laboratory vessels because it is nonporous, generally nonreactive, and has a high fusion point. Porcelain was very expensive during the early part of the eighteenth century, whether it was imported or the product of the few local European manufactories. The increased strength of glass when devitrified by slow heating had been noted for centuries (C.S. Smith, 1969, p. 322).

thinner, and harder) glass scientific vessels (Anderson, 1978, p. 143). On the Continent, similar resources were available. It was mentioned earlier that Tschirnhaus was associated with a glass manufactory at the end of the seventeenth century.

Not surprisingly, glass vessels preserved in museums sometimes show signs of damage (Anderson, 1978, p. 136). They were also recycled. Lavoisier declared that when glass was used as a container for liquid to be distilled,

The best utensils for this purpose are made of the bottoms of glass retorts and matrasses, as their equal thinness renders them more fit than any other kind of glass vessel for bearing a brisk fire and sudden alterations of heat and cold without breaking. (Lavoisier, [1790] 1965, p. 377)

In his search to discover the nature of true porcelain, Réaumur (Fig. 6.4) rediscovered the glass devitrification process and produced "Réaumur's porcelain." He put a glass vessel in a sand bath, slowly raised it to red heat, then allowed it to cool very slowly, after which it could be heated to red heat without damage. During this time, "[T]he various substances of which it is composed are at liberty to exercise their affinities and to crystallize. This makes the vessel lose its glassy structure altogether" (Thomson, [1830–1831] 1975, v. 1, p. 282). Some of this more resistant glass was used in the laboratory. However, when they became available, crucibles and other porcelain ware were preferred. In the thirteenth century, stoneware was imported into England from the Rhineland, which practice continued until the first stoneware was produced in London in the late-seventeenth century. Ceramic apparatus for distillation has survived more frequently than that of glass (Anderson, 2000, p. 13, 21–23).

The use of porcelain has persisted until today. We use porcelain crucibles, evaporating dishes, and the mortar and pestle. Lavoisier mentioned porcelain evaporating dishes (basins), a mortar and pestle, and even an aludel (Lavoisier, [1790] 1965, p. 388). In response to needs of his Lunar Society friends, Josiah Wedgwood began production of porcelain and stoneware utensils in the 1760s and 1770s. Some of the first items were mortars and pestles, suitably deep, fired at high temperature, and of a composition that would not absorb oil (Schofield, 1963, p. 160–161), made for the chemist Joseph Priestley. Wedgwood also sent him "a constant supply of ceramic tubes, dishes, mortars, and crucibles" (Schofield, 1963, p. 161). Soon after, Wedgwood made evaporating pans, funnels, subliming bottles, thermometers set in ceramic, and various sizes of tubing and retorts (Schofield, 1963, p. 161). He published a catalog of laboratory ware in 1772 (Schofield, 1959, p. 183). Thus, through Wedgwood and other ceramics makers, this fertile time for science at the end of the eighteenth century was aided by increasing availability of laboratory ware, with particular attention paid to production of pieces that were resistant to the conditions imposed upon them.

Metal: Iron, Copper, Platinum

Metals were employed for many sorts of implements. The metals used were mainly iron, tin, and copper, although the high temperature of fusibility and the nonreactivity of platinum made it attractive as it became available at the end of the eighteenth century. Iron was relatively inexpensive and commonly available. It had well-known properties, and it could be shaped as desired. It had the drawback of being reactive with both compounds and other metals. Like pottery vessels, iron could interfere with reactions by its reactivity, by releasing small particles, or by the scale that formed upon oxidation of the iron. An iron vessel might be coated with tin to decrease reactivity. Thomson remarked that prior to the work of Réaumur, tin plate was not made in France, being available from Germany, but that Réaumur discovered a workable process (Thomson, [1830–1831] 1975, v. 1, p. 280). Iron vessels remained a valuable addition to the laboratory.

Figure 6.4. René-Antoine Ferchault de Réaumur.

In his plates of laboratory equipment, Gellert illustrated an iron grinding pan, with a grinding hammer, an iron "trevet" to hold a small vial with a few coals underneath it, a pair of tongs, and one of the aforementioned long-handled iron rings to break the necks off of glass vessels. There were also a ladle, flat pans, hammers, files, chisels, and a vise. Clamps, stands, and supports made of iron were common, as were crucibles. There was even an iron wire cage designed to hold ice within a concave mirror in order to determine whether cold, like heat, could be reflected (Anderson, 1978, p. 80).

Copper is less reactive than iron, but is more expensive and more difficult to fabricate. Among the equipment listed by Black when he left Glasgow for Edinburgh was a copper still with a copper head, a copper pan, and a copper Papin's digester (Swinbank, 1982, p. 33). Macquer pointed out that vessels of both iron and copper could be damaged by nearly all saline, oily, and aqueous substances unless they were dilute (Macquer, 1749, p. 275–276). Henry listed "an oval copper boiler, for exhibiting the most important fact respecting latent caloric" (Henry, 1814,

v. 1, p. 335). Clearly, the heat conductivity of copper was recognized, and it was used where appropriate. In fact, Lavoisier said that metal dishes could be used for blowpipe analysis since

if … sufficiently large, they do not melt, because, metals being good conductors of heat, the caloric spreads rapidly through the whole mass, so that none of its parts are very much heated. (Lavoisier, [1790] 1965, p. 476)

Cronstedt noted that platinum, a recent discovery, was well described in the *Acts of the Royal Academy of Sciences* in Stockholm for 1752 (Cronstedt, 1770, p. 178), and listed six of its characters that enabled it to be distinguished from gold. As mentioned above, the high fusion temperature of platinum made it attractive as a container or support in furnaces and when making investigations with the blowpipe. Its lack of reactivity was an advantage in both fusion, where it would not react with the fused substance as ceramic containers might, and in chemical reactions. Both purification and fusion of platinum had been and still were difficult problems, but those difficulties did not preclude its use. The platinum had to be virtually free of arsenic. Otherwise, if used with caustic alkali, the alkali would react with the arsenic rendering the crucible porous (Chaldecott, 1968, p. 33).[7] Much effort was put into fusing platinum in order to purify it, and also to make it malleable. When Hall visited Paris in 1791, he spent time in the company of Lavoisier and others of his circle who were trying to fuse platinum. The Paris goldsmith Marc Étienne Janety claimed a process to make it more malleable. Guyton de Morveau worked with him so that larger crucibles could be made (Smeaton, 2000, p. 216). While in Paris, Hall purchased two cups, two spoons, and some platinum wire for his investigations (Chaldecott, 1968, p. 33–35). With independent incomes, both Hall and Lavoisier were well able to purchase expensive, high-quality equipment for their laboratories. Thomson remarked on the improvements made in chemical analysis by Klaproth when he substituted a silver crucible for those of iron that Bergman had used, and credited Wollaston with the introduction of platinum crucibles (Thomson, [1830–1831] 1975, v. 2, p. 191).

In the first decades of the nineteenth century, Berzelius added more specific instructions for the use of platinum vessels. They could not be used for ignition of caustic alkalis, nitric salts with alkaline earths, alkali sulfides, or alkali sulfates with carbon. If metals were melted in platinum, alloys could be immediately formed, but molten lead or bismuth made holes in the platinum. Other combinations were also destructive to it. The more fragile porcelain crucibles might be heated inside a platinum crucible. Platinum crucibles made in France, being thicker, were said to be both less expensive and more durable than those made in London (Szabadváry, 1966, p. 145). Jameson listed uses for platinum: expansion pyrometers, crucibles, pendulums, reflecting telescopic mirrors, and wheels for the construction of watches (Jameson, 1816, v. 3, p. 4). Platinum's very small coefficient of expansion had been noted, and led to its use as a standard for length.

Balances

Weighing goes back far into antiquity. Multhauf stated that the history of the balance goes back three millennia, although it was not used for theoretical chemistry until the fourth quarter of the eighteenth century (Multhauf, 1962, p. 210). Newman contested the position that alchemy was not quantitative. He contended that balances are pictured more than claimed, more use was made of them, and that they were used for purposes other than assaying and mining, for which he presented convincing evidence (Newman, 2000, p. 39–42). A casual and evidently misleading impression, often repeated in historical literature, is that Lavoisier was the first to take weight differences in chemical reactions seriously. Two things argue against that position. The first is simply the historical record of numerous meticulous workers who included weights, and concerns about weights, in their records.[8] The second is the indisputable existence of balances that were constantly being improved. Lavoisier insisted on excellent balances, but he was not the only inquirer who expressed concern about how accurate and reproducible the values were. Lavoisier's results were indeed excellent, and he displayed great insight and originality both in determining what should be weighed and in interpreting those results.

I will not attempt to write a history of the chemical balance, about which entire books have been written, but instead will discuss how balances were used in this setting of inquiry into earth materials and rock origins.[9] They were occasionally shown in early pictures of pharmaceutical establishments. During their history of being used by alchemists to identify minerals and their components (Newman, 2000, p. 49), balances were much in evidence in depictions of assaying and metallurgy, where the concern was yields from ores or, in coins, the precious metal content. With respect to assaying silver ore, beginning with pulverizing it, Cramer said:

This pulverization must then be done, before the Ore is weighed, because there is always some Part of the Ore adherent to the Mortar, or the Iron Plate on which it is made fine; which Part being lost, the Operation is not exact. (Cramer, [1739] 1741, p. 208)

Weight proportions were important in making coins, and also in making the touch needles referred to in Chapter 2. Agricola discussed a system of standard weights in detail (Agricola, [1556] 1950, p. 260–264). He illustrated three balances, one in a case to protect it from air currents, and each designed for a different range of weights. (See Fig. 6.5.) The balance in the case was the most sensitive, and was used to weigh the buttons of gold or silver from an assay (Agricola, [1556] 1950, p. 265).

---

[7]Caustic alkali refers to the strong hydroxides (Eklund, 1975, p. 23).

[8]Newman wished to correct the impression given by Lundgren (1990) that the balance was not often employed. He also noted that "weight was not given the primacy that it acquired in the works of Lavoisier" (Newman, 2000, p. 49).

[9]For good discussions about them, refer to Stock (1969), Jenemann (1997), and Shannon and Shannon (1999). Encyclopedias include sections on weighing and balances as well.

Figure 6.5. Agricola's three balances (1556, p. 207).

As with any instrument, balance usage in the eighteenth century depended on a number of factors. First, there had to be the recognition that weight was important. It was clearly important in an economic sense, but it had seemed less so in scientific investigations. If it was an important parameter, an instrument to measure it had to be available. Balance making was a skilled craft, and a practitioner needed sufficient economic resources for purchase. All of that may seem obvious, but a close reading of laboratory practice of the time indicates that there were things we now take for granted, including good balances, that were not routinely available. It has been claimed that Joseph Black was the first to employ a balance in virtually all stages of chemical investigation (Anderson, 1978, p. 73). The balance attributed to him in the Playfair collection was accurate to about a grain (65 mg, or 0.065 g). However, in Black's analysis of mineral water, to which we shall return in the next chapter, he must have used more sensitive balances, reporting to either 0.01 or 0.1 grains (Anderson, 1978, p. 74). Black also mentioned weighing a residue of "38 or 39 hundredths of a grain" (Black, 1794, p. 111).

Lavoisier used three balances of different sensitivities (Cadet, 1803, v. 1, p. 377) as well as others for approximate work. It was in the decades toward the end of the eighteenth century that high precision balances appeared, answering the needs of a small group of experimenters who demanded them (Daumas, [1953] 1972, p. 221; Levere, 1994, IV, p. 315). Using the considerable skills of Continental and English clock and instrument makers, the instrument makers identified sources of error and worked to overcome them. Balances alone were not sufficient. In the case of Lavoisier, it was noted that he needed to know not only what

should be weighed, but also how to estimate things that could not be weighed, what adjustments could be made, and what errors were acceptable (Holmes, 1985, p. xviii).

Schmeisser (1795, v. 1, p. 34) was one of the few mineralogists of his time who listed an assay balance among recommended instruments. Sir James Hall regularly recorded his weights for specific gravity measurements with determination to two decimal places (1805a). His meticulous laboratory procedures must have included similar care with weighing, although there are no statements about his balance(s) in his published works that I can find. When Robert Kennedy did the chemical analysis of Hall's whinstones at the same time as Hall was doing the fusion experiments, he noted that none of the rocks scratched the mortar in which they were ground to powder, implying that he was concerned with small amounts. He began with 100 or 200 grains of sample, and recorded half grains. The final results were reported as percentages (parts of 100), and generally did not record tenths.

During this time, there were numerous geologists who did extensive fieldwork and brought samples back to the laboratory for analysis. Included in this group are the Italians Lazzaro Spallanzani and Scipione Breislak (1748–1826), as well as Saussure and Dolomieu, all of whom published field studies backed by laboratory work. They analyzed mineral, rock, and mineral water samples, usually to tenth (0.0) places in parts per hundred. Saussure weighed the results of mineral water analysis to fractions of grains (Saussure, 1803, v. 1, p. 301). The balance was thus used for analysis, which in turn was often used for identification as well as conjectures about rock origin. The work of John Murray (1818), which addressed mineral water analysis, did not mention his balance, but stressed the need for accuracy in identifying very small weights of minor constituents. Berzelius was highly respected for his work in mineralogy and analysis, but historians tend not to discuss how his quantitative results were obtained (for example, see Melhado, 1981). As discussed in Chapter 2, the hydrostatic balance was important in finding specific gravities, and in 1817, Robert Jameson still discussed the required hydrostatic balance at length in his *A treatise on the external, chemical, and physical characters of minerals*. Thus, in a relatively few years, weights and combining weights went from being a matter of little concern in mineralogy and geology to being an integral part of the work of those concerned with geological materials.[10]

Magnifiers and Microscopes

There is no doubt that the Greeks and Romans knew a considerable amount about the optical properties of glass. References to burning mirrors and spheres, and notes on light refraction exist from that time (Wolf, [1935] 1968, v. 1, p. 71; Hogg, 1886, p. 2).[11] The use of magnifying lenses also occurred

---

[10] Just a few others who addressed geological problems in the eighteenth century were Marggraff, Klaproth, and Bergman.

[11] I had access to the eleventh edition of Hogg (1886). The first edition was printed in 1854.

early on. For precise work, a supply of good optical glass was necessary, a need sufficiently addressed only in the middle of the eighteenth century (Daumas, [1953] 1972, p. 158). The inherent problems of spherical and chromatic aberration were minimized by the use of the simple microscope, with a single lens. At the beginning of the seventeenth century, that microscope was much used by scientists to see fine detail, especially by transmitted light through transparent objects. The compound microscope has tentatively been credited to Zacharias Jansen of Holland ~1590 (Wolf, [1935] 1968, v. 1, p. 72), although it didn't become popular until after Robert Hooke published his *Micrographia* in 1665 (Turner, 1998b, p. 388). Chromatic and spherical aberration were handled with increasing success at the end of the eighteenth and the beginning of the nineteenth centuries.

At that time mineralogists and geologists continued using both magnifiers and microscopes. The use of a hand lens was as common then as it is now. A triple-lens magnifier was included in von Engeström's pocket laboratory, as discussed in his 1788 translation of Cronstedt's treatise on mineralogy. When the lenses were appropriately combined, it could produce seven different powers of magnification "the better to distinguish the structure and metallic parts of ores, and the minute particles of native gold, whenever they contain that metal" (Cronstedt, 1788, v. 2, p. 987). In his list of instruments necessary to identify minerals, Schmeisser listed a "microscope or magnifying glass" (Schmeisser, 1795, v. 1, p. 34). Saussure used a microscope to observe analysis products that didn't otherwise appear to show regular forms (Saussure, 1803, v. 1, p. 301), while Breislak used the instrument for natural minerals as did Pinkerton. For sal ammoniac adhering to cellular lava, Pinkerton noted that "small articulations are clearly distinguished, composed of octaedra [*sic*], placed one on the other" (Pinkerton, 1811, v. 2, p. 410). Haidinger used the microscope to assist in accurate specific gravity determinations. In order to assure the homogeneity of his samples, he pulverized minerals until "even by the assistance of the microscope, we can no longer detect a want of continuity in the fragments" (Haidinger, 1825, v. 1, p. 309).

While magnification was a normal procedure for many years, hindsight shows us three stages in its use, two of which occurred after the period discussed in this book. The first was Pierre-Louis-Antoine Cordier's (1777–1861) separation of mineral grains and their close observation. The second was William Nicol's (1768–1851) fabrication of the thin section in ~1815, followed by his invention of the polarizing microscope in 1828 (Hurlbut and Klein, 1977, p. 6). The third, in 1851, was Henry Clifton Sorby's (1826–1908) application of thin section technique and the polarizing microscope to natural rocks. It is interesting that this work of both Cordier and Nicol was not immediately applied to problems in geology in its full power.

The study of small mineral grains collected from rivers in order to determine the kinds of country rock was not unusual (Dolomieu, 1791a, p. 313). Cordier, the French mining engineer and geologist, initiated the study of the separated mineral grains

of basalt with the help of a microscope.[12] At this time, the basalt controversy had been raging (with most of the rage in Britain) for several decades. Some sources, particularly those in English, claimed that Sir James Hall's slow cooling of fused basalts to a somewhat crystalline texture "proved" their igneous origin. However, there were immediate objections and counterclaims, one of which was the suggestion of several well-known Neptunists that the change in texture was due to driving off volatile alkalis by heating. Individual crystals in basalt were too small to be seen, but Cordier found a way to first separate the minerals and then examine them microscopically, both before and after fusion with a blowpipe. He crushed the rocks, initially separated the minerals by their densities by flotation and/or with the magnet, and then manually under the microscope. He could identify mineral characteristics both before and after fusion. Cordier applied this laborious process to volcanic rocks of many kinds from many locations, with profound results for geology (Zittel, 1901; Ellenberger, 1984).[13]

The use of thin slices of minerals or rocks brought the advantage of transparency. In 1763, the French naturalist Jean-Étienne Guettard (1715–1786) remarked on observation of thin slices of petrified wood under the microscope by a naturalist named Moll (Ellenberger, 1984, p. 45). William Nicol, a lecturer in natural philosophy at the University of Edinburgh whom we associate with the invention of the polarizing microscope in 1829, also made thin sections of fossilized wood and reported on the fine structure seen (Geikie, [1905] 1962, p. 463). After a period in which that method received little attention, an independent researcher, Sorby of Sheffield, applied it to making thin sections of natural rocks, following which the optical properties of minerals were increasingly investigated (Dawson, 1992).

Collateral Equipment

Much of the ordinary equipment used for chemical determinations in the laboratory by chemists and geologists of the late eighteenth century has been discussed. It is common to find books of the time and journal articles with carefully drawn and described illustrations of this equipment (see Fig. 6.6). With respect to the varieties of apparatus used to collect, react, and analyze gases, there must have been almost as many ways of doing this as there were investigators. Breislak even devised a structure made of reeds leading to a dome of inert material to collect corrosive gases emitted from fumaroles (Breislak, 1792, p. 37). There were also efforts to protect workers from noxious air and gases. Agricola offered a quite lengthy discussion of mine ventilation (Agricola, [1556] 1950, p. 200–211). It was recognized that good ventilation was needed in laboratories also, and efforts were made to protect workers from harmful fumes by covering the tops of stills with hoods (Greenberg, 2000, p. 43).

---

[12]There is a good summary of use of the microscope in geology from the beginning until that of Sorby, emphasizing Cordier's work, in Ellenberger (1984).

[13]Cordier's work will be further discussed in Chapter 10, which addresses the topic of basalt.

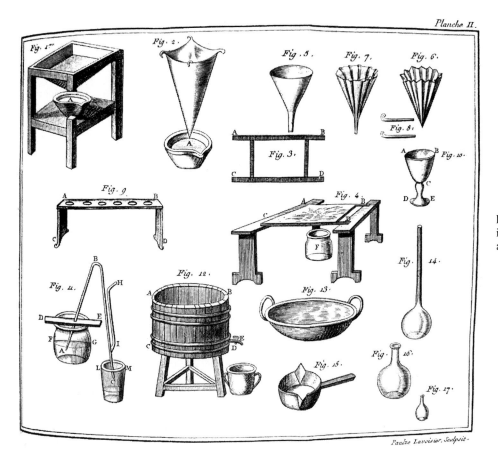

Figure 6.6. Lavoisier's (1789) Planche II, illustrating various sorts of equipment and filtering arrangements.

Nevertheless, laboratories were often still cold, hot, damp, or evil smelling, and generally unpleasant places.

Filtration was apparently performed from the earliest time. Mercury was squeezed through a leather bag to purify it in the sixteenth century and perhaps earlier (Greenberg, 2000, p. 17). Paper folded into a cone was described early, and continued in use. Sieves were made of hair-cloth, silk gauze, or parchment depending on their intended use (Lavoisier, [1790] 1965, p. 361). Mortars and pestles made of various materials were employed from antiquity. A flat table might also be used as a grinding surface. I have mentioned stands, to which apparatus could be attached by clamps, screws, and rings.

In short, a twenty-first-century geochemist would not be at a complete loss if transported to a laboratory from 250 or more years ago. All this equipment was used in procedures and reactions employed to investigate rock and mineral behavior and composition. Those procedures and reactions are the subject of the next chapter.

CHAPTER 7

Chemistry in the service of geology: Procedures, reagents, reactions

With such aid [galvinism], chemical experiment, conducted by such men as
Sir H. Davy, and other able chemists of the present day, may work a complete
reformation in science: it is impossible even to conjecture where discovery will stop.
—Phillips ([1816] 1978, p. 27)

INTRODUCTION

By the final decades of the eighteenth century, the equipment and instrumentation for laboratory work in geology, mineralogy, and crystallography existed, thanks to innovative experimenters and skilled instrument makers. Reactions with geologic materials were familiar from metallurgy and industrial processes, as well as from alchemy and the practical arts. Lime, cement, and gunpowder were produced, as were metals and alloys. Soap, dyes, glass and enamels, wine and beer, and salts for various purposes were all routinely made. Reagents had been developed for industrial, medicinal, and alchemical purposes. Despite sometimes fanciful terminology, the effects of acids, bases, and indicators were often predictable and repeatable. In order to aid that predictability, procedures were carefully specified and contaminants guarded against, although those efforts were not always successful.

Alessandro Volta (1745–1827), the "electrician" from Austrian Lombardy, devised his "pile" in 1800. With the advent of this first battery, electrical separation of active elements was possible early in the nineteenth century, as investigated by the gifted Humphry Davy. Argenville had written:

Les Chymistes conviennent cependant que malgré leurs découvertes, ils ignorent encore les premiers principes des corps, & que ceux qui se rendent sensibles, ne sont que secondaires par rapport à ces premiers principes, dont les élémens sont composés. (Argenville, 1755, p. vii)[1]

In 1789, Lavoisier had published his list of "simple substances" (including light and caloric) that were taken to be the end point of analysis, a list similar to that of Boyle (Partington, 1962, v. 3, p. 485), which list was also directly indebted to the work of Bergman (Oldroyd, 1973b).[2] Davy's work soon altered the list. This was also the period when the elemental composition of many gases and water was identified. It isn't surprising that prominent scientists, whether we now label them as "geologist," "chemist," "mineralogist," or "physicist," should have been

intrigued by the new discoveries and should have joined in the inquiries with the new tools available. Despite the scale of earth processes, laboratory investigation of them was a constant theme. Then and now, geology is a rather "unruly" science, gathering in a plethora of concerns, from the physical through the chemical into the biological. Various authors have also noted differences in approach between French, German, Italian, Spanish, Swedish, Norwegian, etc., work, and that in Britain (Schimkat, 1998; Laudan, 1987; Guntau, 1984). Not surprisingly, a good deal of space is given by secondary authors to the exertions of a number of German chemists who were associated with the Bergakademie at Freiberg. However, in reading the eighteenth-century literature, there is unity both in the sorts of questions thought worthy of investigation and also in the methods applied, regardless of the broader theoretical models that might be espoused.

There might be questions as to the reason this chapter is subtitled "Procedures, reagents, reactions" when it's part of a book pertaining to the history of geology, particularly when it is concerned with the eighteenth century. The answer is basic to the motivation for writing this book: This is the evidence that the analytical procedures were part of what prevented geology from making what we may now see as the obvious and correct choice for at least some rock origin. Confronted with a large rock with interlocking crystals of several minerals, we immediately say "igneous," and scoff at the idea that one could put it in water, even broken up, and see it dissolve. Thus, when we read that an aqueous origin was suggested and/or accepted for basalt, granite, and other rocks, we might think, among other things, that the people espousing that view were ignorant, bad observers, and/or poor geologists. However, we will see that origin from solution was actually a more plausible explanation than the alternative which involved an enormous heat source of subterranean location and origin. Solution and precipitation were demonstrable processes in the laboratory (Newcomb, 1986; Laudan, 1987). In a sense, origin from solution was also observable in the natural world, because the source of at least one class of rocks, those we call sedimentary, was not hard to extrapolate from present inspection of drying sediments left behind by floods.[3]

---

[1]"However, chemists acknowledge that in spite of their discoveries, they are still unaware of the first principles of bodies, and those that make themselves perceptible are only secondary in relation to the first principles of which elements are composed" (my translation).

[2]In that paper, Oldroyd gave a brief synopsis of the first time each of the earths was recognized as a separate entity.

[3]For extensive and important background see Laudan (1987) and the papers in Oldroyd (1998).

PROCEDURES

Many of the procedures carried out on rocks and minerals in the late eighteenth century had their roots in antiquity. Partington (1989, p. 23) identified processes known from Greek sources as "fusion, calcination, solution, filtration, crystallization, sublimation, and especially distillation." Lists of the procedures such as distillation, filtration, precipitation, etc., followed in laboratories for the investigation of minerals, their purification or manipulation, didn't change much from the seventeenth through the eighteenth centuries. In the thirteenth century, in his book on minerals, Albertus Magnus described their properties along with his ideas about their origin. For processes to attempt to change base to precious metals, he listed calcination, sublimation, and distillation as means of having the "elixir" penetrate the metal (Magnus, [~1262] 1967, p. 171). It can be argued that there is only a fine point of difference between recipes and procedures. Much alchemical knowledge was passed on by means of recipes—in other words, reliable methods for obtaining a desired result. To this day, many K–12 school and undergraduate chemistry laboratory exercises employ methods more akin to recipes than to true experiment.

As we've seen, implements for distilling, filtering, dissolution, ignition (burning), precipitation, digestion, and heating to various temperatures were necessary in laboratories of the sixteenth and seventeenth centuries (Ferchl and Süssenguth, 1936, p. 95–103). In Becher's time, in the seventeenth century, raw materials might be treated in three different ways: ground and washed, resulting in a sludge; treated with a liquid and crystallized or distilled; or dried and fused. These processes were related to the ways in which the elements earth, water, and fire supposedly behaved (P.H. Smith, 1994, p. 237). The laboratory was divided into areas that had the tools appropriate for each process. Cramer noted that assaying was designed to separate the parts of minerals from each other, particularly the metals. Most chemical operations were required at some time in assaying (Cramer, [1739] 1741, p. 182–183). Under "dry" procedures, he listed fusion, vitrification, scorification, "coppelling" (cupellation), reduction, amalgamation, sublimation, cementation, roasting, calcination, eliquation, and precipitation under fusion. Under "moist," he put elutriation, edulcoration, quartation, and moist precipitation (Cramer, [1739] 1741, p. 184).[4]

Gellert wrote:

Those performances which, by means of chymical agents and dissolvent menstrua [solvents], produce an intended alteration of certain bodies; that is, by which these bodies are either separated or compounded, are called *chymical operations.* (Gellert, 1776, p. 178)

His categories are somewhat difficult to understand. First, he remarked that those desired alterations could only be effected by both a "chymical" agent and a dissolvent menstruum. But in his five categories of operations—by fire, air, water, earth, and dissolvent-menstrua—there were many operations that appear to belong to more than one category (as we understand them), although each operation was listed only once under a single category. Many required heat. Among those listed under "Fire" were fusion, burning, digestion, and decomposition, which latter had ten subcategories including roasting, calcining, distilling, evaporation, chrystallization (as the translator spelled it), "dephlegmation," and concentration (Gellert, 1776, p. 179).[5] The seven categories listed under "Air" could be investigated without heating. They included (with some subcategories) liquefaction, impregnation and exhalation, fermentation, putrefaction, and gradation. Under "Water," were recorded washing, elixivation, and solution, which had the subcategory edulcoration.[6] "Earth" as an operation contained only fixation, with the subcategories "partly" and "wholly" (Gellert, 1776, p. 180). The final operation, Dissolvant-Menstra, had three parts, the first of which was amalgamation. The second was solution in the dry way, which included glass making, uniting in fusion, parting in the dry way (precipitation, scorification, working upon the test), reduction of metalline cakes into metal, and volatilizing in the dry way (sublimation). The third operation, solution in the liquid way, included precipitating, extracting, cementing, and volatilizing in the liquid way (distillation) (Gellert, 1776, p. 181). There was no dearth of means to alter mineral materials.

Fourcroy remarked: "La chimie, … , est une science dont l'objet est de reconnoître la nature & les propriétés de tous les corps par leurs analyses & leurs combinaisons" (Fourcroy, 1789, v. 1, p. 2)[7]. Rock and mineral discussion and analysis took up a good part of the first volume of his five-volume work. At this beginning period for the new chemistry, Fourcroy spoke of pulverization, porphyrification, the action of lime and the leas of wine, and use of the chisel (Fourcroy, 1789, vol. 1, p. 51).[8] As opposed to the early mineralogical treatises, Fourcroy gave his instructions in the framework of chemical theory, and spoke of the need to overcome the attractive forces between the parts of substances. In exploring rocks and minerals, he observed the effects of air and water vapor on them, as well as of lixiviation by hot and cold water (Fourcroy, 1789, v. 1, p. 381).[9]

---

[4]Elutriation or eleutriation is defined as separation and purification of a mixture of granular solids with water by (a) decanting, (b) straining, or (c) washing. Edulcoration is defined as the washing of a solid, often a precipitate, with water to free it from soluble impurities such as salts and acids (Eklund, 1975, p. 26). The meaning of eliquation is not so easy to find. Since it is under "dry" reactions, I assume it involves heat. Quartation is the process of combining gold and silver in the ratio of one to three. That combination is dissolved in nitric acid, and the pure gold separated (Eklund, 1975, p. 35–36).

[5]Dephlegmation means "to remove water from a solution, usually one of an acid or alcohol. There is a sense of refining or purifying about the term, as opposed to simple concentration" (Eklund, 1975, p. 25).
[6]Eklund defined "elixation" as "the action of boiling or stewing" (1975, p. 26). The difference in spelling is not surprising.
[7]"Chemistry … is a science the object of which is to examine nature and the properties of all the substances by their analyses and their combinations" (my translation).
[8]As nearly as I can tell, porphyrication refers to the use of a slab of porphyry, a hard rock, as a surface upon which to grind samples. My only reference, a dictionary (the unabridged *Random House dictionary of the English language,* 1950, Stein, J., ed., New York: Random House), has that listed as one of the uses of the rock.
[9]Lixiviation is defined as "separation of soluble from unsoluble solid substances by soaking the mixture of solids and removing the resulting solution which contained the soluble material" (Eklund, 1975, p. 30).

These operations are familiar to chemists today and many are still in use, particularly in sample preparation for our modern instrumentation. Some, such as precipitation, will be mentioned frequently. Others, discussed in the next section, were used for sample preparation and/or influencing the speed of the reaction. They might also be used for the purification, analysis, or characterization of the materials of Earth. The division into categories is not always clear, as the same process may be used at several stages, but our interest is in how geologists applied these operations, with what equipment, and how they interpreted the results.

Solution

Solution could be called one of the most important processes, because it enables reactions that would not occur in the solid state to proceed. We use the term "solvent" for that which does the dissolving, and "solute" for what is dissolved. In the eighteenth century, a solvent was called a "menstruum." Dissolution in water was recognized as a physical process by some workers, because the solute could be recovered on evaporation. For example, Agricola noted that common salt had been produced by evaporation since antiquity. He also reported variable solubilities for different kinds of salts (Agricola, [1546] 1955, p. 29). Other solvents, such as oils or alcohol, might be used. Some solvents reacted with the solute, such as acids with metals. Since in this case the original solute could not be recovered, the process of dissolution was known to be a chemical change. Boerhaave was said to equate solution and chemical reaction, as the fine particles of the solvent penetrated between the particles of the solute (Leicester, [1956] 1971, p. 124). Solution was aided by varying degrees of heat, and by division of solids into fine particles. The debate about whether solution was a physical or chemical process continued among chemists, both positions supported by evidence.

Cramer included a chapter on the "Virtues of the Menstrua" in his work on assaying (Cramer, [1739] 1741). He wrote:

We call solutions moist, when a Body is distributed through the very minutest Particles of an aqueous Fluid, in such Manner that both may turn into a Fluid to Appearance homogeneous, which goes through all Filters without being detained in them, and the smallest Part of which contains in it a proportionable Quantity of both the dissolvent [solvent] and the dissolved Body. (p. 194)

He emphasized the use of different solvents for different materials and the need to have the correct proportions, as well as the advantages of having the solute finely divided. A true solution would pass through all filters. The process of solution known as edulcoration applied when a soluble salt was mixed with an insoluble material. The salt was dissolved away, and the solution was separated from the insoluble part by filtration. The process was quicker if the initial material was pounded into small particles so all could be "reached" by the solvent, which might also be heated to boiling (Cramer, [1739] 1741, p. 195). Cramer thought that "moist" solution could include elutriation, or washing with plain water. This was applied to two different, insoluble substances, one of which was notably lighter than the other. The heavy substance would stay at the bottom of the flask, while the light one could be poured off with the water. Elixation (or elixivation) might also aid solution, as it referred to the act of boiling or stewing (Eklund, 1975, p. 26).

Considering solvents other than water, reasonably pure alcohol was probably not known in Europe until several hundred years after the Chinese had distilled it from wine in about the fourth century A.D., following their ability to increase its concentration by freezing the wine (Brock, 1993, p. 24). The solution powers of oils had also been noted, although in general this was not important with rocks and minerals. Bituminous substances were observed to act as a solvent on a few things as well.

Many mineralogists used solution as part of the sequence of reactions used to identify minerals. For example, Kirwan used a change in solubility as one of his criteria for ascertaining whether a "chymical union" had taken place (Kirwan, 1784, p. 17). His entire conception about the origin of crystalline rocks rested on what he knew of solution, some of it closely and accurately, and other parts being imaginary or fanciful. Procedures for putting rocks into solution in order to achieve quantitative analysis were well known. This will be discussed later in this chapter. William Withering (1741–1799), the chemist, botanist, and physician of Birmingham, said that for his rock analyses, he only used water that had been distilled in glass vessels or from a tin refrigeratory (Withering, [1782] 1809, p. 291).[10] Black demonstrated the presence of silica in mineral water, although it was recognized that other stronger solvents such as the alkalis were needed to dissolve rocks. Kennedy repeatedly put the basalts of Sir James Hall into water solution, and then precipitated portions in his analysis of them (Kennedy, 1805). Kirwan claimed that the fact "that siliceous earth is soluble in simple water, when sufficiently comminuted, appears from various observations" (Kirwan, 1799a, p. 111). By 1814, in the third American edition of Henry's chemistry book, it was noted that "the term SOLUTION comprehends an extensive class of phenomena" (Henry, 1814, v. 1, p. 40). Listed were solids with liquids, and also liquids soluble in each other and gases with both. Solution was indicated by a perfect transparency, and was the result of chemical affinities between the bodies. Saturated solutions were defined as they are today (Henry, 1814, v. 1, p. 40).

Digestion

Digestion was an aid to solution, or to chemical reactions. It consisted of heating, usually for a lengthy period, to a temperature less than the boiling point. It had been a valuable tool for the alchemists of the fourteenth and fifteenth centuries, and continued in use later with modernized glassware (Ferchl and Süssenguth,

---

[10]A refrigeratory (or refrigoratory) is a container filled or surrounded with cold water at the head (top) of a still in order that all vapors can be condensed (Eklund, 1975, p. 36).

1936, p. 45). A sand bath was a popular way to transmit the heat, or a special digestion furnace was used that would maintain a steady, low heat. Papin's digester was mentioned previously, wherein heat was applied under pressure. A somewhat different pressure apparatus, still designed to supply moderate heat under pressure, was in use by the middle of the nineteenth century in pioneering organic reactions (Russell, 2000, p. 316–318).

Distillation

A procedure handed down from antiquity was distillation, a major activity among alchemists, which exploited different boiling points of liquids.[11] The original mixture was heated in a flask or retort, and the vapors directed upward and through some sort of distillation head or "pelican" in which they were cooled. If a solid was dissolved in the liquid, it would be left behind in the flask. Ihde ([1964] 1984, p. 140) included an illustration of the early evolution of still heads from a simple lid to a lid that might be curled under at the edge to catch the condensate, and then development of a tube leading to a separate receiver. By the sixteenth century, a double condenser with a countercurrent flow of cold water and steam distillation were reported (Ihde, [1964] 1984, p. 19). Cooling of the vapors might be aided by covering the collector with a blind head through which water could be poured (Greenberg, 2000, p. 21), or sometimes by surrounding it with a cloth kept moist with cold water or ice. The condensate then accumulated in a receiver. Gellert mentioned three kinds of distillation, over the helmet, sideways, and downward, each of which would have its own equipment (Gellert, 1776, p. 179).[12]

Ideally, the higher boiling point component was left behind. Sometimes mixtures of solids in solution were distilled over long periods, and volatile products collected. As we saw in Chapter 6, vessels for this were usually made of glass, sometimes covered with lute or tin. It was relatively easy to get a separation of, say, "spirits of wine," or alcohol, which has a boiling point of 78.5 °C, from water which has a boiling point of 100 °C, in this manner. Repeated distillation was often employed to improve separation. The form of the equipment might be different according to the properties of the substances to be distilled, i.e., a longer neck for a flask if the vapors were more difficult to separate.

Although solid rock cannot be distilled in the usual sense of the word, the process was most useful in preparing reagents that were used to dissolve the rock as indicators, or for reactions useful in identification. Early examples were the acids used by assayers and alchemists, the preparation of which required distillation. Agricola gave ten "recipes" for making what he called *aqua valens*, or strong water, of which four were particularly powerful (Agricola, [1556] 1950, p. 439–441).[13] His purpose was to dissolve and thus purify gold. His distillation mixtures, which were basically pulverized solids combined with water contained in a luted glass ampula, were heated in a furnace for as much as 24 hours, and the drops collected in another glass receiver (Agricola, [1556] 1950, p. 443). Cramer ([1739] 1741) discussed preparation of both *aqua regia* (a mixture of hydrochloric and nitric acids) and *aqua fortis* (concentrated nitric acid) by distillation. Because preparation was difficult, he gave a method for distilling *aqua fortis* out of a solution which had dissolved metal so that it could be used again (Cramer, [1739] 1741, p. 37). The mercury used to dissolve gold could also be distilled off and reused.

By the late eighteenth century, the use of distilled water, often distilled in glass, was standard for analytical work, as noted by Withering ([1782] 1809, p. 291).[14] Black employed it in his analyses of mineral waters, and also used distillation as a separation step. Kennedy used distillation to remove uncombined acid in one step of his basalt analysis. In the final decades of the eighteenth century, Guyton de Morveau "explained at some length the value of distilled water in practical chemistry" (Smeaton, 2000, p. 224). It is not surprising that at a time of continuing interest in mineral water analysis, the presence of salts in rain and surface water was noted. Thus, distillation was a standard purification or separation technique that required only standard equipment to which appropriate improvements for new usages could be easily made. We often think the more important component is that which is volatilized and collected, but in reading accounts of both alchemists and assayers/metallurgists, I have sometimes wondered if the substances remaining in the heated flask were occasionally what were sought. In that case, the process would be more akin to digestion. Biringuccio discussed extracting water or oils from a variety of substances such as roots, fruit, flowers, as well as minerals (Biringuccio, [1540] 1990, p. 340). Evaporation, which is heating without trying to collect the fluid portion, was practiced since antiquity as a way to provide common salt (Agricola, [1556] 1950, Book XII).

Precipitation

Precipitation is the production of an insoluble substance, generally by the chemical interaction of solutions of two soluble substances. Ideally, and usually, the other compounds formed will stay in solution, allowing the insoluble material to be filtered out. Sometimes a more active element will substitute in the solution for one less active, allowing the latter (an element, as opposed to a compound) to precipitate out. A good demonstration of pre-

---

[11]In their translation of Agricola's *De re metallica*, the Hoovers gave a capsule history of evidence for distillation, beginning with Aristotle's statement that sweet water can be produced from salt water by condensation of vapors from heated salt water. Mercury was also distilled early, and there is evidence of appropriate apparatus from the first to the sixth centuries ([1556] 1950, p. 441 fn).

[12]Anderson described distillation apparatus and its uses from the time of Greek antiquity in a most interesting paper titled "The archaeology of chemistry" (Anderson, 2000).

[13]The term meant "powerful water." From the ingredients and the way they were treated, hydrochloric, sulfuric, and nitric acids might all have been produced, along with *aqua regia*, the latter consisting of one part nitric to three or four parts of hydrochloric acid (Agricola, [1556] 1950, p. 440).

[14]I have discussed distillation only as employed by people who related it to geological questions.

cipitation is to mix solutions of lead nitrate and potassium iodide, both colorless, causing brilliant yellow, insoluble lead iodide to form, with colorless potassium nitrate remaining in solution.[15] Solubility varies with both temperature and pH, and not always in the expected direction. We associate an increase of solubility with increase of temperature, and this is true for sugar and many salts. However, there are some exceptions to this generalization, such as cesium sulfate, whose solubility decreases with increasing temperature. There are also wide variations in both absolute solubility, often expressed as grams of salt per 100 g of water, and in change of solubility with change in temperature.

Ursula Klein (1994) has written a luminous article about the origin of the concept of a chemical compound, which included a discussion of the meaning attached to precipitation in the sixteenth and seventeenth centuries. Gold, silver, and other metals in turn could be separated, or precipitated out, with the proper combinations of solution and precipitation. The reactions of a range of substances with acids and salts were known empirically, and were presented in table form by Étienne François Geoffroy (1672–1731), for a time lecturer in chemistry at the Jardin du Roi (Partington, 1962, v. 3, p. 50), in 1718 (Klein, 1994, p. 165). We can read into the metal reactions carried out in the sixteenth century our knowledge of the sequence of relative reactivities, as desirable metals and compounds were produced. Cramer described precipitation as follows:

Precipitation is called moist, in German Fallung, when a Body, which has been dissolved the moist way, is again driven out of the Dissolvent, so as either to swim in the Menstruum, or to sink to the bottom of it: which most commonly looks like Powder; but the Separation of the Menstruum from the Body precipitated, is afterwards performed by either decanting, or filtering. (Cramer, [1739] 1741, p. 196)

By the mid-eighteenth century, such reactions were applied to mineral analysis, as exemplified by Wallerius in his analysis of phosphoric gypsum. After calcining this substance, he dissolved it in water and found that a black precipitate was present when the solution was treated with solutions that contained lead, silver, or (probably) iron sulfate, and also effervesced with all the acids. Clearly, it isn't easy to determine what the actual reactions were, particularly as his "phosphoric gypsum" undoubtedly contained a number of impurities. Wallerius concluded that the mineral contained alkaline and sulfurous substances, as well as arsenic, and had a taste and odor resembling those of a mixture of orpiment (As_2S_3) with quicklime and water (Wallerius, [1747] 1753, v. 1, p. 109).

In his analysis of mineral water from Iceland (Fig. 7.1), Joseph Black demonstrated his thorough familiarity with solution and precipitation. In his preliminary investigations, he saw that a solution of sal saturni (lead acetate) made the water white and muddy (which would have been lead carbonate), but vinegar dissolved nearly all of the precipitate, making the water

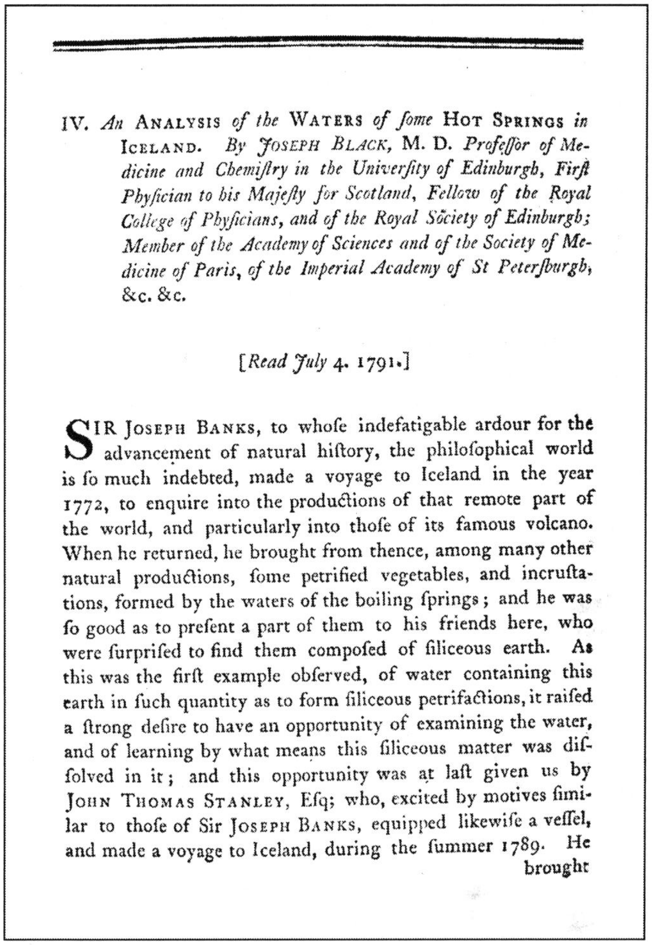

IV. *An* ANALYSIS *of some* HOT SPRINGS *in* ICELAND. *By* JOSEPH BLACK, M. D. *Professor of Medicine and Chemistry in the University of Edinburgh, First Physician to his Majesty for Scotland, Fellow of the Royal College of Physicians, and of the Royal Society of Edinburgh; Member of the Academy of Sciences and of the Society of Medicine of Paris, of the Imperial Academy of St Petersburgh,* &c. &c.

[*Read July 4.* 1791.]

SIR JOSEPH BANKS, to whose indefatigable ardour for the advancement of natural history, the philosophical world is so much indebted, made a voyage to Iceland in the year 1772, to enquire into the productions of that remote part of the world, and particularly into those of its famous volcano. When he returned, he brought from thence, among many other natural productions, some petrified vegetables, and incrustations, formed by the waters of the boiling springs; and he was so good as to present a part of them to his friends here, who were surprised to find them composed of siliceous earth. As this was the first example observed, of water containing this earth in such quantity as to form siliceous petrifactions, it raised a strong desire to have an opportunity of examining the water, and of learning by what means this siliceous matter was dissolved in it; and this opportunity was at last given us by JOHN THOMAS STANLEY, Esq; who, excited by motives similar to those of Sir JOSEPH BANKS, equipped likewise a vessel, and made a voyage to Iceland, during the summer 1789. He brought

Figure 7.1. The title page of Black's (1794) influential paper on the analysis of mineral waters.

almost clear (Black, 1794, p. 97).[16] There were numerous other instances of his using precipitation both to give him an idea of how to proceed, and to separate components for quantitative analysis. Saussure remarked on "un précipité d'une belle couleur orangée," which was produced when he tested a residue he believed to be fixed alkali (sodium carbonate) in a mineral water analysis (Saussure, 1803, v. 1, p. 299).[17] Kennedy (1805) used precipitation at several points in his analysis of Hall's lavas and whins, which revealed probably a little of what he called muriatic acid, some calcium carbonate, and silicon dioxide among other things.[18]

---

[15]It should be noted that in the real world there are intermediates to "soluble" and "insoluble."

[16]Addition of the soluble lead acetate could precipitate insoluble lead chloride, -sulfate, and -carbonate from the contents of Black's mineral waters (see his analyses, 1794, p. 119). Some lead salts, however, are soluble to some degree in excess acetate, which the vinegar would have provided (Latimer and Hildebrand, 1951, p. 353).

[17]He used "corrosive sublimate" or mercuric chloride ($HgCl_2$) to effect the precipitation. Some mercury salts are orange, although mercuric carbonate does not exist alone (Latimer and Hildebrand, 1951, p. 146).

[18]By this time, procedures for dissolving silicate rocks and proceeding with analysis were fairly standard (see Newcomb, 1986).

Titration

Titration is now understood as a quantitative means of determining the concentration of one solution by means of a measured volume of another solution of known concentration. Its practice arose from the need to determine quickly if the chemicals used in necessary industrial processes, such as making textiles and gunpowder, were of proper strength (Szabadváry, 1966, p. 198). While Joseph Louis Gay-Lussac (1778–1850), the French chemist who discovered the law of combining volumes, has been credited with the invention of the method in 1835 as a volumetric means to determine the purity of silver, one can find numerous instances of the concept in use from the mid-eighteenth century (Szabadváry, 1966, p. 197). For instance, in 1756 what was called the "value" of pearl ashes, purified potassium carbonate, which was used in bleaching, was determined by how much nitric acid could be added before effervescence stopped. Similar reactions had been used even earlier to determine the strength of vinegar and nitric acids using potassium carbonate (Ihde, [1964] 1984, p. 288–289). Measured volume increments might be added, drops counted, or the solutions weighed before and after use. Both change in color of an indicator or cessation of effervescence might be used to determine when an alkali or an acid was used up. For volumetric measuring devices, there was a progression through the graduated cylinder, a transfer pipette, calibrated glass tubes, and, finally, the burette (Ihde, [1964] 1984, p. 289).

An approximate measure of the acidity or alkalinity of mineral water was determined by additions of known amounts of acid or base and, again, the value indicated by the end of effervescence, or an indicator color change. A number of indicators were known at the time, and will be discussed later. It should be noted that the volumetric methods of the late eighteenth century were not accurate, since the indicators then available generally did not provide precise end points (Brock, 1993, p. 183–184). They could, however, indicate whether a solution was saturated with acid or base, and thus produce conditions most favorable for precipitation of particular components. Szabadváry (1966, p. 200) discussed the difficulty of finding a "first" for the institution of titrimetric procedures as the methods came into use for diverse reasons in diverse places, many outside "academic" chemistry. W.A. Campbell provided a good summary of the history of titrimetric analysis with references that gave more detail (Campbell, 1985a, p. 180–182).

Black, Saussure, Kirwan, Breislak, and Kennedy all used indicators, but not quantitative methods, to give a preliminary idea of the properties of substances found in mineral waters.[19] Indeed, knowing whether a natural water was acidic or basic was one of the first concerns of the many people who analyzed them, as it would tell them immediately what sorts of things were likely to be dissolved in the water, and how to approach the analysis.

Breislak also dissolved volcanic gases in water and tested the resultant solutions. He was curious to know if both hepatic (H_2S) and carbonic (CO_2) "airs" could be present, as this was a matter of some confusion. He found it necessary to investigate the properties of litmus solutions which might sometimes regain their original color after a first color change, so he added mineral acid drop by drop to such a solution until he again just effected a color change (Breislak, 1792, p. 91). In this case, he applied the method of titration to clarify the nature of the indicator itself in order to further investigate volcanic gases. This was just a small part of his investigation of those gases.

Another person we know primarily as a geologist, but who in his time was known as one of the most innovative analytical chemists, was Richard Kirwan (Fig. 7.2). He is credited with developing a method for the determination of iron that was based on precipitation-titration. For it, he used a standard solution of phlogisticated alkali, or potassium ferrocyanide. The standard solution, a solution with a known weight of reacting reagent in a known volume, was just coming into use (Szabadváry, 1966, p. 207). Szabadváry (1966) provides an excellent account of the development of volumetric chemistry, which was applied for geological or mineralogical knowledge.

Filtration

Unlike the more complex procedures of precipitation and titration, filtration is a simple mechanical process of separating a solid from a solution. As noted above, Cramer used what he called edulcoration to separate a saline substance from another solid. However, because the solid was usually in very fine particles and it had to be recovered quantitatively if an analysis was being done, care had to be taken to recover all of it. Since precipitation had been commonly used as a preparation for many years, means of separation, although not completely quantitative, were not novel. There was disagreement among practitioners as to whether filtration or decantation was better for quantitative work. With decantation, some of the precipitate might inadvertently be poured off with the supernatant solution, or the residue was left very wet. However, use of filter paper meant that the paper had to be very carefully washed to remove all the precipitate. If the paper was ignited in the case of a non-fusing precipitate, the ash would add to the weight of the residue, and sometimes to the amounts of substances such as silica in the analysis, but this could be accounted for. In both cases, the precipitate itself had to be washed with distilled water to remove all soluble material, and dried before weighing. Unlike Black, Withering did not use filter paper. He allowed solutions containing precipitate to settle for days or weeks (Withering, [1782] 1809, p. 291).

Geologists/chemists were well aware of these niceties. Black dealt with very small quantities of precipitates when analyzing Icelandic mineral waters, and developed a method of recovering them, which he set out in detail. He first allowed the solution with precipitate to settle for a long while so that most of the

---

[19] In the seventeenth century, Robert Boyle was known to use violets, cornflowers, cochineal, and litmus to indicate alkalinity or acidity (Szabadváry, 1966, p. 36). I have generally discussed people who contributed to the geology of the eighteenth century.

Figure 7.2. Richard Kirwan.

one of the finest of mineral analysts, was able to order special filter paper, made in the winter with long fibers, that was allowed to freeze when wet to expand the fibers (Szabadváry, 1966, p. 145). An act as simple as filtration was given close attention.

Pulverization

The large store of practical knowledge extant in the late eighteenth century included the recognition that most reactions and solution occurred far more readily if the substance under investigation was finely divided. Describing edulcoration, Cramer said:

> It is proper in the Operation to inlarge the Surface of the Body to be edulcorated, by pounding of it; that the Solution may have a speedier Success: for which Purpose indeed, they moreover break it in Pieces, … That all Particles of the Body … may on all Parts be contiguous to the Particles of the washing Fluid … (Cramer, [1739] 1741, p. 195)

When he pulverized his samples, Withering used either "those excellent … [mortars] made by Mr. Wedgewood" or "a steel mortar tempered so hard that it will bear the grinding of enamel in it without discolouration" (Withering, [1782] 1809, p. 291). Kennedy reduced his lava samples to fragments before attempting solution (Kennedy, 1805, p. 78, 83, 92). A number of these chemist-geologists mentioned weighing the mortar and pestle before and after grinding rock or mineral samples to ensure that material from them did not add to the mass in the analysis of samples.[20]

REAGENTS

There were a large number of substances available to effect reactions. As noted above, useful materials had been gathered or produced since early human history, and were increasingly employed in the production of necessities. There is no "beginning" to this story. Most early technology could be effected with salt, sal ammoniac (ammonium chloride), saltpetre (potassium nitrate), copperas (iron sulfate), lime, alum, sand, and fat (Campbell, 1985b, p. 239). By the ninth and tenth centuries, an alchemical text mentioned metals, vitriols, boraxes, salts, stones, and sal ammoniac, the latter prepared by distillation of hair with salt and urine (Brock, 1993, p. 22). Sulfur and its compounds were popular. The whole of metallurgy employed reagents for what we recognize as oxidation-reduction reactions. Agricola's and others' mid-sixteenth-century texts detailed their preparation and isolation. In the early seventeenth century Johann Rudolph Glauber (1604–1670), the Bavarian chemist who settled in Amsterdam, and who has been credited with being one of the earliest industrial chemists, manufactured acids, bases, and salts, and did much experimentation on the chemistry of mercury and arsenic (Leicester, [1956] 1971, p. 103).

liquid could be decanted. The remaining material was agitated thoroughly and poured into a carefully prepared four-inch filter paper, which was first folded, then reopened and warmed. The margin was coated with melted tallow or beeswax to a width of an inch, the center remaining nonwaxed. The paper was refolded while still warm and allowed to cool before use. The impervious border allowed all adhering precipitate to be washed off and collected in the center. After drying, the paper was spread flat, the waxed border carefully cut off, and the whole weighed. Black would then take two pieces of the same filter paper, cut them out in the exact shape of the one that held the precipitate, weigh them several times, and average the weights so determined. Subtracted from the weight of the filter paper plus precipitate, the best value for the weight of the precipitate was obtained (Black, 1794, p. 124–125 [124 is misnumbered as 224]).

Such care in analysis was not unusual, although Black was recognized as a notable chemist. Scientists who presented their work in the forums that Black did would be expected to be similarly fastidious about their procedures. We are again reminded of how often the scientists of the eighteenth century read the work of others, often in multiple languages. Breislak, Spallanzani, Saussure, and Dolomieu were just a few of the seminal thinkers in geology who included the details of their analytical work in their geological observations, and frequently referred to that of others. By the early nineteenth century, Berzelius, who was recognized as

---

[20]There are few examples of mortars and pestles extant. Earlier examples made from metal gave rise to fear of contamination, which was the impetus for Wedgwood to produce a kind of stoneware that would not abrade into the sample (Anderson, 1985, p. 224).

Eklund's dictionary of terms and reagents in use in Britain in the eighteenth century included roughly 450 of the latter, some being duplicates or near relatives (Eklund, 1975, p. 20–45). They were of both organic and inorganic origin, and were both natural and man-made. A number of histories of chemistry give a good account of the origin and use of reagents.[21] While their compositions were not known, their action was known and predictable.

The following are just a few examples of reagents found necessary to investigate rocks, minerals, and ores in the eighteenth century. For assaying, Cramer included a chapter titled "Of the docimastical *Menstrua*, and the preparations of them,"[22] divided into those of metals and their products, of semi-metals, of pure sulfureous or oily bodies, of salts, of sulfur, of cements, and of the simpler and reductive fluxes (Cramer, [1739] 1741). He discussed the menstrua before the laboratory instruments because the latter had to contain the former and their reactions. His comment about them was:

Those Bodies, which being applied to others according to certain Rules, dissolve them so as to adhere themselves in a State of Division to the Particles of the Body dissolved, and cannot separate from them again of their own accord, are called Menstrua. (Cramer, [1739] 1741, p. 14)

Metals and metal compounds answered that description, as he illustrated with many examples. The effects of sulfur and its compounds were also noted, both alone and combined with salts. The acids utilized included acid of vinegar, acid of vitriol, *aqua fortis* (the same as spirit of niter, nitric acid), and *aqua regia* (spirit of salt plus spirit of niter). The effect of the first upon some classes of earths and stones was recorded as "violent Ebullition," although Cramer recorded that there was no reaction with flints, sand, or many other materials (Cramer, [1739] 1741, p. 30).

Cramer also noted the action of borax in dissolving earths and stones into glass when both were well pulverized together. He advocated rubbing borax on the inside of crucibles that were to be used when melting precious metals, covering the crucible with a thin crust that filled cavities, so that all the metal could be poured out cleanly (Cramer, [1739] 1741, p. 41). He pointed out that because of this use, borax was erroneously listed among the reducing bodies. The error came about because the borax did not reduce the ore to metal, but only helped to coagulate it (Cramer, [1739] 1741, p. 42).

In the 1776 translation of Gellert's *Metallurgic chymistry*, symbols were given for, among other things, acid of vegetables (essentially acetic acid), distilled vinegar (the same), acid or spirit of common salt (HCl), acid of niter (HNO_3), acid of vitriol (H_2SO_4), and fixed alkaline salt (solid potassium carbonate). Altogether there were 63 symbols, for compounds and metals, as well as for fire, air, water, and earth. A few examples are: Δ for fire (which we still use for heat applied), O for alum, X for zinc, and W for bismuth (Gellert, 1776, p. 183). It must be noted that the symbols assigned to both elements and compounds changed enormously over the eighteenth century, and the same symbol might represent another element, compound, or procedure in different lists. However, this metallurgical chemistry clearly delineated the properties of the substances later used specifically to test rock composition or origin.

A list of "fluid tests and re-agents" was included in the necessities for the humid, as opposed to the dry, or fusion, laboratory (Cronstedt, 1788, v. 2, p. 989–991). Besides those more common, some of the 38 reagents listed were the muriates (chlorides) of barium and calcium, acid of sugar (oxalic acid), oil of tartar *per deliquim* (potassium carbonate dissolved in the water it attracts through deliquescence), and phlogisticated alkali (potassium ferrocyanide). This was for the traveling wet, or humid, laboratory, the case of which was necessarily twice as deep as that of the dry laboratory in order to hold all the bottles of solutions. For the use of those reagents, readers were referred to Bergman, Kirwan, some *Memoirs* from Dijon, and Fourcroy's lectures (Cronstedt, 1788, v. 2, p. 991).[23]

In his investigation of mineral waters, Black made use of various indicators, plus limewater, acid of sugar, corrosive sublimate (mercuric chloride), *sal saturni* (lead acetate), a solution of barites in muriatic acid (barium chloride), and silver chloride. We easily see that some of those reagents were designed to put a material into solution by forming a soluble salt (an acetate or a chloride), so that substances of interest could then be precipitated out and weighed. Breislak's account of the investigation of the mineral waters of Solfatara included those and others, such as the oxide of arsenic, the addition of which would induce turbidity in the presence of hydrogen sulfide (Breislak, 1792, p. 41–42). Saussure referred to "sugar of Saturn," another name for lead acetate (Saussure, 1803, v. 1, p. 296). He also mentioned a preparation of galls, which preparation would have been gallic acid. Galls were a parasitic growth on oaks that when dried, ground, and dissolved, was used to detect iron (Eklund, 1975, p. 28). Thus, the analyzers of minerals, rocks, and mineral water had an extensive armamentarium of reagents to call upon, applied in a sequence that became more refined as wet chemical analysis was itself refined.

Acids and Bases

The preceding section makes it clear that acids and bases and their reactions were well known by the time investigations of geological materials were undertaken with a view to clarifying rock origin. Vinegar, fruit juices, and hydrolyzed salts were the first acids in use (Ihde, [1964] 1984, p. 13). Mineral acids were mentioned as early as A.D. 1200, and were probably first prepared by Europeans in the thirteenth century (Multhauf, 1966, p. 141; Partington, 1989, p. 31; Brock, 1993, p. 23).[24] They were carried in the portable, or pocket, laboratories of eighteenth-century min-

---

[21]For example, see Brock (1993), Ihde ([1964] 1984), Partington (1989), and Szabadváry (1966).

[22]These would be the solvents used in assaying, which was also known as docimacy.

[23]There are several lists giving meanings for archaic terms, but I mainly rely on J. Eklund's *The incompleat chymist* (1975).

[24]N.G. Coley and J.R. Partington, among others, have reminded us of how indebted early chemistry was to the alchemists (Coley, 1985, p. 53).

eralogists. The humid laboratory referred to by von Engeström in the appendix to Cronstedt (1788) included concentrated vitriolic acid, more and less concentrated marine acid (HCl), *aqua regia* in several strengths, and "nitrous acid," the nitric acid of modern chemistry (Cronstedt, 1788, v. 2, p. 989). The acids Kirwan looked for in mineral waters were sulfurous, sulfuric, muriatic, boracic, and nitrous (Kirwan, 1799b, p. 13–14).

Indicators

Indicators were available to determine the strengths of some industrial chemicals from before the seventeenth century. In Boyle's work on colors, he investigated many substances that would alter color when changed from basic to acidic solution, or vice versa. A partial list of indicators in use in the eighteenth century would include infusions of red rose leaves, syrup of violets, solutions of turmeric, ground galls, and berry juices, as well as infusions of different woods. A little kitchen chemistry shows that Concord grape juice and red cabbage have the same activity, the latter showing multiple color changes. Then as now, a popular item was litmus, made from the plant lichen *Crozophora tinctoria*. This was often given the name turnsole, which name was spelled in a variety of ways (Eklund, 1975, p. 43; Brock, 1993, p. 162).[25]

Breislak employed a *teinture de tournesol* (a solution made from sunflowers) to investigate the properties of volcanic gases that had been absorbed in water (Breislak, 1792, p. 89). In his analysis of Iceland's mineral water, Black used "paper stained blue with the March violet" and "cambric stained to a bluish color with infusion of litmus" (Black, 1794, p. 97). Results with both showed an alkaline solution. Saussure tested waters issuing from a volcano with syrup of violets, which turned green and indicated the waters were alkaline (Saussure, 1803, v. 1, p. 297).[26] Although he did not mention using an indicator, Kennedy may have used one when he discussed putting an excess of sulfuric acid into a solution from his basalt analysis, which he then neutralized with carbonate of soda (Kennedy, 1805, p. 80). As noted in the titration section above, volumetric titration was beginning to be used, but more often, particularly in the field or with samples brought back to the laboratory, indicators were used for a general characterization that might show acidity or alkalinity.

Salts

Salts were known in great variety from antiquity onward. In many languages, their names often reflected a property, characteristic, or origin.[27] We know thousands of them as the combination of a cation, often metallic, and an anion, usually nonmetallic.

They figured largely in alchemical investigations, and had many practical uses. Hundreds of them were known by the end of the eighteenth century, and early in the nineteenth century, thanks to Volta's pile and improvements to it—and the work of Davy, Berzelius, and many others—the electrical nature of compounds and their combination was clarified.[28] Their relative solubility or insolubility in different combinations was the basis of the various steps in assaying and mineral analysis.

Pott had used fusible salts mixed with the different earths as the second step of his investigation of the fusibility of earths (Pott, [1746] 1753, p. 10). As mentioned in Chapter 3, which addressed fusion, the salts borax (hydrated sodium tetraborate), mineral alkali or sal soda (sodium carbonate), and microcosmic salt (hydrated sodium ammonium phosphate) served as fluxes in blowpipe analysis. In the humid laboratory of von Engeström, a number of salt solutions were listed, many of them as being the nitrate, as well as several salts in solid form (Cronstedt, 1788, v. 2, p. 990). The usual solubility of nitrates and acetates was put to good use. Later reports of mineral and mineral water investigations in the field routinely include them.

METHODS AND REACTIONS

We will look at the methods and reactions of four kinds of geological material analysis: that of minerals, rocks, mineral waters, and gases. By definition, analysis is designed to ascertain constituents. In chemistry, synthesis is often considered proof of analysis: If the material can be put back together from the constituent parts, the analysis is correct. This approach was not foreign to eighteenth-century workers. Black is credited with being one of the first to recover the original mass of a reactant, namely *magnesia alba* ($MgCO_3$), after a series of reactions (Anderson, 1982, p. 8). The recovery of minerals after a series of reactions led to proper synthesis, but often the lack, or lack of understanding, of appropriate pressure and temperature conditions precluded reconstitution of geological materials.

Mineral Analysis

In the simplest sense, mineral analysis had been pursued since antiquity. Substances could be recognized not only as something that might effect a desired change, but also as a source of a desirable material. The latter was exemplified by ores that yielded metals. The identification and characterization of the "earths" was important in the seventeenth and following centuries, and that search has engaged the attention of historians of geology.[29] Work with the earths was pivotal to mineralogy, and it

---

[25]A further point of confusion about his indicator, besides the variety of spellings in a variety of languages, is that the same name is used for a class of flowers ("sunflowers" in English) that turn toward the sun.

[26]Boyle found that when put on white paper, blue syrup of violets turns green with alkalis and red with acids (Partington, 1961, v. 2, p. 534).

[27]For example, "sugar of lead" (lead acetate, which tastes sweet); "sedative salt" (sometimes sodium tetraborate); "marine salt" (sodium chloride); and "marine alkali" (sodium carbonate).

[28]Eklund noted that in the sixteenth and seventeenth centuries, salts were identified as being a group of solid, soluble, noninflammable substances, with characteristic tastes. In the eighteenth century, they were increasingly thought of as being the result of processes, such as the reaction of an acid and base (Eklund, 1975, p. 38).

[29]For an illuminating series of papers on the evolution of understanding about earths, see Oldroyd (1973a, 1973b, 1974a, 1974b, 1974c), many of which are included in his 1998 book. Also see Laudan (1987).

provided the beginnings of the use of chemical analysis for mineral classification. Chemical analysis as we know it could well be said to have originated with Black's work in determining that the mineral, magnesium carbonate ($MgCO_3$), consisting in part of an imponderable fluid (carbon dioxide), could be recovered quantitatively after dissociation (Anderson, 1982, p. 8).

Characterization of minerals by their composition could not be successful until chemical procedures were reliable, and methods were developed to isolate and recover their constituents. There were many reasons for problems: (1) Constituents might be present in such small quantities that they were difficult to detect; (2) elements that were present might not yet have been identified, there being only 54 by 1830;[30] (3) elements might be so active that it was difficult to isolate them, as with the metals in some kinds of metal oxides before Volta's pile; and (4) the form of the constituent could be resistant to separation methods. With respect to the last, anyone who has precipitated silica or some metal hydroxides knows that they are flocculent solids, not easily amenable to filtration. It is not enough to have some knowledge, via early tests, that a particular constituent might be present. A sequence must be followed that allows serial separation. Along with all of that, it is necessary to have pure reagents and distilled water, as well as equipment that does not deteriorate or break during the necessary processes.

Oldroyd has reviewed the identification of the "earths" as it occurred over time. Calcareous earth, or calcium carbonate, had been known since antiquity. Magnesia, magnesium carbonate, was probably recognized by the German medical chemist Friedrich Hoffmann (1660–1742) when he found a different product upon treating both magnesia and calcareous earth with the same reagents (Oldroyd, 1973b, p. 337 [1998, VIII, p. 1]). These were followed by ponderous (barium oxide), gypseous,[31] argillaceous (clay, or Al_2O_3), and siliceous (silicon dioxide) earths. Oldroyd pointed out the necessity of proper recognition and characterization of the earths in order to have a chemical means of classifying minerals. It is also instructive to look at the classification schemes he has extracted from the work of mid- and earlier eighteenth-century workers, including Henckel, Pott, and Wallerius, whose work is cited often in this volume. We now use the categories "minerals" and "rocks." Major categories then might have included earths, stones, minerals, and concretions, with what we now designate simply a "mineral" appearing in more than one category (Oldroyd, 1974c, p. 288 [1998, IV, p. 288]). For example, Kirwan's analysis scheme was titled "Of the chemical analysis of earths and stones," not "minerals and rocks" as we would expect (Kirwan, 1794a, v. 2, p. 459).

Thus, in all studies of chemical composition, it is imperative to have some knowledge of categories as well as of possible constituents so as to know where to begin. Minerals may be easier to analyze than rocks because they have fewer constituents. Although generally correct chemical formulae are now given for minerals, in the natural world there are unlimited opportunities for anions and cations of appropriate size and/or charge to substitute in crystal structure. As mentioned in Chapter 2, which addressed characters, it was recognized very early that trace amounts of metals could color minerals. Minerals were characterized by external characters, and then by the blowpipe via heat alone, followed by heat plus the standard fluxes. Cronstedt used this sequence (Oldroyd, 1974a, p. 510 [1998, VI, p. 510]). For the next development, Thomson credited the Swedish chemist Torbern Bergman: "The first person who attempted to lay down rules for the regular analysis of minerals, and to reduce these rules to practice, was Bergman" (Thomson, [1830–1831] 1975, v. 2, p. 190).

Bergman demonstrated that most of the minerals he studied could be dissolved in muriatic acid (HCl) after being reduced to a fine powder and heated strongly, or after fusing with an alkali carbonate. Earthenware crucibles would be damaged by the strong alkali, so Bergman used iron crucibles, which, however, were susceptible to corrosion at high temperatures. He purified his reagents, used glass containers whenever possible, and washed precipitates until the washing water showed no reaction. Finally, precipitates were filtered through a weighed filter paper (Szabadváry, 1966, p. 77).

Bergman used the Papin digester to heat samples to a state of incandescence, and cautioned readers to use pure distilled water with no fixed air or salts dissolved in it to obtain good results (Bergman, 1784a, p. 19). The reaction to putting silica into solution via fusion with an alkali had been long known from glass manufacturers (Oldroyd, 1973b, p. 337 [1998, VIII, p. 2]). Bergman exploited it to put minerals into solution for what is recognized as the first scheme for sequential wet chemical analysis of mineral components. While his actual results were not accurate, Bergman's methods showed the way to wet analysis, which quickly became a powerful analytical tool in the hands of Kirwan, Vauquelin, and, especially, Klaproth (Oldroyd, 1973b, p. 338 [1998, VIII, p. 4]). As more constituents were identified, often by discrepancies in mass within analyses, analytical reaction sequences became necessarily more complex.

Procedures

Kirwan's methods for mineral analysis were often cited. He began by classifying materials with respect to their reaction with chemical agents, which gave him seven classes: (1) those wholly or partially soluble with effervescence in nitrous acid (nitric acid) of higher than 1.4 specific gravity; (2) those insoluble in 1.4 nitrous acid, but soluble in 1.25 nitrous acid at a temperature of 60°; (3) those insoluble in the former, but soluble or partially soluble with effervescence in 1.10 spirit of niter (dilute nitric acid); (4) those soluble at atmospheric temperature in spirit of niter, but

---

[30]It is instructive to look at the dates of discovery for each element. Ihde ([1964] 1984, p. 747–749) included such a list. Surprisingly, molybdenum, tellurium, and tungsten were isolated in the early 1780s, while silicon, aluminum, and bromine were not identified as elements until the middle of the 1820s. A number of other elements were first identified in the oxide form.

[31]"Gypseous earth" sometimes referred to gypsum (hydrated calcium sulfate), or to calcium oxide (Eklund, 1975, p. 28).

without effervescence; (5) those soluble in spirit of niter without effervescence, but only at 150 °F to 180 °F or higher; (6) those that don't effervesce with spirit of niter, but do so slightly with concentrated vitriolic acid (sulfuric acid); and (7) those soluble in neither acid, or only a little, with little or no effervescence (Kirwan, 1794a, v. 2, p. 459–460).

That last class of minerals was very difficult to get into solution, and included corundum, adamantine spar, and zircon. Thomson reported that Klaproth first ground those minerals into a very fine powder, digested them in a solution of caustic potash (potassium hydroxide) until all the water had evaporated, and then fused the residue in a silver crucible. It was then soluble. Most silicate minerals could be put into solution "after having been kept for some time in a state of ignition with twice their weight of carbonate of soda." The silica combined with the soda, and released carbonic acid (Thomson, [1830–1831] 1975, v. 2, p. 200). We would write it as:

$$SiO_2 + Na_2CO_3 \xrightarrow{\Delta} Na_2SiO_3 + CO_2,$$

the sodium silicate being soluble in acidic solution.[32] If the soda did not suffice, caustic soda, or sodium hydroxide, was substituted. However, that reaction might be said to be just the beginning. If sodium or potassium was among the components being sought, other methods had to be used, but their presence often went unsuspected.

Kirwan advocated first powdering earths and stones, then boiling them in 15 times their weight in water, so that saline substances were dissolved. Then the dried residue was heated to find volatile constituents, followed by heating to red and white heat to see changes of color and loss of weight. In a procedure I wouldn't be comfortable performing, both residues were then placed on niter (potassium nitrate) and heated to red heat to see if they had a carbonaceous component. Kirwan felt he needed to account for nine earths, five acids, and five metallic substances in his analyses, although he stated that it was unlikely to find all of them in any sample (Kirwan, 1794a, v. 2, p. 460).[33] Additionally, he listed the reagents needed for such work.

It is possible to write the reactions in modern notation for single steps in much of this analytical work. What is not so clear is the cumulative, sequential effect of a series of reagents on the varied contents of natural minerals. Trying to follow procedures with my "enlightened" knowledge was not as easy as I first assumed, and there was always the problem of nomenclature. For example, the term "calx" might have different meanings in different circumstances, and might be a term for the result of a procedure (heating to dryness) or for a specific chemical entity.

Figure 7.3. Martin Heinrich Klaproth.

There are many such terminology problems. The same compound might have as many as eight or ten different names in different accounts. Conversely, a single name could refer to a single compound, a class of compounds, or several closely or only tenuously related compounds.[34] If there is a good description of a reaction, it is sometimes possible, with the help of modern knowledge, to figure out what were the products and the reactants. Solubility product constants can help with precipitate identification.[35] In addition, there is a massive primary literature about mineral analysis. I have tried to limit my comments to those early workers who were themselves active in geological thought, or who were often cited by those who were.

Leucite

Among the myriad possibilities for an example of mineral analysis at this seminal time, we may discuss an analysis of leucite, commonly found in lava, done by the master analyst Martin H. Klaproth.[36] It demonstrated both the sophistication and the problems of analysis of the time. Klaproth (Fig. 7.3) is

---

[32]For conditions of solution and equations in modern notation, see Newcomb (1986, p. 91).

[33]Melhado pointed out that Cronstedt listed nine earths, but Andreas Sigismund Marggraf demonstrated only three that were not compounds of each other, namely, alumina, silica, and chalk. Bergman added magnesia and barites (Melhado, 1981, p. 53).

[34]Potassium carbonate is an example of the first; in another context, "vitriole" might be used for a number of different sulfates without modifiers.

[35]This is an equilibrium constant related to the equilibrium between a solid salt and its ions in solution. It gives a quantitative measure of the solubility of the salt. If a salt is barely soluble in its water solution, it is likely to be the precipitate.

[36]I used the copy of Klaproth available to me, an English translation published in 1801, of the original 1795 German edition.

credited with introducing silver crucibles, although those were shortly supplanted by those of platinum, and refining analytical methods (Thomson, [1830–1831] 1975, v. 2, p. 191). His methods were meticulous, and his report on leucite fairly represented the puzzlement felt by a competent chemist in response to inexplicable results. I will discuss the analysis in some detail.

Klaproth began with remarks about the occurrence of leucite and a description of its crystals. We now know it as a potassium aluminum silicate, or $KAlSi_2O_6$. Klaproth was sure of the identity of his leucite samples, and took pains to see that they had not been altered by heat or decomposition. It was infusible alone. When 100 grains were heated strongly to red heat for an hour, only one-eighth of a grain was lost (Klaproth, 1801, v. 1, p. 350). In his humid analysis, he first digested 100 grains of powdered leucite with muriatic acid (HCl), which he said dissolved a substantial portion of the sample. The solid remainder was 54 grains of siliceous earth. A schematic for that reaction is:

leucite 100 gr.   hydrochloric acid Δ silicon dioxide, 54 gr.

$KAlSi_2O_6$ +   HCl $\rightarrow$ solution a + SiO_2

He then twice ignited that residue with caustic alkali, dissolved the product in water, added hydrochloric acid to excess, re-precipitated, and then dried and weighed the residue. He found thus "a trifling decrease of weight" (Klaproth, 1801, v. 1, p. 350). He treated the acid solution with "prussiate of pot-ash" and after heating found a precipitate representing one-eighth of a grain of iron oxide.[37]

solution a + potassium ferricyanide(aq.) $\rightarrow$ blue precipitate

$Fe^{+2,+3}$ $K_3Fe(CN)_6$ $\rightarrow$ $KFeFe(CN)_6 \cdot 3H_2O$

                                                             Δ

blue precipitate $\rightarrow$ iron as oxide[38]

This was not added to the total in the analysis because he thought it represented contamination of the leucite with a small amount of hornblende (Klaproth, 1801, v. 1, p. 350), and we note that iron was also part of the reagent.

Klaproth then treated the solution from the first step with ammonia solution and got a precipitate. There was no further reaction upon adding sodium carbonate to the remaining solution.

solution a   + NH_4OH $\rightarrow$ precipitate
solution a remainder + Na_2CO_3 $\rightarrow$ no reaction

The precipitate was digested with vinegar, neutralized with ammonia to re-precipitation, dried, and weighed:

precipitate + $\begin{array}{l} CH_3COOH \text{ (vinegar), neutralized with} \\ NH_4OH \text{ (ammonia), and re-precipitated} \end{array}$

To further test what it was, he dissolved the precipitate in sulfuric acid, producing alum as shown:

                    sulfuric acid   alum
precipitate + H_2SO_4 $\rightarrow$ $KAl(SO_4)_2 \cdot 12H_2O$
               (which would have been in solution,
               and needing to be subsequently crystallized)

This convinced him the precipitate was "alumine" ("alumina" or aluminum oxide). By treating the various washing waters, he found another half grain of siliceous earth. And here was his quandary: From his initial 100 grains he had produced 54.50 grains of silex, and 24.50 grains of alumine, leaving a loss of 21 grains. But a quick, twenty-first-century calculation shows that his values for silicon dioxide and aluminum oxide were within a couple of percentage points of what would be expected.[39]

The loss required Klaproth to repeat the analysis. This time he followed a slightly different method of separating the silica, but ended with essentially the same weight of it. On repetition of the method for recovering alum made from the aluminous earth, again, the mass was the same. He admitted to uncertainty as to where to look for the lost component, but was now convinced that there was a part that had so far escaped notice. He was accustomed to some loss in analysis, but ~21 percent was far too high to be accounted for by loss of air and water. He had determined in the initial tests that leucite contained little or no aerial acid (carbon dioxide) or water. At this juncture, he looked for phosphoric, fluoric, and boracic acids, but found none (Klaproth, 1801, v. 1, p. 353). However, he thought he had identified potash, or potassium carbonate.

Klaproth's success in finding that missing component is surprising, considering that one way of separating silica in mineral and rock analysis was to fuse with caustic potash, or potassium hydroxide (KOH), thus dissolving the silica, which was then precipitated with acid. To avoid initial use of the potash, he returned to the first method of analysis and digested finely ground leucite with muriatic acid (HCl), with 109 of his initial 200 grains of leucite remaining as an insoluble residue. He evaporated the acid solution to what he called the consistency of honey. At that point, the surface was covered with a saline crust. On cooling, it appeared to be a clear yellow oil, full of crystals—some cubical, some tabular. After pouring off the oil, he rinsed the crystals with alcohol, then evaporated the rinsings to produce a little more of the salt. The dried salt weighed 70 grains. He dissolved it in water, added some ammonia which separated some aluminous earth, and then let the solution dry and crystallize, which produced only cubical crystals.

Leaving those crystals for the time, he diluted the yellow "oil" with water, decomposed it with carbonated soda (sodium

---

[37]The precipitate would have been Prussian blue, ferric ferrocyanide, $Fe_4[Fe(CN)_6]_3$. Klaproth's comment was then that the amount was "of so small a quantity, as hardly to indicate one-eighth of a grain of *oxyded iron*" (Klaproth, 1801, v. 1, p. 350).

[38]Chemists may be slightly uncomfortable with this chemistry. The interactions of ferro- and ferricyanides are not simple, particularly with a component such as "solution a," which is the acidic remainder of a rock digestion. Klaproth was probably correct in assigning the iron to a category of a minor contaminant.

[39]This was on the basis of percent composition of leucite with no cation substitution.

carbonate) to precipitate 47.25 grains of aluminous earth. To prove its composition he treated it with concentrated sulfuric acid, producing alum. He also purified the 109 grains of insoluble residue, most of which was silica, and from the washings isolated another 1.5 grains of alumine, leaving 107.5 grains of silica. His totals were now 156.75 (107.5, silica; 47.25 + 1.5 grains of aluminous earth) of his original 200 grains, which is 78.4% of the total. His total of the two from the first analyses was 79%. Clearly, he was a consistent analyst, and equally clearly, something was missing, namely, whatever was in the 70 grains of crystals mentioned earlier. The crystal shape was the same as "muriated potash" or potassium chloride (KCl). Its solution did not change either blue or red litmus, nor did it react with sodium carbonate or caustic (concentrated) ammonia (NH_4OH). After treating three parts of the salt with two parts of concentrated sulfuric acid, evaporating the excess acid, redissolving in water and again drying, he obtained "sulphat of pot-ash (vitriolated tartar)," or what we would call potassium sulfate, in crystals that he recognized. By treating the remainder of the salt with acid of tartar (tartaric acid, $H_2C_4H_4O_6$), he produced cream of tartar, which on further treatment and testing gave a remainder of carbonate of potash (potassium carbonate, K_2CO_3), itself changed to "prismatic nitre, nitrate of potash" (potassium nitrate, KNO_3) on treatment with nitric acid (HNO_3) (Klaproth, 1801, v. 1, p. 356).

At this point, Klaproth announced that therefore his missing 70 grains consisted of what he called vegetable alkali (Klaproth, 1801, v. 1, p. 356). This compound, as well as potash, can be identified as potassium carbonate (K_2CO_3). And there a twenty-first-century reader becomes rather unsettled. Note that leucite has no carbon in it. Carbon and oxygen are introduced via the tartaric acid. Klaproth's final percent composition for leucite is: silex, 53.750; alumine, 24.625; and potash, 21.350, for a total of 99.725 (Klaproth, 1801, v. 1, p. 357). A modern calculation of weight percent for the constituents of leucite, using modern atomic weights, yields, roughly: silicon dioxide, 57.0%; aluminum oxide, 24.3%; and potassium (K), 18.6%. Klaproth's value of 21.35% would seem to indicate that he had indeed gotten a reasonable value for just potassium.[40] He remarked on the presence of this substance, which he also quotes Bergman as calling "the alkaline basis of muriated potash" (Klaproth, 1801, v. 1, p. 357). Until his analyses, it was thought to exist only in plants, and he suggested that a name other than vegetable alkali would now be appropriate (Klaproth, 1801, v. 1, p. 367).[41]

In 1789, Lavoisier listed 17 known metallic substances, which list contained neither alkali nor alkaline earth metals (Lavoisier, [1790] 1965, p. 160). His table of simple substances contained those 17 metals as well as the five earths—lime (including

calcareous earth), magnesia, barites, argil, and silex (Lavoisier, [1790] 1965, p. 175–176). Potash was clearly not a simple substance (an element). The most active metals were not to be isolated and characterized for another several years. However, mineral analysis definitely pointed the way to more complete knowledge of the natural world.

Rock Analysis

As with so many things in history, we again need to be conscious of changes in terminology. By definition for us, rocks, with a few exceptions, are made of several minerals. We also expect the natural variability mentioned above, because most rocks were formed in open systems, and may well have continued to be in open systems for thousands or millions of years. Also, in the eighteenth century, decomposition products were beginning to be recognized. So what a rock consisted of was and is a matter of even more variability than was true for minerals. However, because they are generally formed of combinations of minerals, their analysis was an extension of mineral analysis. Physical properties of rocks, particularly granite and basalt, had been investigated, as discussed in earlier chapters. But because of recognition that rock behavior depended upon their constituents, and there was great interest in how rocks were formed and emplaced, their composition was a natural subject of investigation, being an extension of mineral analysis.

A survey of the many mineralogy books of the eighteenth century shows that rocks were often included in those reports. As with the minerals, most often it was the properties of color, hardness, fracture, specific gravity, and behavior under the blowpipe that were given. But there were also attempts to determine constituents. Some rocks, such as a large-grained granite or porphyry, might have individual crystals broken out and analyzed. Not until Cordier was there a concentrated attempt to separate out and identify minerals from finer-grained rocks such as basalt (Cordier, [1816] 1868). However, what we would call whole-rock analysis for chemical constituents began quite early. Again, because of the plethora of examples, I will discuss only a few.

Serpentinite

One of the first quantitative rock analyses I've seen was of what was called "serpentine." Like limestone and marble, serpentinite is a rock mass that consists mainly of one mineral, a magnesian silicate, although especially in the case of serpentinite, there are accessory minerals. And while in the eighteenth century, the mass was called "serpentine," as if it were just that mineral, modern usage in North America designates it "serpentinite." In this section, I will use the word "serpentine" as did the primary reference authors. The first analysis, at least the one most often referred to, was that of the Berlin chemist Andreas Sigismund Marggraf (1709–1782), who was probably the first to use the flame test to distinguish between potassium and sodium salts (Ihde, [1964] 1984, p. 233). Marggraf's analysis of serpentine turned out to be rather significant. He had previously been

---

[40]My calculation was for just potassium, and not the oxide. We note that, IF it was the carbonate, the carbon (C) could have entered the reaction via the tartaric acid.

[41]At this time, potassium had not yet been separated as an element. In a personal communication of 2008, David Oldroyd has questioned how Klaproth knew how to proceed with his analyses, finding them nearly miraculous for his time. I agree.

able to distinguish magnesia from lime, which confirmed earlier work. He is credited also with determining that the serpentine of Saxony did not contain alumina (Partington, 1961, v. 2, p. 728).[42] However, modern analysis shows the presence of aluminum substituting in the crystal lattice, so it is a little difficult to second-guess the analyses from our time. Marggraf did use a microscope to help in identifying the crystalline products of chemical analysis (Meyer, [1891] 1975, p. 113).

Pierre Bayen (1725–1798), who was active in chemistry and pharmacy in France, repeated Marggraf's experiments. Heating a quantity of pulverized serpentine resulted in the formation of a small quantity of acid water. Further testing revealed "sel marin," or sodium chloride. He repeated the experiment on the serpentine from Germany four times, obtaining comparable results, and also tested serpentines from France and Corsica. For the rest, he noted the difficulty there had been in distinguishing Glauber's salt (sodium sulfate) from the "sel amer" (magnesium sulfate) (Bayen, 1797, p. 131). After paying close attention to reactions and crystal form, he reported: 20/48 (41.6%) "cristaux talqueux,"[43] 5/48 (10.4%) argil, 1/48 (2.1%) iron, 16/48 (33.3%) magnesia, and 6/48 (12.5%) water (totaling 48/48), but there was also that small quantity of marine acid (HCl).[44] Bayen determined that Marggraf had been substantially correct in his identities. Bergman reported Marggraf's and Bayen's work with serpentine, and recorded basically the same analysis, although he again noted problems with determining precisely which minerals were present (Bergman, 1784a, p. 114). In the 1792 annotated La Métherie translation of Bergman, the same analysis was reported, this time with results stated in parts per hundred, but with the water omitted so that it totaled only 86%.

By the time English geologist William Phillips (1775–1828) wrote his *Outline* (Fig. 7.4), he put serpentine in a footnote stating: "Serpentine is composed of about 45 parts of silex, 30 of magnesia, 15 of alumine, with some oxide of iron, and water" (Phillips, [1816] 1978, p. 127). That same year, Parker Cleaveland reported chemist Robert Chenevix's (1774–1830) analysis of serpentine as silex 28.0; magnesia 34.5; alumine 23.0; lime 0.5; water 10.5; and oxide of iron 4.5, with magnesia varying from 23 to 36%. This equaled 101%. However, he noted that Klaproth had found no alumine in serpentine from Saxony, so perhaps magnesia and silex were the only necessary ingredients (Cleaveland, [1816] 1978, p. 347). It is clear that the natural variability of rocks, the sometimes confounding effects of the reagents used, and the difficulty of dealing with small quantities combined to confuse the most expert analysts.

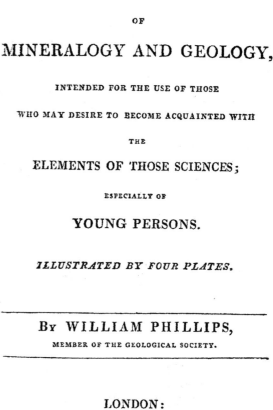

AN

OUTLINE

OF

MINERALOGY AND GEOLOGY,

INTENDED FOR THE USE OF THOSE

WHO MAY DESIRE TO BECOME ACQUAINTED WITH

THE

ELEMENTS OF THOSE SCIENCES;

ESPECIALLY OF

YOUNG PERSONS.

ILLUSTRATED BY FOUR PLATES.

By WILLIAM PHILLIPS,

MEMBER OF THE GEOLOGICAL SOCIETY.

LONDON:

PRINTED, AND SOLD BY WILLIAM PHILLIPS,
GEORGE YARD, LOMBARD STREET.

1815.

Figure 7.4. The title page of Phillips's *Outline* (1815).

Basalt and Lava

Comments about basalt were frequently included in mineralogical works of the eighteenth and early nineteenth centuries. With the usual caveats about whether their "basalt" and our "basalt" are the same thing, we will look at a few of its analyses. For a number of people, before and even after Guettard's and the French geologists Barthélemy Faujas de Saint-Fond's (1741–1819) and Nicolas Desmarest's (1725–1815) interpretations of the rocks of the Auvergne region of France, basalt was exclusively regarded as a recent extrusive rock.[45] Other deposits,

---

[42]Partington also noted that Marggraf apparently did not know of Black's work (Partington, 1961, v. 2, p. 728).

[43]Talc is a secondary mineral formed by alteration of magnesium silicates.

[44]At this point, I'm not sure of the date of Bayen's analysis, or of its publication. McDonald wrote that Bayen burned his manuscripts during the Terror of the French Revolution (1970, p. 529). The work on serpentine must have been published because it was mentioned by Bergman in 1782, which work was translated by both Mongez (1784a) and La Métherie (1792), all well before the 1797 date of the collected works of Bayen that I reference. Further library sleuthing is in order.

---

[45]The Neptunist position was that most basalt, other than that obviously associated with an active volcano, which could be called "lava," was a crystalline rock deposited from water solution or suspension. As is usual in history, categorizing geologists as Vulcanists and/or Huttonians, or Neptunists, is not an easy matter. Guettard had recognized old volcanoes in central France. Both Faujas and Desmarest positively connected basalt layers with the volcanoes (Challinor, 1971, p. 548; K.L. Taylor, 1971, p. 70).

such as those around Edinburgh variously named "toad-stone," "Rowley Rag," and "whinstone," were not then recognized as the same sort of rock, as they subsequently were.

In beginning his analysis of Rowley-Rag stone, Withering remarked that "when exposed to the weather, [it] gets an ochry colour on the outside; strikes fire with steel; cuts glass; melts, though not easily, under the blowpipe" (Withering, [1782] 1809, p. 292). Withering broke up his sample on a steel plate with a steel hammer, after which he ground it to a fine powder in one of Wedgwood's mortars, which "lost only 1/3 of a grain weight during this operation" (Withering, [1782] 1809, p. 292). One hundred parts of the powder were repeatedly boiled with marine acid (HCl), at first showing a slight effervescence. The acid and washings were decanted off, and there remained 80 1/4 parts. The liquid was treated with phlogisticated fixed alkali (potassium ferrocyanide) until no more Prussian blue was precipitated. That precipitate weighed 47 parts. The remaining powder was mixed with twice its weight of fossil fixed alkali (sodium carbonate) and heated to redness for 2 hours, after which steps similar to mineral analyses, beginning with silica, were followed. Of note was Withering's comment that of the total iron precipitated, he needed to subtract the calculated amount of iron already present in his weighed amount of phlogisticated alkali. From his 100 parts he ended with:

Pure siliceous earth, 47½, (SiO_2)
Pure clay, free from fixable air, 32½ (Al_2O_3)
Iron in the calciform (oxide) state, 20, (Fe_2O_3) which sums to 100.
(Withering, [1782] 1809, p. 293)

Despite his neat total, he didn't account for the loss due to the slight effervescence at the beginning.

Withering followed a similar sequence in the analysis of toad-stone, which he described as dark brownish-gray, with a granular texture and several cavities filled with crystallized spar. It didn't strike fire with steel, but did fuse to a black glass. His results:

Siliceous earth, 56
More ditto, 7
Calciform iron (iron as oxide), 16
Calcareous earth, 7 and 5/10
Earth of alum, 14 and 8/10, which sums to 101 and 3/10.
(Withering, [1782] 1809, p. 294)

Withering felt that the excess was due to components that could combine with fixed air, not being fully saturated with it at the beginning of the analysis. It should be noted that those two examples would not easily be identified as the same rock on the basis of analysis alone. Withering's analyses were often cited in the next decades, including by Fourcroy, who also enumerated the analyses of Bergman for basalt and a less compact lava "improprement appellées pierres-ponces" [or pumice] denoted cellular lava (Fourcroy, 1789, v. 1, p. 374).

Dolomieu found that compact lavas contained ~60% siliceous earth, 25%–30% argilaceous earth (alumina), 8%–10% iron, and a small amount of magnesia (Dolomieu, 1788, p. 189). The silica content might be lower when the iron content was higher. Breislak, like others, noted that he had scrupulously followed Kirwan's method in analyzing the lavas of Solfatara (a volcanic region near Naples), but didn't want to bother the reader with chemical details (Breislak, 1792, p. 119). He had noted that the lavas appeared decomposed, particularly those exposed to acid vapors. His investigations of the contents of fumarole gases and water collected nearby gave him much information about what might effect the changes in what he called the "réargillisation" of the lavas. He recognized that the process occurred far more slowly than could be observed in the time that he was able to observe the changes (Breislak, 1792, p. 119). Two of Breislak's analyses were:

Fresh lava: 48% silica   Altered lava: 59% silica
35% alumina   19% alumina
3% lime   1% lime
5% iron   3% magnesia
9% lost = 100%   1% iron
                       10% lost = 93%
                    (Breislak, 1792, p. 120)

These numbers illustrate the folly of easy hopes for numerical comparisons. I had hoped both to compare Breislak's values with those of other analysts, and to look for confirmation of his claims about how exposure to volcanic gases might change the lavas. Even with a "lost" category, which would presumably lead to an ability to sum to 100%, the second list, duplicated as printed, does not do that. Another analysis with the same categories sums to 103%. I don't know if the discrepancies were due to analytical, printing, or other errors. A slightly later pair of analyses of both unaltered and altered lavas was given, and although one totaled 94% and the other 90% (the "lost" category was not included), one can compare the amounts of silica and alumina and from those two see that Breislak's comments that alteration did not appear to change the proportions of silica and alumina were supported (Breislak, 1792, p. 122–123). It may be that his analytical skills were not as great at Klaproth's, as judged by these results and their respective reputations.

Other original basalt and lava analyses were done by Robert Kennedy at the behest of Sir James Hall, on the samples used in Hall's fusion experiments. The paper was read in 1798. Kennedy recorded that his first sample, basalt of Staffa (an island in the Inner Hebrides, site of Fingals Cave), was not attracted to a magnet, had a specific gravity of 2.872, and lost 5% of its weight after exposure to both low and high heat. He used weighed crucibles made of the porcelain clay of Cornwall that had been baked at 100 °WW. Kennedy followed a sequence similar to Withering's, but also tested for manganese, barium, strontium, and calcium compounds. His process was essentially the same as that used for decades to put siliceous rocks into solution. The basalt of Staffa, boiled with an excess of HCl, gave a solution and a residue. The residue was fused with caustic potash (KOH) to a glassy solid, which was dissolved with hot water, and supersaturated with HCl, which then, with heat, gave a gelatinous residue. The residue was diluted, digested, and filtered. This silex remained on the filter. Kennedy tested the silex to see if it was contaminated with other earths, which it was not (Kennedy, 1805, p. 79–80). The solution from the first part of the investigation revealed calcium carbonate, argil, and iron oxide.

The following traces the steps of his analysis, with major results in bold type:

1000 gr. basalt + 1200 gr. HCl $\xrightarrow{\Delta}$ Solution(1) + residue

$\rightarrow$ Residue(1) + KOH $\xrightarrow{\Delta}$

$\rightarrow$ Glassy solid + hot water + HCl $\rightarrow$ Solution

$\rightarrow$ Gelatinous residue, after which heat and filtration gave silica, or **SiO$_2$**

Solution(1) + NH$_4$OH $\rightarrow$ precipitate(1) + Solution(2)
yellow brown   dirty green,   + H$_2$SO$_4$ $\rightarrow$
                                 to brown in air   **no reaction(NR)**
                                                   so no barytes or
                                                   strontium (or lead)

Solution(2) [reduced in volume] + (NH$_4$)$_2$CO$_3$ $\rightarrow$ **CaCO$_3$**

Ppt.(1) + KOH $\rightarrow$ spongy brown solid (SBS) + solution(3)

Solution(3) + excess H$_2$SO$_4$ neutralized with Na$_2$CO$_3$ $\rightarrow$ **argil**, or clay (most are hydrated aluminum silicates) + H$_2$SO$_4$; resulted in alum, proving the argil

SBS + KOH $\rightarrow$ N.R.

SBS + HNO$_3$ $\rightarrow$ solution, re-precipitated $\rightarrow$ **Fe$_2$O$_3$**

The totals were:

Silex	48
Argil	16
Oxide of iron	16
Lime	9
Moisture and volatile matter	5
	94

That total, and other low totals in similar analyses, led Kennedy "to suspect that some saline substances existed in these stone" (Kennedy, 1805, p. 82). He had also observed that Sir James Hall's "artificial crystallites"—the result of his fusion followed by slow cooling—showed an efflorescence on their surfaces with a salty taste (Kennedy, 1805, p. 83). He took another sample and removed the silex, the argil, and the lime. After treatment, the residue appeared to be sodium sulfate, which he identified by six different tests. The identification of soda in rocks of this kind was both important and influential. Among other things, it was one solution to the conundrum of Breislak's missing material, as was Kennedy's discovery of muriatic acid in basalts.[46]

Kennedy began with 400 grains of the Staffa basalt, mixed it with 1200 grains of sulfuric acid, and heated it slowly to dryness.

This residue was boiled with water, and the insoluble part, the silica, was filtered out. The silica was washed and again treated with acid. The two resultant solutions (the original and the silica washings) were evaporated to dryness, and what Kennedy called the "saline mass" was heated at red heat for an hour. When cool it was brick red. Kennedy felt that sulfates of iron and argil would have been decomposed by the red heat, and he found very little of "earthy matter" in the remainder. Here he tested for sodium sulfate by the six tests, and found the tests conclusive (Kennedy, 1805, p. 85). He adjusted his method with other samples to improve the separation of components from each other. He identified the muriatic acid by digesting the rock with nitric acid, then distilling off the remaining acid. The remaining liquor was tested with nitrate of silver to show the presence of chlorine.

Thus, in a total analysis of the basalt of Staffa, Kennedy found, besides the earlier contents, ~4% soda and 1% HCl, and had a total of 99% (Kennedy, 1805, p. 89). He remarked that these discoveries could explain why Dolomieu and others found soda in the vicinity of volcanoes (Kennedy, 1805, p. 94). It also led Kennedy and other analysts to look for it in rocks of different kinds. So, with Klaproth's discovery of potash in rocks, both soda and potash could be sought in rock analysis (Kennedy, 1805, p. 96).[47] These discoveries are but one instance in which careful analysis led to further possibilities for geology.

In 1799, Kirwan compared the results of eight analyses of trap and basalt done by four other analysts (Kirwan, 1799a, p. 197–198). He attempted to identify rocks by the variations in their contents, which effort we will pursue further in the chapter on rock origins. Klaproth also had reported the results of a basalt analysis, which he felt differed little from one of Bergman's, with the exception of 2% soda. He said that he had not seen Kennedy's analysis when he did his own. However, after seeing it, he noticed the report of muriatic acid, searched for it, and found it, although it was a very small amount (Klaproth, 1801, v. 2, p. 203). Jean-François Daubuisson de Voisins (1769–1841) compared Klaproth's and Kennedy's basalt analyses, and showed their substantial agreement (Daubuisson, 1814, p. 14).

Klaproth found what he called "pumice-stone" interesting. He mentioned that Bergman and others had considered it as an "*asbest* changed in its mixture by volcanic fire" (Klaproth, 1801, v. 1, p. 368). The others had cited the fact that pumice contained magnesian earth as one reason for that belief. Klaproth failed to find it in his analyses and thus didn't agree with the others (Bergman, Cartheuser, and Spallanzani) about its origin. Klaproth described his samples as "common grey-white, fibrous pumice-stone, swimming on water, and procured from Lipari" (Klaproth, 1801, v. 1, p. 369). When boiled in water, the water gave a small indication of muriatic acid with silver nitrate.

One hundred grains of the stone were gently ignited, pulverized, and then heated to redness with twice its weight of caustic

---

[46]He did indeed identify that portion of the volatile substances as muriatic acid, our HCl.

[47]We remember that salts of sodium and potassium were not identified as precipitates, because all of their salts are soluble. Instead, solutions of the salts were crystallized, and they were dried to crystals and then identified by their characteristic shapes or by further reactions.

alkali. The fused mass was bright green, which to Klaproth indicated some manganese. The color changed to brown on addition of water, which was followed by digestion with dilute muriatic acid. This left a remainder of 77.5 grains. Caustic ammonia was added to the acid solution, and the brownish precipitate was filtered out. The precipitate was dissolved with hot caustic alkali, leaving behind 1.75 gr. of iron oxide. Klaproth then added carbonated alkali to the solution, and aluminous earth precipitated. This was proved by the formation of alum after it was redissolved with sulfuric acid and liquid acetated potash (a solution of potassium acetate) was added. Klaproth's values for the contents of pumice-stone were:

Silex	77.50
Alumine	17.50
Oxyd of iron	1.75
Besides a trace of manganese	
Total of	96.75

(Klaproth, 1801, v. 1, p. 370)

He remarked that boiling with acid had had little effect on the pumice, but made no conjectures about the missing 3.25%. He had heated the stone before pulverizing for his sample, so one cannot assume that moisture was the cause.

There are a number of other rocks whose analysis we could follow over three or four decades. However, I will let these examples suffice. Chemical analysis played a most important part in learning about the natural world, specifically the materials of Earth. The chapters on rock origin will build on this and discuss it further.

Mineral Waters

Both mineral water and gas analysis are enormous subjects, about which papers and books have been written for several hundred years. I shall not attempt more than a few statements about them, but such analyses cannot be ignored, as both were put in the service of attempts to puzzle out questions of rock formation or origin. While gases were not well characterized until the last quarter of the eighteenth century, solution analysis, with real knowledge of at least some components, began well before that time. Both mineral water and gases could be seen as products of earth processes, as well as agents of change for earth materials. Here are a few examples of work that directly impinges on rock analysis or origin.

John Woodward (1665–1728), an English fossil collector ("mineralist") and believer in the Deluge, recognized that water carried all manner of substances to locations within the crust (Woodward, [1695] 1978, p. 190). From being a convenient agent of deposition, mineral water advanced to being perceived as an agent for good health, as well as a reflector of geological processes.[48] People in a region could hope to attract users to its

spas with analyses of mineral water that might cure ailments and support health. The course of analytical chemistry was most certainly advanced by the desire to prove the ingredients in natural waters, or to duplicate them in artificial mineral waters. Methods evolved constantly through the eighteenth and nineteenth centuries (Hamlin, 1990).

Bergman is sometimes credited with originating the first systematic method of "humid" analysis in his *De analysi aquarum* of 1778. No water found on Earth was pure, which might be due to the seasons, the climate, or other factors (Bergman, 1784b, v. 1, p. 106). He included a history of water analysis that began with Hippocrates and Pliny (Bergman, 1784b, v. 1, p. 92). Quantitative methods were not seen in water analysis until the seventeenth century with the work of Boyle and others. Bergman detailed the reagents and reactions in use, and noted the very small amounts of material that needed to be detected, perhaps one in 50,000 or 100,000 parts (Bergman, 1784b, v. 1, p. 109). He also listed the components that might be found. He noted that siliceous earth was incontrovertibly found in the waters at Geyser in Iceland. While attempts at dissolving siliceous earth in boiling water had not been successful, Bergman suggested that it might be tried with the Papin's digester, wherein higher temperatures could be attained. He thought that when water boiled below the Earth's surface, the heat was much more intense and, "in caverns closed up by very thick strata of stones," it might form "an apparatus far more effective than Papin's digester" (Bergman, 1784b, v. 2, p. 51). Here was a direct analogy from the laboratory to geological processes, enabled by chemical analysis.

Like all good chemists, Fourcroy included an entire section on mineral water analysis in his *Elémens* of 1789. He divided waters into classes depending on their contents (acid, sulfurous, etc.) and the temperatures at which they'd been found, hot or cool. He also performed initial tests (specific gravity, taste, color) before employing the reagents and indicators we have discussed, as well as distillation and evaporation. He saw fabrication of artificial mineral water as a great good.

Breislak reported investigations done on the waters of Solfatara by chemists associated with the hospital for incurables, and listed ten initial steps to determine the contents (Breislak, 1792, p. 40). In his fieldwork, he made the direct connection between deposits from volcanic exhalations and the contents of the water, some of them being soluble salts such as sodium and ammonium sulfates, carbonates and chlorides, and he also found alumina, iron, and sulfur. Black joined the debate about the solubility of silica with his painstaking analysis of hot-spring water from Iceland (Newcomb, 1986). Detecting the silica was not difficult for two reasons. The first was that the solid material deposited by those springs had been sampled and analyzed. The second reason was that the analysis itself, while requiring much dexterity and care in the laboratory, was, after all, the reverse of the process used to analyze rocks for silica contents, where one begins with the solid. Silica in the water could be precipitated, dried, and weighed. Black's analyses

---

[48]Its importance is underscored by C. Hamlin's book about water analysis in Britain in the nineteenth century (1990).

revealed the silica contents, as well as caustic fossil alkali, argillaceous earth, common salt, and Glauber's salt (sodium sulfate) (Black, 1794, p. 118).

Kirwan's essay on mineral water analysis (1799b) was extensive, containing in excess of 277 pages, with methods for detecting a large range of possible solutes. He had already published a number of articles that led to his high reputation as a chemist.[49] The analysis of mineral waters had two goals: discovery of the contents, and their weight (Kirwan, 1799b, p. 175). He discussed both earlier methods and his own new and improved procedures. Kirwan began by determining the gases present and their quantities. He discussed the general classes of substances that might be found dissolved in water: aerial fluids (gases); acids; alkalis and earths; neutral salts; and what he called "associated salts." He spent several pages discussing lime and its solubility characteristics, including with respect to the presence or absence of fixed air (carbon dioxide). He considered siliceous earth, and its relation to alkali content. For his many constituents, he provided a test, frequently quantitative, to determine each of them. The second part of the paper dealt with the actual sequence to be followed to determine what was in a sample of water. He acknowledged his debts to many analysts before him, but still suggested many improvements.

A careful reading of these and some of Kirwan's other papers gives a different picture of him from the one sometimes suggested, that of a convinced Neptunist indifferent to evidence (see Gillispie, [1951] 1959). Instead, one sees mastery of methods of detecting and quantifying instances of mineral behavior in real-world conditions.

This tradition of mineral water analysis continued. For Klaproth, the presence of silica was still to be remarked upon:

This earth [silica] has always been considered as a substance, by itself, absolutely insoluble in water. It was, therefore, totally neglected in *hydrologico-chemical* inquiries, … until *Bergmann* [*sic*] directed the attention of chemists to its solubility in simple water, and demonstrated that it exists in a state of solution in the Geyser, and other boiling springs of Iceland. (Klaproth, 1801, v. 1, p. 399)

Klaproth found that silica was present in cooler springs as well. Upon discovering the report of Black's analysis of Iceland spring water, he compared his own analysis with Black's, and found they substantially agreed (Klaproth, 1801, v. 1, p. 404). However, he disagreed with Black about the necessary presence of alkali in order to keep silica in solution, feeling that it had not been demonstrated, while the absence of alkali had been.

Saussure and Dolomieu were additional geologists who took mineral water analysis to the field. Their goals were varied: to determine if a spring had medicinal properties; to determine if the immediately surrounding area was safe, by determining if dangerous gases were present;[50] or, simply to detail what was in the water. By the time John Murray wrote his treatise on mineral water analysis, he could apply the new French chemistry in a sophisticated fashion, discuss the direct and indirect methods of analysis, and determine weight relations within salts (Murray, 1818).[51] The interest in water continues, literally through today, and for many of the same reasons. Recently in our Washington, D.C., area newspaper, there have been two stories about water. One was concerned with adequate water supply in normal and drought conditions, and thus discussed water sources, both surface and underground. The other detailed an instance of a public water supply that was suspected of causing miscarriages because of the combination of natural organic substances with the chlorine used for bacterial control. To a geologist, both stories underline the relation of water to the rocks and sediments through which it flows.

Gases

The final decades of the eighteenth century were the time of great discoveries about the identity and properties of gases, and the newfound knowledge was quickly applied to geological questions. Laboratory equipment to collect and confine gases was being perfected. There were ingenious solutions to the problem of equipment that would function in the field, such as Breislak's assemblage of tubes made of reed with a central covering dome of pumice that would collect fumarole gases (Breislak, 1792, p. 36–37). In a later 1801 publication, there was an illustration (Fig. 7.5). Metal tubes would react with the gases, while glass was too fragile. The hot, moist gases condensed so that he could test the water for its contents, mixing it with various reagents such as chalk, litmus, potash, sulfuric acid, copper sulfate, oxalic acid, barium chloride, silver nitrate, and arsenic oxide. As he approached fumaroles, he noted that silver blackened, small animals sometimes died, and a candle flame went out. Water could absorb the fumarole gases, giving a nauseating taste and the odor of hydrogen sulfide. Breislak also tested the comments that Kirwan and La Métherie had made about the effect of "hepatic" gas (hydrogen sulfide) and fixed air (carbon dioxide) on each other, when both were dissolved in water. Not surprisingly, he found that carbon dioxide could both enter water and be driven out of it, and noted that the hydrogen sulfide reacted with the oxygen in the atmosphere to form sulfuric acid (Breislak, 1792, p. 94). Breislak regretted being unable to have a proper pneumatic trough in the field to collect gases

---

[49]Among them, "Experiments and observations on the specific gravity and attractive powers of various saline substances," 1781, with continuations in 1782 and 1783, in *Philosophical Transactions of the Royal Society*; and "On the strength of acids and the proportion of ingredients in neutral salts," 1790, in *Transactions of the Royal Irish Academy*; and "Additional observations on the proportion of real acid in the three ancient known mineral acids, and on the ingredients in various neutral salts and other compounds," 1797, in *Transactions of the Royal Irish Academy*.

[50]Even nontoxic gases such as carbon dioxide are dangerous if their presence is not suspected. Carbon dioxide has no odor, but people lose consciousness or worse, if it replaces too much oxygen.

[51]The direct method involved evaporation to dry salts, followed by use of solvents and/or precipitants. The indirect method applied reagents directly to the water, which was sometimes concentrated in order to show the effects of the reagents better (Murray, 1818, p. 260).

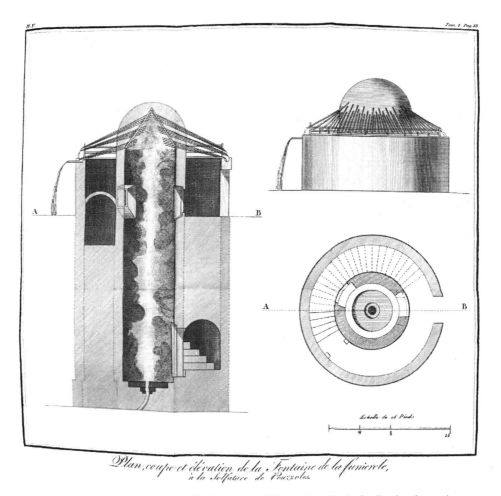

Figure 7.5. Breislak's (1801, v. 2, p. 88) illustration of his novel method of collecting fumarole gases.

more quantitatively, but he could characterize them well. These field tests were not that unusual, although Breislak's were more thorough than most. He cited the work of Priestley, Berthollet, and others in support of his identification and characterizations of the volcanic gases (Breislak, 1792, p. 146–147).

Kirwan listed the gases frequently found in mineral waters. He included oxygen, fixed air, heavy inflammable air or carbonated hydrogen,[52] mephitic air (Kirwan adds, azotic air of the French),[53] common air, hepatic air, and sulfureous air that seems to be sulfuric acid (Kirwan, 1799b, p. 8–13). But as noted, despite glossaries and references, the identity of the gases was sometimes not clear. In the section on earths found in mineral waters, he made an extensive investigation of the interactions of fixed air and water, noting that a sufficient pressure of fixed air could keep lime in solution, forming what we call "calcium bicarbonate." Later, in the section called "Tests," he discussed how to identify the gases in mineral water.

CONCLUSION

From the discussion above, it is clear that by the end of the eighteenth century, there was detailed knowledge of the natural world from samples collected in the field and sent to laboratory workers, those collected and analyzed by the same person, and from analysis done while in the field. It was normal to have two or three trials of the same procedure, and to have other workers repeat analyses to check results. Niceties such as neutralizing excess alkali if it had been necessary to use it to put a sample into solution were mentioned (Murray, 1818, p. 279). As we will see, this specific sort of knowledge was later directly applied to geological questions about rock origin or processes. The connection between earth processes and earth materials was becoming clearer, as we will see as the story continues in Chapter 8. Analysis of composition helped separate the possible from the unlikely.

---

[52]The difficulty in figuring out what that gas might be is explained by Eklund who tells us that this gas might be carbon monoxide, water gas (hydrogen and carbon monoxide combined), or methane (Eklund, 1975, p. 29). All are much heavier than hydrogen.

[53]This is another puzzle. According to Eklund, mephitic meant noxious. He then said mephitic air was carbon dioxide. This could be termed noxious as it doesn't support life. However, azote is nitrogen (Eklund, 1975, p. 22, 31).

CHAPTER 8

Other investigative methods

INTRODUCTION

This chapter deals with phenomena such as gravity, magnetism, electricity, and occurrences such as earth tremors, which excited curiosity and engendered observations and instrumentation long before the phenomena were understood or predictable.[1] Some of those areas of study, such as magnetism, the early thoughts about seismology,[2] and gravimetry, continued in use, gaining in sophistication and becoming a major source of information, both locally and on an extremely large scale. The ubiquitous electricity was employed, as were new technologies developed in mining. Other methods proved to not be based on science, might be unreliable, and only exist today, if at all, as folk tools. In some cases, it is difficult to tell where folk "wisdom" ends and science begins. Known and quantifiable properties, such as response to water vapor, might be applied in a measurable scientific way as in a hygroscope, or in the controversial dowsing rod.

The relations between gravity, magnetism, and electricity were tantalizing. Earth and/or its minerals were clearly somehow a repository of all three, but their nature and possible connections remained mysterious. All three relations had been noted and investigated since antiquity. Their treatment in this chapter will by no means be exhaustive, nor is the reference list complete.[3] Most of these fields are now encompassed by what we call "geophysics," but that designation didn't come into general use until the mid-nineteenth century (Buntebarth, 1998a; Good, 1991). The lack of a specific term was not, however, an obstacle to eighteenth- and nineteenth-century scientists who were not discipline-bound. Important truths about Earth were discovered by physical methods, although more complete understanding did not come until the last quarter of the nineteenth century. Instruments to measure electricity, magnetism, and gravity have evolved steadily, and the interaction of understanding and instrumentation is worthy of the extensive attention given to it. This chapter will discuss what was known in studies of Earth as earth science was becoming a discipline, with some discussion of the instrumentation available during that earlier time.

Much of this investigation in the eighteenth century was more observational than experimental, as, indeed, is much of the work recounted in this book. But the properties of, particularly, gravity and terrestrial magnetism manifested themselves in ways that could not be ignored, and were the subject of extensive studies of Earth in the eighteenth century.

GRAVITY

As scientists in the modern scientific age, we take so many things for granted that perhaps it is difficult to think about a time when concepts of mass, current, and magnetic fields didn't exist as part of the background of our reasoning. Caneva (2001) has pointed out how our mind-set, and those of earlier generations, alters our perception of the thought processes and accomplishments of earlier workers. Understanding of Earth's shape and of gravity itself came in increments over time, but the fact that things fall toward the center of Earth was and is inescapable. I will be generally concerned with effects seen on our Earth, and not the larger solar system.

For hundreds of years after Aristotle's time, it was assumed that heavy objects fell faster than light ones.[4] Not until the sixteenth century did Galileo Galilei (1564–1642) argue that objects fell at the same rate if air resistance was nullified, although his experimental demonstration of this is in doubt. Robert Hooke (1635–1703) and Isaac Newton (1643–1727) further described gravity in their interactions and writings in the last decades of the seventeenth century. Despite our association of Newton with the formulation of gravitational force relations, there is evidence of Hooke's contribution to the issue, or that he may have had a role in suggesting salient features (Drake, 1996, p. 78–79; Greenberg, 1995, p. 623–624).

In the eighteenth century, investigations of gravity continued. We see a complex interaction of theory, observation, instrumentation, and even economics, which characterized the science of the time. Newton's contributions had been incomparable:

He not only discovered the principle but also formulated the quantitative laws of gravity. His laws of motion and the laws of gravity make it possible to arrive at a logical explanation of tides, the motions of comets, precession of the Earth's axis and the motion of any body in a gravitational field. (Gondhalekar, 2001, p. 116–117)

The shape of Earth had been investigated for centuries. Pythagoras (fl. second part of sixth century B.C.) described Earth

---

[1]This chapter is not the long-awaited general history of geophysics, even for the eighteenth century. It is simply a short recital of how those methods were applied to earth knowledge at the time.

[2]Oldroyd et al. (2007) discussed the study of earthquakes during the 100 years after the great Lisbon earthquake of 1755.

[3]For an excellent introduction to some of the classic references, I suggest the end-of-entry references in Bud and Warner (1998) and Good (1998a).

[4]Actually, the state of science education in the United States is such that a general gathering, if asked if heavier objects fall faster, will provide a few positive responses.

as a sphere, and Eratosthenes of Alexandria (ca. 275–194 B.C.) estimated the length of its radius and circumference (Lenzen and Multhauf, 1966, p. 306). The flattening of Jupiter at the poles had been observed by English astronomer John Flamsteed (1646–1719) and the Italian-born astronomer Giovanni Domenico Cassini (1625–1712), who worked in France after 1665 (Greenberg, 1995, p. 1). By 1722, Jacques Cassini (1677–1756), son of the latter, published the results of 50 years of work regarding the length of degrees of latitude as measured along a meridian passing through Paris (Greenberg, 1995, p. 15). Lenzen and Multhauf (1966) have ably discussed the sequence of events that led to convincing evidence that Earth is flattened at the poles, and in a book of considerable complexity, Greenberg (1995) has set forth in detail the mathematical innovations involved, which were made by a series of extremely talented mathematicians, who developed new modes to express curves, motions, and forces, and the interaction of gravity and centripetal force. Newton and Huygens realized that Earth is an ellipsoid of revolution, as opposed to a sphere, while variations in gravity and pendulum period were noted as early as 1671 (Pledge, 1939, p. 71; Torge, 1989, p. 4).[5] The whole question of the shape of Earth and the flattening at the poles, which produces a larger gravitational attraction there (because of the shorter distance to the center), as well as the mathematics developed to describe it, is one of the enduring stories in the history of science.[6] We are concerned here with attempts to demonstrate differences in gravity.

Besides his role in introducing the concepts of centripetal and centrifugal forces to celestial mechanics, Hooke used several methods, including a pendulum, to attempt to measure gravity (Drake, 1996, p. 32). He suggested the use of a carefully constructed spring balance as well, and said:

The Scale I contrived in order to examine the gravitation of bodies towards the Center of the Earth, viz, to examine whether bodies at a further distance from the Center of the Earth did not lose somewhat of their power or tendency towards it. (Hooke, 1678, *in* Boynton, 1948, p. 92)

At equilibrium, the spring balance balances the force of gravity with an equal pull of its own, represented by the recorded weight so that extension serves as a measure of the weight. In 1678, Hooke had stated the law of elasticity, that elongation of a body in tension, in this case his spring, is directly proportional to the stretching force, gravity (Torge, 1989, p. 11).[7] He tried to find a difference in gravitational attraction at the top of the tower of Saint Paul's Cathedral in London, and at the top and bottom of Westminster Abbey, but could not find any significant difference. The same attempt was made in what he called deep mines, but the instrument used was judged inferior, and no difference was detected (Boynton, 1948, p. 92).

The pendulum was the other instrument of choice for gravity measurements from the fifteenth through the nineteenth and into the twentieth centuries. A simple pendulum is a mass on the end of a cord or string. Put into motion by raising the mass to the side, the mass is attracted by Earth, so that it completes its arc by rising to the same height on the opposite side, and then returns to the original height. There are obvious problems of friction, air resistance, humidity, and so on, all of which the early scientists worked to minimize. Galileo used a pendulum to measure periods of time, while Huygens invented and patented a pendulum clock in 1657 (Lenzen and Multhauf, 1966, p. 305). Shortly before these efforts, Marin Mersenne (1588–1648) had determined the length of a pendulum that oscillated in the period of a second (Lenzen and Multhauf, 1966, p. 305), which length varies according to the local attraction of gravity. The slowing of a pendulum clock denoting less attraction by gravity at lower latitudes had been observed.

Members of the famous expeditions to Peru and Lapland to determine the length of degrees of latitude took pendulum clocks with them. Every effort was made to establish the pendulum length initially required for a second's beat, and to protect the pendulums from as many changes in conditions as possible. Very fine suspending wires, temperature control, and eventually a vacuum chamber in which the pendulum could swing all aided the efforts.[8] The clock pendulum gave way to a free pendulum compared to the clock, and eventually progressed to two pendulums, one to determine absolute gravity and another to compare site gravity with that of a base station (Agnew, 1998, p. 453). Anomalies in gravity were noticed, such as the attraction of large mountain masses. Nearly a century after Newton and Hooke, Astronomer Royal Nevil Maskelyne (1732–1811) followed Newton's suggestion that it would be useful to know the density of Earth in order to calculate the masses of Earth, Sun, and other bodies, for which he used a plumb line (Boynton, 1948, p. 93).[9] Subsequently, these deviations in gravity were frequently investigated.

---

[5]Galileo Galilei and Simon Stevin (or Stevinus) (1548–1620) had discovered that the period of a pendulum depends only on pendulum length (Torge, 1989, p. 4). Corrections for temperature, expansion, and air resistance were made later. Because of the lesser attraction of gravity at lower compared with upper latitudes, a pendulum's period is longer.

[6]For an introduction, and references to the very large literature on this subject, see Pintus (1998, p. 399–403); Lenzen and Multhauf (1966); and Greenberg (1995).

[7]Hooke's own words were:

Take then a quantity of even-drawn wire, either Steel, Iron, or Brass, and coyl it on an even Cylinder into a Helix of what length or number of turns you please, then turn the ends of the Wire into Loops, by one of which suspend this coyl upon a nail, and by the other sustain the weight that you would have to extend it, and hanging on several Weights observe exactly to what length each of the weights do extend it beyond the length that its own weight doth stretch it to, and shall find that if one ounce, or one pound, or one certain weight doth lengthen it one line, or one inch, or one certain length, then two ounces, two pounds, or two weights will extend it two lines, two inches, or two lengths; and three ounces, pounds, or weights, three lines, inches, or lengths; and so forward. And this is the Rule or Law of Nature, upon which all manner of Restituent of Springing motion doth proceed, whether it be of Rarefaction, or Extension, or Condensation and Compression. (Boynton, 1948, p. 88–89, from Hooke's 1678 *De potentia restitutiva* as reprinted in Gunther, 1931)

[8]Lenzen and Multhauf (1966) include a long discussion of methods and instruments designed to make pendulums more reliable, as well as Huygens's compound pendulum.

[9]His attempt utilized positions on either side of the granitic mass of Schiehallion in Perthshire, where the pull of the mass approximated the pull of gravity. His method is explained in Wolf ([1952] 1961, v. 1, p. 111–112).

Henry Cavendish (1731–1810) refined Maskelyne's density determination for Earth by means of a torsion balance. The torsion balance was also used to determine forces in electricity and magnetism.[10] To this point, gravity had been expressed as a proportionality relation because the value of the gravitational constant was not well known, but Cavendish could use relations between known weights. The expression for the force of gravity, as we now state it, is $F_{grav} = Gm_1m_2/R^2$, where m is mass and R is the distance between masses. G can be solved for $G = F_{grav}R^2/m_1m_2$, if the force can be measured via the torsion balance, with the masses and distance known (Holton, 1973, p. 147). The apparatus was constantly refined after Cavendish's work during 1798 (Khan, 1998, p. 394), and values were updated well into the twentieth century. Gravity anomalies subsequently became useful in exploration geology, for example, in the discovery of the lower density of salt domes associated with oil deposits, or the higher density of metallic ores.

MAGNETISM

While magnetism and static electricity were known since antiquity, the interrelation between them was not investigated until the nineteenth century. Magnetism is another of those topics about which many books have been written, both by scientists who attempted and attempt to delineate its parameters, and by historians tracing the evolution of knowledge about it and the instruments used to detect it. A property of Earth itself like gravity, magnetism differs from it by both attracting and repelling magnetic objects. And while the phenomenon of magnetism was put to use early for direction-finding, the source of Earth's magnetism remained a matter of conjecture until the mid-twentieth century, and to some extent even today, remains inaccessible to empirical investigation.[11] This discussion will focus on what was known in the late eighteenth century and how it was known, with a brief discussion of periods prior to and a little after that time.

The property of magnetism and its source were much conjectured about in the eighteenth century. The natural magnet, lodestone,[12] was recognized, probably by about 500 B.C., in China and the West. Little was done in the West, while in China there was considerable development. Pumfrey has related that in China:

[S]outh-pointing spoons of loadstone were in use by the first century C.E. By 1100 the magnetization of iron, the polarity of loadstones, the use of compasses, and even the declination of magnetic from geographical north were established, a century before they became known, probably via trade routes, to Western scholars. (Pumfrey, 1998, p. 345)[13]

It was difficult to carve the lodestone so that it showed the correct direction, as the act of carving itself disturbed its magnetization. Later, a light, flat, magnetized iron "fish," which could be floated on water, was employed. Investigations were made for the best way to mount a magnetized needle. A "turtle" with a magnetic needle inside it was utilized so that the needle's movement was not impeded. Soon, a disk was added that was marked with 24 directions. These innovations were all done in China as navigation needs increased in the eleventh and twelfth centuries (Lin Wenzhao, 1983).

In those early times, magnetic force was probably not understood in the sense that we use the word "geomagnetism." Biringuccio recounted that Albertus Magnus had said there were lodestones that would attract all manner of things. But he himself described the common lodestone that he knew, "that sailors use in their voyages in order to return the enfeebled compass to the sign for observing our pole, from which some say this particular virtue [force] proceeds" (Biringuccio, [1540] 1990, p. 115). Alchemists noted magnetic force as a metaphor for natural operations and a supposed link between parts of the universe (Findlen, [1994] 1996).

For several hundred years, philosophers continued to investigate the properties of the lodestone, compared them to those of artificial magnets, and tried to determine their strength.[14] When the property was put to use in the West, despite the efforts of a few philosophers, magnetism was employed chiefly as an empirical aid to direction-finding. The variation from true north (first known as variation, then as declination) was noted early, probably in the thirteenth century.[15] By the sixteenth century, the dip of a magnetic needle (or variation from the horizontal) was regularly noted.[16] At first, to observe those properties, Western

---

[10]Wolf discussed the apparatus:

two small leaden balls hanging by short wires from the ends of a wooden beam suspended at its midpoint by a torsion wire of copper, the whole protected from air currents by being enclosed in a chamber…. Two massive leaden balls were brought up outside the chamber close to the suspended balls, one on one side of the beam and one on the other, so as by their attractions to turn the rod through an angle which could be measured. Each large ball was then transferred to the other side of the beam so as to turn it in the opposite direction, and the readings were again taken. Half the difference between the two sets of readings gave the mean deflection due to the attracting masses. (Wolf, [1952] 1961, v. 1, p. 112–113)

[11]Le Grand stated:

For most geologists and geophysicists rock magnetism was in the 1930s no more than a curiosity. There was no agreed theory as to the origin of the earth's magnetic field, the supposed cause of paleomagnetism; the methods used to measure rock magnetism were quite crude; igneous rocks were of less interest to geologists than fossiliferous sedimentary rocks but the remanent magnetism of the latter was much weaker and possibly less stable; and, even for igneous rocks, there was no reason to assume that their magnetism should remain unchanged over millions of years or that this would be unaffected by folding, elevation, erosion, and other geological processes. Finally, there were some rocks which had a magnetization almost directly contrary to the earth's present field: this seemed to be evidence for the instability and unreliability of paleomagnetism itself. (Le Grand, 1989, p. 55)

[12]Two spellings are used for the mineral: lodestone and loadstone. I use the spelling that appears in the reference cited.

[13]Pumfrey referenced Joseph Needham's "monumental work" (see Needham, 1959).

[14]An example of many similar reports is "An account of some magnetical experiments, shewed before the Royal Society by Mr. Gowan Knight, on Thursday the 15th of November, 1744" (Knight, 1746, p. 161–166). In the account, he brought different poles of various magnets of various sizes together to try to measure attraction and repulsion, and documented other properties.

[15]Lin claims that people of twelfth-century China had noted the "angle between the magnetic north-south and the geographical north-south" (Lin Wenzhao, 1983, p. 157).

[16]Landmark works in the history of magnetism include the work of Petrus Peregrinus (fl. ca. 1269) and William Gilbert's classic *De Magnete* (1600). The latter has many editions/translations, including the 1956 edition from Dover.

experimenters such as Robert Norman (fl. ca. 1590), had floated magnetic needles in water, under the surface (by density adjustment), or mounted them so they rotated freely (Good, 1991, p. 155). Careful work showed both dip and declination, as well as daily variations, to be intrinsic to magnetic measurement and not purely from bad instruments. By that time there was demand for well-made compasses, which were supplied by increasingly sophisticated instrument makers who employed ingenious methods to determine declination and dip reliably.[17]

Early experimental investigation of lodestone properties was done by Petrus Peregrinus, as reported in his *Epistola de magnete* of 1269 (Wolf, [1935] 1968, p. 291).[18] William Gilbert (1544–1603), the English court physician, is generally credited with recognizing that Earth itself is a magnet. He noted the attraction and/or repulsion of the poles of lodestone, and made an analogy with Earth from his spherical lodestone, the terrella (Good, 1991, p. 157). But the concept of two magnetic poles, as opposed to four, was not easily attained. For us, it is so much a part of our perception of Earth that it seems overwhelmingly obvious. Good has traced the history of ideas about magnetic poles, and the evidence that backed them, just before and throughout the nineteenth century. The great polar expeditions of the nineteenth century, and on into the early years of the twentieth, were undertaken to locate them. The latter expeditions also attempted to determine the movements of the magnetic poles.[19]

Many instruments used by geologists needed to be sturdy due to their use in the field or on shipboard, as well as in the laboratory. Instrument makers in England, and some on the Continent, supplied that condition. The requirement to avoid close proximity to other magnets or magnetizable metal (basically iron or steel) was also recognized. On land, special rooms were sometimes built to avoid magnetic interference. The problem was more difficult on shipboard, and it became worse as more and more steel was used in ship construction.[20] For both declination and dip, reliable producers of good needles were sought, and proper balancing of the needle was required. Early in the seventeenth century, there were compasses with dip needles added to the conventional needle (Multhauf and Good, 1987, p. 8). Later the needle might be balanced on a fine pivot point or suspended on a thread, both of which were anticipated by the Chinese.[21] The instruments became more and more complex. In order to solve the navigation problem of longitude, compasses were combined with methods to determine true north astronomically, and apply

known variations (declination). World maps of lines of equal magnetic variation and of equal intensity resulted from long expeditions (Multhauf and Good, 1987).

Whereas many of the early instruments had been used outside the laboratory for navigation, in the later eighteenth century scientists contributed to instrument development in their quest for more precision in measurement (Multhauf and Good, 1987, p. 5). At this point there was much theoretical conjecture, but no clear theory about the magnetism of Earth. Charles Coulomb's (1736–1806) torsion balance was used to measure both electrical and magnetic forces, for which he discovered an instantiation of the inverse square law (see Figs. 8.1 and 8.2). For this balance, the magnetic needle was suspended by a fine silk thread, and "the angle of the twisted thread was taken to be proportional to the elastic force with which it strained against the magnetic force to unwind itself" (Dörries, 1998, p. 626), a general adaptation of Hooke's law of elasticity. More and more data of increasingly greater precision and accuracy, as far as was known, were collected, but it took discoveries relating electricity and magnetism in the early nineteenth century, and the novel mathematical approach of Carl Friedrich Gauss (1777–1855) in 1839, to provide a new foundation for investigations (Good, 1998b, p. 351).[22]

Saussure is a good example of the scientists who investigated magnetism because of the variations that so vexed navigators. In 1779, he published the results of his investigation of magnetism during his travels in the Alps. Saussure noted the investigations that had been made into variations in declination, inclination, and intensity (Saussure, 1779, v. 1, p. 375). As his extensive reports demonstrate, the field was truly his laboratory. He found it advantageous to study the strength and direction, as well as the daily variations, of magnetism in the field. Saussure wrote:

L'idée de ces recherches me vint premièrement par rapport aux montagnes. Il me parut intéressant d'éprouver, si la direction de l'Aiman ne seroit point différente sur leurs cimes & si la force attractive ne diminueroit point comme la gravité, & peut-être plus rapidement encore, en s'eloignant de la surface de la Terre. (Saussure, 1779, v. 1, p. 376)[23]

He determined direction very exactly at the base of the mountain, and from there determined a point at the summit. When carried to that summit, and oriented in the same direction, the compass showed the same direction. When he did find differences, they were due to iron deposits on one or the other side of

---

[17]Multhauf and Good (1987) provided an excellent introduction and discussion of a variety of magnetic instruments in the collections of the (then-named) National Museum of American History. They included a brief history of geomagnetism.

[18]Wolf ([1935] 1968) used the English translation by Silvanus P. Thompson, London, 1902, *Epistle concerning the magnet*.

[19]See Good (1991), "Follow the needle: Seeking the magnetic poles." He discussed the invaluable contributions of Gauss and Edmond Halley (1656–1742), as well as the details of many polar expeditions.

[20]There is an excellent discussion of how these problems were approached and solved in Multhauf and Good (1987).

[21]Good has pointed out that further scholarship is needed on the Chinese record (G.A. Good, 2003, personal commun.).

[22]By 1800, there were two major ways of approaching magnetic theory: the first was based on an imponderable magnetic fluid, which imponderable fluid approach had been successful in the eighteenth century for electricity, heat, and light; the second was based on the increasing knowledge of electromagnetic fields as investigated by Ampère, Faraday, and others. "Gauss suggested that discussions based on magnetic fluids or on electromagnetism could produce the same appearances" (Good, 1998b, p. 351, 352).

[23]"The idea of these investigations came to me first in relation to the mountains. It appeared interesting to me to test if the direction of the magnet would not be different at their summits, and if the attractive force would not diminish as gravity does, and perhaps still more rapidly, when removed from the surface of the Earth" (my translation).

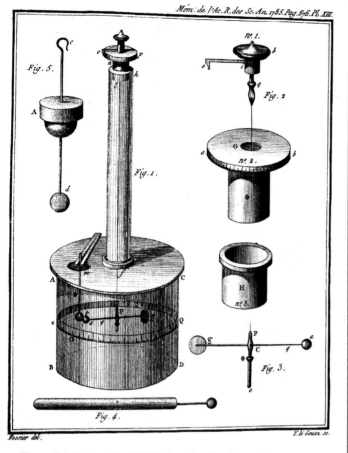

Figure 8.1. Coulomb's (1788) plate showing his torsion balance.

Mém: de l'Ac.R.des Sc.An. 1785.Pag.576.Pl.XIII.

Rossier del. *Y. le Gouaz. sc.*

DES SCIENCES. 569

PREMIER MÉMOIRE

SUR
L'ÉLECTRICITÉ ET LE MAGNÉTISME.
Par M. COULOMB.

*Construction & usage d'une Balance électrique,
fondée sur la propriété qu'ont les Fils de métal,
d'avoir une force de réaction de Torsion propor-
tionnelle à l'angle de Torsion.*

*Détermination expérimentale de la loi suivant laquelle les
élémens des Corps électrisés du même genre d'Électricité,
se repoussent mutuellement.*

DANS un Mémoire donné à l'Académie, en 1784,
j'ai déterminé, d'après l'expérience, les loix de la
force de torsion d'un fil de métal, & j'ai trouvé que cette
force étoit, en raison composée de l'angle de torsion, de
la quatrième puissance du diamètre du fil de suspension
& de l'inverse de sa longueur, en multipliant le tout par
un coefficient constant qui dépend de la nature du métal,
& qui est facile à déterminer par l'expérience.

J'ai fait voir dans le même Mémoire, qu'au moyen de
cette force de torsion, il étoit possible de mesurer avec
précision des forces très-peu considérables, comme, par
exemple, un dix millième de grain. J'ai donné dans le même
Mémoire une première application de cette théorie, en
cherchant à évaluer la force constante attribuée à l'adhé-
rence dans la formule qui exprime le frottement de la
surface d'un corps solide en mouvement dans un fluide.

Je mets aujourd'hui sous les yeux de l'Académie, une
balance électrique construite d'après les mêmes principes;
Mém. 1785. Cccc

Figure 8.2. The first page of Coulomb's "Mémoire" in *Academie des Sciences* (1789), where he introduced the torsion balance.

the line connecting the two stations (Saussure, 1779, v. 1, p. 376–377). The direction of the force was easier to measure than its attractive strength, but there were differences in both even at the same time and place (Saussure, 1779, v. 1, p. 377). For attractive strength he had a well-mounted needle, but, among other things, he felt that dust in the groove that supported the needle was a cause of inaccuracy. Saussure then designed an instrument that would better suit his purpose. He mounted a small piece of iron on a delicate spring attached to a magnetized needle. He thought he could measure the differences in the attraction of the iron to the spring, varying as Earth's magnetism did. But "cette idée ne me satisfit pas" (this idea did not satisfy me) (Saussure, 1779, v. 1, p. 378). The spring itself entered into the results. But Saussure by no means gave up.

To solve this problem, Saussure looked at means of measuring gravity, itself not constant, but following well-known laws that enabled it to be estimated. Applying the pendulum in a way to measure differences in magnetic attraction, he placed a ball of iron on a long, light attachment (as a pendulum), and very accurately measured differences in the arc it described in relation to another magnet. He had an instrument maker construct two of these instruments, each with the pendulum long enough to be effective, but short enough that the instrument was portable. It

exceeded his expectations. He used a level with a bubble of air to position it exactly in a vertical position, and enclosed it in glass so that air currents wouldn't affect it. Saussure used the instruments for five years, and determined that heat was the principal reason for the variation in magnetic attraction, the force decreasing as temperature increased, shown by a change of just half a degree of Réaumur's thermometer (Saussure, 1779, v. 1, p. 378–380). He called the instrument a "magnétromètre" (Saussure, 1779, v. 1, p. 380). Saussure remarked that the relation of magnetic force to distance was still unknown, but promised to further the calculations in the second volume of his work. It was shortly after this that Coulomb discovered the inverse square law of attraction for both electricity and magnetism, which has been called the greatest achievement of the eighteenth century in magnetism (Wolf, [1952] 1961, v. 1, p. 269), although it had been predicted earlier.[24]

Pinkerton noted that even geologists who recognized the common presence of "ferruginous or siderous" earths[25]

---

[24]This law is very tricky to prove (G.A. Good, 2003, personal commun.).
[25]These are earths that contain iron.

didn't commonly associate them with the ideas of those who wrote about magnetism, who had suggested that Earth was a great mass of iron (Pinkerton, 1811, v. 1, p. xl). A little later, Bakewell used an analogy to observations of Watt (1804a) with respect to the interaction—in Watt's slow cooling experiment with basalt—of heat and magnetic particles, to suggest that this interaction might be "illustrative of the cause and variation of magnetic polarity in the earth" (Bakewell, 1813, p. 323). Volcanic heat might be sufficient to orient or reorient the magnetic particles to produce the effect.

Because of its importance in navigation, the work in determining direction and strength of magnetism continued well into the twentieth century with expeditions mounted by most developed nations to determine them over the globe. However, the basic premise of instruments designed to detect and measure them didn't change a great deal, although there were many improvements (see Good, 1998c, p. 368–370). There was constant effort to use properly magnetized needles, to produce good artificial magnets, and to reduce or eliminate friction and all magnetic attraction other than the natural field. More was learned about the magnetic field, and theories as to its cause were advanced. There were attempts in the nineteenth century to use a compass needle for mining exploration. And much later, Earth's magnetism proved to be a key to understanding the conundrum of the moving, solid crust of Earth, and aided in constructing theories for the origin of the magnetic field of Earth (Le Grand, 1998, p. 652).

ELECTRICITY

Joseph Priestley had this to say about electricity in 1767:

The electric fluid is no local or occasional agent in the theatre of the world. Late discoveries show that its presence and effect are everywhere, and that it acts a principal part in the grandest and most interesting scenes of nature. It is not, like magnetism, confined to one kind of bodies, but everything we know is a conductor or non-conductor of electricity. (Priestley [1767] 1948, p. 281)

Boynton noted four kinds of electrical phenomena that had been recorded since antiquity: lightning, the shock from an electric eel, the attraction for light materials that amber and some other substances acquired after being rubbed, and the glow of St. Elmo's fire, sometimes seen at the ends of metal fittings on a ship (Boynton, 1948, p. 281–282). There has probably been more written about the history of studies and knowledge about electricity than about magnetism. Unlike gravity and magnetism, it does not have such an immediately obvious relation with Earth, although, besides the electrical properties of minerals, it was applied to broader questions. Lightning and static electricity, like gravity, were hard to miss, but very difficult to investigate. With the boundless curiosity and the inquiring spirit of the eighteenth century, it is understandable that electricity would be part of field observations as well as laboratory inquiry. In Chapter 2, I noted that electrical properties of minerals were investigated as part of mineral characterization, identified as vitreous or resinous. Mohs

wrote, "Several minerals produce electric phenomena; some of them by friction, others by pressure, others by communication, and others by heat" (Mohs, 1825, v. 1, p. 310).

In the early- to mid-eighteenth century, several Englishmen recorded the electrical conductivity of some rocks and minerals as well as Earth itself, and credit for the independent discovery of terrestrial conductivity is given to William Watson (1715–1787) in 1746 (Spies, 1998, p. 199). One of those who looked at rock and mineral conductivity was the English "electrician" Stephen Gray (c. 1666–1736), who was credited with being the first to discover the property of conductivity in various materials in a series of experiments begun in 1729 (Boynton, 1948, p. 294). He first found that he could "communicate" electricity from a rubbed flint-glass tube through, among other things, a piece of wire. Over several years, he extended his investigations to other materials at increasing distances, eventually finding that "flint-stone," "sand-stone," "load-stone," bricks, tiles, and chalk could have "Elektrick Vertue" communicated to them (Gray, 1733, p. 22). In the process, he discovered there were many materials that could conduct electricity from place to place, as well as those that would not.

The ability to produce an electric current, as opposed to the static electricity created by a friction machine mechanically rubbing a sphere of sulfur or glass as done in the eighteenth century, also added to knowledge of terrestrial materials. Volta's "pile" was the first electric battery (Fig. 8.3). It consisted of metal disks separated by moist disks of paper or a liquid such as lye or brine. It was a dependable source of electric current, as Volta announced to the Royal Society in 1800. This means to supply a current was immediately hailed in London, France, and the Netherlands, and experiments were quickly designed for its use.[26] An immediate result for geology was a new theory advanced by Davy on the possible cause of volcanism. Davy was able to isolate some of the alkali and alkaline earth metals, which had previously been known only in compounds. Their extreme reactivity in water with generation of heat induced Davy to hypothesize such reactions as a possible cause of volcanism, because compounds of some of them had been shown to be present in volcanic rocks. That might provide the means to produce heat enough to vaporize water, thus causing volcanic explosions.[27]

The electric nature of some atmospheric phenomena was also demonstrated artificially. One example was the Dutch Martinus van Marum's (1750–1837) 1787 demonstration with "artificial clouds" in a lecture hall, one of which contained a mixture of hydrogen and oxygen, with a brass chain construction that enabled a spark to detonate the "cloud" (Roberts, 1999, p. 713). Benjamin Franklin's (1706–1790) kite experiments (~1750) to demonstrate

---

[26]To trace the fascinating story of Volta's discovery, his experiments, and his relations to the scientific communities in Italy, London, Paris, the Netherlands, and elsewhere, see Giuliano Pancaldi's excellent book, *Volta: Science and culture in the age of Enlightenment* ([2003] 2005). For a thorough treatment of electricity, see Heilbron (1979).

[27]Pancaldi's ([2003] 2005) Chapter 7 details the first investigations of various chemical effects made possibly by the "pile." It is worth noting that Dr. William Cruikshank (d. 1810 or 1811) was one of several people who improved the design of the "pile" so it could be easily reproduced.

Figure 8.3. Alessandro Volta (1745–1827).

a cap so that it could be used in rain or snow, as well as in fog or dew (Saussure, 1786, v. 2, p. 203). The pith balls were set to be very sensitive, and he added a scale so that he could tell how far they diverged when charged, enabling him to estimate intensity. He found that atmospheric electricity is subject, like the sea, to a flux and reflux, growing and decreasing twice in 24 hours (Saussure, 1786, v. 2, p. 222). He remarked:

Car j'ai vu avec cet instrument que cet état varie, & par rapport à l'intensité absolue de l'electricité, & par rapport à la distance de la terre à laquelle cette électricité commence à se faire sentir. (Saussure, 1786, v. 2, p. 204)[28]

He also noted that atmospheric electricity increases a great deal around an active volcano, and suggested a theory as to why this might be. Water that contributed to burning of the volcanoes, compressed by the weight of the air in subterranean caverns, would fall into the burning furnaces and probably would attain a degree of heat greater than that attained in experiments. He tested to see if electricity would be produced by, essentially, boiling water by splashing it on extremely hot iron, and found that it was. His further proofs and investigations occupy many pages of *Voyages* (Saussure, 1786, v. 2, p. 228–244). Saussure and other researchers amassed a great deal of information about the electrical properties of Earth. At this time—the end of the eighteenth century—electrometers or electroscopes, both the static machine and Volta's new battery, Leyden jars to store electricity, and soon galvanometers, which cause deflection by a current shown by a magnetic needle, were all employed to study electrical properties.

In further work, changes in electric currents around ore deposits were explored as early as 1830 (Spies, 1998, p. 199). This was soon put to use in mines by the English scientist Robert Fox (1789–1877), as reported by Thomson:

His galvanometer consisted of a magnetic needle contained in a box 4 inches square and 1 in depth, round which a copper wire, covered with silk, was passed twenty-five times; small copper discs were placed in contact with the minerals in the mine, and these, by means of copper wires (sometimes several hundred feet long), were brought in contact with the two poles of the galvanometer. (Thomson, 1836, v. 2, p. 227)

After determining in which mineral veins current would flow, Fox attempted to relate direction of current flow in various veins to their orientation (north to south, east to west, dipping in different directions), but he obtained no definitive results.

SEISMOLOGY

Earthquakes are of a different nature from gravity, electricity, and magnetism, being episodic and not an intrinsic property of matter. With better delineation of the makeup of the crust of Earth from magnetic and other data in the 1960s, the sometimes

atmospheric electricity, and show it was the same as the electricity in a Leyden jar, were well known. Franklin applied his findings to recognition that buildings could be protected from lightning if the charge was conducted to the ground by wires attached to a metal lightning rod. Saussure is credited with installing a lightning conductor in his house in Geneva in the early 1770s, following talks with Franklin, and after quelling his neighbors' fears of disaster (Freshfield and Montagnier, 1920, p. 130). This identity of atmospheric electricity with that produced in the laboratory also provided new avenues for conjectures about the cause of earthquakes, a "hot topic" because of earthquakes in Lima, Peru, Italy, and supposedly aseismic England in the middle of the eighteenth century, followed by the destruction of Lisbon in 1755 (Guidoboni, 1998a, p. 208). It was conjectured that earthquakes might be caused when a cloud with electric charge discharges against Earth, also producing noise. Like magnetism, electricity was invoked as the cause of many phenomena.

From the seventeenth century onward, electricity was detected by what we call an "electroscope." At first threads enclosed in a glass bottle were used. Two suspended pith balls, or two pieces of gold leaf, proved better. The correspondence between the amount of divergence and the strength of the electric charge was noted (Hackmann, 1998b, p. 219–220). Credited by some as the first to devise a true electrometer, Saussure employed it to record atmospheric electricity during his travels. A metal hook conveyed electricity inside the glass. He furnished it with

---

[28]"For I have seen with this instrument that this state changes and makes itself felt in proportion to the absolute intensity of the electricity, and in relation to the distance of the earth from which this electricity begins to be perceived" (my translation).

suggested possibility of plate tectonics gained recognition. Earthquakes themselves provide data about deep and shallow earth structure. They are possibly the natural occurrences that did, and still do, arouse the most fear among people subjected to them. Despite efforts to codify conditions of weather, animal responses, well levels, or other portents, there was little of predictive value, other than their frequent association with volcanic phenomena. In nonvolcanic regions, there was nothing of much use. Guidoboni has distinguished what she calls "historical seismology" from seismology itself, and has defined the former as "that area of seismology which uses historical data to analyze seismic activity over the long period in order to arrive at estimates of seismic hazard and risk" (Guidoboni, 1998b, p. 749). Those compilations and catalogs of the earthquakes of various regions and countries have existed since more than 1000 years before the Christian era, and continue in our computerized databases today. As would be expected, countries such as Italy, which experience many earthquakes, produce most of these kinds of investigations.

Historical Records

Instrumental efforts to detect, and later, to record, earthquakes began (famously) in China, with the instrument of Chang Heng (72–132), which showed not the magnitude, but the direction of seismic movement (Ferrari, 1998b, p. 530). Oldroyd has given a succinct description of the instrument:

[I]t is thought to have consisted of a heavy pendulum within a large jar. There were eight mobile arms inside the jar which, according to the direction of the seismic impulse, would release mechanically one of eight balls held in the outside of the jar. The falling ball was supposed to be caught in the mouth of one of the eight model frogs, appropriately arranged around the jar. (Oldroyd, 1996, p. 340)

There have been a number of attempts to construct or illustrate such an instrument. A thorough history was supplied by Needham (1959). Trees, suspended objects, and the behavior of liquids in various vessels were all recognized as detectors of earthquake motions (Ferrari, 1998a, p. 468), and led to ideas for instruments.[29]

In the seventeenth century, Hooke noted that the effects of earthquakes on Earth could be seen long after the event (Drake, 1996, p. 314). Due to extensive lists of earthquakes (see Davison, [1927] 1978), by the middle of the eighteenth century it was recognized that earthquakes had occurred over much of the earth, although there were regions that were far more active than others. Buffon recognized two kinds of earthquakes: those related to volcanoes, and those that apparently were not (Davison, [1927] 1978, p. 5). By the middle of the eighteenth century and later, new attempts were made to measure earthquake parameters. Mercury flowing into and out of different channels, pendulum instruments making tracing on sand or other media, and vari-

ous different pendulum arrangements were eventually utilized to record direction and, sometimes, strength. The attempts to find reliable indicators led to increased knowledge of the behavior and structure of Earth, and efforts were made to determine how far earth disturbances were propagated, and what their strength was. However, recognition of the different kinds of waves, and methods for detecting and describing them as well as estimating their strength, were not truly implemented until the beginning of the twentieth century.

Meanwhile, ideas about the causes of earthquakes didn't change a great deal, at least after the realization that they were generated by natural rather than animistic causes, such as the idea that the world was on the back of a large tortoise that shrugged. Hooke likened earthquakes to the effect of gunpowder or lightning, which aided rapid expansion (Drake, 1996, p. 359).[30] Volcanism was, of course, closely associated. It demonstrated large, if misunderstood, reservoirs of heat. The effect of heat on water and the power of the subsequent volume increase were familiar. Throughout the seventeenth and eighteenth centuries, there was much conjecture about circumstances that could produce such effects (Guidoboni, 1998a). Fires and electricity were both invoked as possible sources for the necessary energy. Guillaume Amonton (1663–1705) suggested that compressed air trapped below Earth could produce earthquakes (Guidoboni, 1998a, p. 209). Some experimentalists favored explosions and/or underground winds. There were conjectures about the known reactions between iron filings, sulfur, and water, or of strong acids, water, and iron, as sources of heat and disturbance. Rock fracture was not seriously considered until well into the nineteenth century (Howell, [1986] 1990, p. 118).[31]

John Michell (1724?–1793) exemplified the thoughtful eighteenth-century scientist. He was known as an astronomer and for his work on artificial magnets, but like many of his peers, he was intrigued by devastating earthquakes in Lisbon and elsewhere in the middle of the century. He investigated the locations of earthquakes, noting that they tended to reappear in the same localities, and attempted to time the velocity of propagation of earthquake waves as well. He explained earthquakes as a combination of the interaction of fires from inflammable materials, the subsequent volcanoes, and the relation of confining cavities and overlying strata, and attempted to corroborate those suppositions with surface observations. As more specific studies of earthquakes continued, compilations of seismic activity continued to be made and, beginning with the work of Robert Mallet (1810–1881), were used to show areas of seismic activity as well as for efforts at prediction (Guidoboni, 1998b, p. 751).[32]

---

[29]Ferrari (1998a) gave a detailed summary of methods attempted during the eighteenth century and later.

[30]David Oldroyd reminded me that Hooke's *Discourse on earthquakes* led Hooke to an entire theory of Earth, for which see Hooke in Drake (1996) and Oldroyd (1972, 1998).

[31]Oldroyd reviewed "explanations" for earthquakes in the Greek and Roman literature in his 1979/1998 paper about the poem "Aetna."

[32]Oldroyd (1996) has given a good summary of the history of seismology in his chapter titled, "Thinking with instruments: Earthquakes, early seismology and the Earth's hidden interior." Also, the entries about seismology in the *Sciences of the Earth* encyclopedia (Oldroyd, 1998) provide good reference lists.

Despite ingenious attempts to detect and quantify earthquake motions, the development of accurate seismic instruments lay in the future.[33]

HYDROSCOPE-METALLOSCOPE: DOWSING AND DIVINING

Electricity, magnetism, and gravity continued to be mysterious. Their causes, and those of earthquakes, were veiled, despite the successful mathematical description of gravitational attraction. Quantitative treatment for these forces and reliable instrumentation were just at their beginnings, or were still in the future. In such a situation, it is perhaps not surprising that natural philosophers, as well as practical miners, seriously investigated the efficacy of forked sticks or rods in trying to locate underground water or ore deposits. What is somewhat surprising, after many unfavorable findings and reports from the sixteenth century onward, is that people did and still have a belief in such methods. In the eighteenth century and later, some believers made use of the extended interest in electricity and magnetism to explain the apparent successes of some dowsing trials. Interestingly, the debate can be used as an indicator of what was perceived to be scientific as opposed to unscientific in the eighteenth century, as will be discussed in the concluding chapter to this book.

Biringuccio noted surface staining, some kinds of mineralization, and the flavor of groundwater arising from ores. He even said that evaporating the waters might show the minerals (Biringuccio, [1540] 1990). Agricola was an exemplar of the practical man. He described what he saw and how things worked. While demons or goblins were not entirely ruled out as causes of ill fortune in mines (Agricola, [1556] 1950, p. 217–218), for finding ore deposits he generally relied on simple, objective methods, such as outcrops, soil alteration, or the appearance of streams. Agricola scoffed at dowsing as a means to find ore, and said that if a miner was "prudent and skilled in the natural signs, he understands that a forked stick is of no use to him" (Agricola, [1556] 1950, p. 41). Nevertheless, he illustrated the technique (Fig. 8.4). He also reported the assertions of supporters who said that if the stick was not the right size or shape, or if it were held incorrectly, it wouldn't work. Agricola averred that the use of natural indicators such as surface staining, etc., saved much useless labor that might be expended in digging where a forked twig might indicate (Agricola, [1556] 1950, p. 41).

To find ore, Agricola cited color, small bits of ore or minerals washed out of veins, the tastes (of six possible kinds) of springs draining veins, the color of plants, and even the effect of frost on vegetation near veins—all were "indicators."[34] Although Martine de Beausoleil (died ca. 1643) possessed divining rods, she listed rules for finding ores that depended on the same sorts of surface observations, and the jury is still out on whether she put faith in

Attamen uenam dilatatam raro labor hominum aperit, fed plerumcp uis aliqua, interdum uero uenæ profundæ puteus aut cuniculus. Venæ autem inuentæ

Figure 8.4. Despite his skepticism, Agricola (1556) illustrated dowsing for ore.

dowsing (Kölbl-Ebert, 2003). There was a tradition of using dowsing to make "all kinds of discoveries" as seen in Figure 8.5. As the centuries progressed, the miners' methods were expanded and quantified, but recognized the same reality. Now plants are analyzed for minute amounts of metals, which is a more reliable indicator than a shift in their leaf color, but which depends on the same principles. However, belief in dowsing persists, even today.[35]

As the years passed, tales of divining or dowsing seem to have been unduly influenced by the mind-set of the reporter. Objectivity is apparently in shorter supply than in other scientific reports.[36] However, there is a large literature on the subject. Divining rods themselves might be forked sticks from several favored trees or shrubs, or made of other materials such as metals. They were distinguished from the divining rods of antiquity or later times, which people hoped would predict human affairs.

---

[33]To carry the story forward, see Oldroyd et al. (2007).

[34]Air issuing from veins was thought to be warmer, and thus frost would have less effect on the area (Agricola, [1556] 1950, p. 37).

[35]It is interesting to note differences in the views of historians of science, which may represent their access to or interest in primary literature. In a footnote by the Hoovers in their Agricola translation, they state that 100 years after Agricola, Robert Boyle "was convinced of the genuineness of the divining rod" (Agricola, [1556] 1950, p. 38 fn). But in their book, Barrett and Besterman argued that Boyle couldn't come to an opinion about it (Barrett and Besterman, [1926] 1968, p. 13).

[36]In a slight twist on this topic, I refer readers to the article by Commander titled "Dowsing" in *Sciences of the Earth* (Good, 1998a, p. 172–173). In the very brief entry, he remarked that no new historical information had been published since Ellis (1917), for which see reference list. Actually the literature is quite large, and the 2003 Kölbl-Ebert paper (see fn. 37) is a good place to start, as is a subject search in library databases. A real dissertation on the matter is Barrett and Besterman ([1926] 1968), a reprint and updating of an earlier work, *The divining rod, an experimental and psychological investigation*.

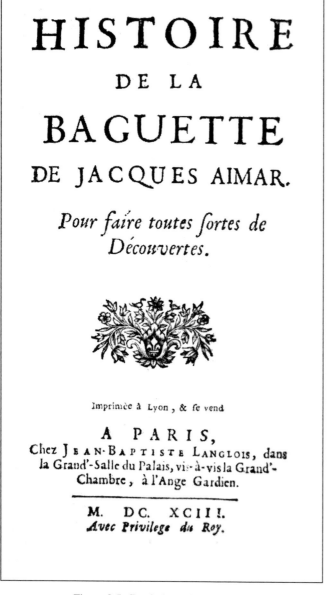

HISTOIRE

DE LA

BAGUETTE

DE JACQUES AIMAR.

Pour faire toutes fortes de
Découvertes.

Imprimée à Lyon, & ſe vend

A PARIS,
Chez JEAN-BAPTISTE LANGLOIS, dans
la Grand'-Salle du Palais, vis-à-vis la Grand'-
Chambre, à l'Ange Gardien.

M.   DC.   XCIII.
Avec Privilege du Roy.

Figure 8.5. Garnier's (1693) title page.

partly associated with her advocacy of a series of "metallic and hydraulic" divining rods. The politics of the France of her time, as well as de Bertereau's self-confident pronouncements, entered into the circumstances that led to her imprisonment, as well as that of her husband and daughter. She died in prison.[37]

In his survey on changing attitudes toward dowsing from the beginning of the eighteenth century through the later Enlightenment years, Lynn (2001) has noted the change in attitude toward it. There was more effort to find a scientific explanation for (or against) it as the eighteenth century progressed. Trials at the end of the seventeenth century to see if criminals could be identified by dowsing (Lynn, 2001, p. 39–44) had been supplanted by more specific tests involving water and minerals. Lynn discussed the celebrated case of the French peasant Barthélémy Bléton born in the 1740s. He was said to have been very successful at detecting underground water and metals, but when he was brought to Paris in the 1780s, there was no clear demonstration of his powers. Tests of his ability to detect buried metals and to trace water lines were inconclusive. Bléton continued, however, to locate springs in the countryside. The perception of what scientific proof was, and who should have a voice in determining it, had changed enormously during the century. Scientific journals were much more widely disseminated, and many more people felt they could judge what constituted scientific proof. Even so, there was no plausible explanation for apparent dowsing successes.

Naturally, those who studied Earth were interested in these questions, and both the investigations and their interpretation remained complex, as they are today. One of our major protagonists, Sir James Hall, visited Paris during the trials of Bléton's abilities. Hall couldn't believe that reputable scientists could accept the "cursed nonsense" of dowsing (Eyles, 1963, p. 166). Despite Bléton's endorsement by highly regarded French scientists, Spallanzani was not convinced by those and other demonstrations, as he stated in letters to Italian naturalist Alberto Fortis (1741–1803), who thought the demonstrations damaged the reputation of some French scientists (Fortis, 1802, v. 2, p. 138–216). He felt that the subject of dowsing, also known as rhabdomancy, was insufficiently explored in light of the new knowledge about electricity and magnetism. There were many variables in conditions, and personal animosities and jealousy were sometimes involved. But Fortis did not condemn it absolutely. It is clear that dowsing did and does exist at a very interesting intersection of "pure" science, folk psychology, empirical testing, and belief.

DRILLING AND BORING

Drilling, digging, and boring or tunneling through Earth generally don't qualify as experiments, although in the twentieth century onward, a number of efforts have been preceded by

Very early on, it was recognized that observable motion could emanate from the holder of the rod and not from the substance searched for. The action was attributed to everything from purposeful motion by the holder, to corpuscular motion, or to any newly characterized portion of science (e.g., electricity, magnetism) that could be brought to bear on it.

Reception of reports of dowsing was problematical. The career of Martine de Bertereau (Madame de Beausoleil) is a cautionary one, at the intersection of sorcery and objectivity. As pointed out by Kölbl-Ebert (2003), opinion has not been finalized as to whether she was a rational, objective person who was demonstrably experienced in mining as well as other sciences, or a sorceress, or at the least, a charlatan. The latter view was

---

[37]Kölbl-Ebert (2003) has assessed literature from the early eighteenth century until the present on the efficacy of Madame de Beausoleil as a mineral explorer, and has come to the conclusion that reports about her represent as much the changes in historical "fashion" as they do objective assessments of her ability.

hypotheses about what might be found in test holes and what they might mean. Certainly the increasing ability to penetrate Earth's surface was of enormous help in unraveling stratigraphy and finding mineral resources. The earliest efforts to go below the surface were concerned with water, brines, and metal ores. Usually, the depth of drilling or digging for water and salt or brines far exceeded that done for metal ores and, later, coal and oil. Materials for tools progressed from the Paleolithic wood, bone, antler, or stone, to copper, and then iron (Singer, 1954, v. 1).[38]

Water

There is a large literature about water wells, some of them dating from as long ago as the eighth century B.C., written by hydrologists and others interested in the history of understanding about groundwater.[39] In general, the ancient wells were dug wells, despite the considerable depth of some of them excavated with pick and spade, the dirt being removed with sacks carried to the surface by men or brought up by a winch. The Chinese may have drilled a water well 1500 meters deep, but generally such wells were for salt brine (Corapcioglu, 1998, p. 835). Important as water was—as attested to by the elaborate supply systems for transporting salt made in antiquity, particularly in arid regions—a larger impetus for improved drilling technology came from the search for mineral resources that were not available on the surface.

Salt

The salt sodium chloride is a ubiquitous necessity for animal life, with much written about the history of salt sources and their exploitation.[40] Knowledge of geology was definitely increased by the search for subsurface salt, whether solid or as brine. The necessity of drilling for brine sources provided the impetus for technological advances in drilling. As Needham (1959) pointed out, Chinese drilling technology was far advanced compared to that in the West. A more recent publication (Chinese Academy of Sciences, 1983), has detailed its history. Wide brine wells, sunk by digging, were common several centuries B.C. In 300 A.D., some Chinese wells reached 100 meters. At that time, the Chinese began to exploit the common occurrence of natural gas associated with salt, sinking gas wells next to the brine wells, and conducting both through bamboo pipes to processing areas where the gas was ignited to concentrate the brine (Yang Wenheng, 1983, p. 266). The well shafts became much narrower as drilling techniques improved, and the wells were cased with bamboo. Yang noted how drilling was done for small and deep wells in the mid-eleventh century:

The drill and bits then used were made of an iron shaft and round iron blades, which could reach hundreds of metres below while the well was but 20 centimeters in diameter. Thick bamboo was used as an outer piping to stop fresh water seeping in. Smaller bamboo sections were used as containers to haul up the brine. The hauling was done with the aid of an animal-drawn wheel. The iron shaft and round blades, indispensable to deep-well drilling, were the forerunners of modern drills. (Yang Wenheng, 1983, p. 267)

The Chinese continued drilling, reaching oil, and putting it to use 300 years before the famous oil well in Pennsylvania. In the mid-nineteenth century, they completed the first 1000 meter natural gas well with the same simple technology (Yang Wenheng, 1983, p. 268).

Rather surprisingly, Yang did not mention another Chinese advance in drilling. The Chinese developed percussion drilling in the mid-eleventh century. Kurlansky has described the process:

A hole about four inches in diameter was dug by dropping a heavy eight-foot rod with a sharp iron bit, guided through a bamboo tube so [that] it kept pounding the same spot. The worker stood on a wooden lever, his weight counterbalancing the eight-foot rod on the other end. He rode the lever up and down, seesaw-like, causing the bit to drop over and over again. After three to five years, a well several hundred feet deep would strike brine. (Kurlansky, 2002, p. 27–28)

Note that percussion drilling was not (re)invented in the United States until the end of the eighteenth century (Kurlansky, 2002, p. 250).

The technology for the exploitation of salt sources throughout the world did not change appreciably except for minor innovations for hundreds of years. What did increase was knowledge of their geology. The frequent association of brines with natural gas was generally recognized, and later, the association with oil. Solid beds of rock salt were seen to extend hundreds of miles in some cases, and this aided regional synthesis. In England in the late seventeenth century, boiling brine to produce salt led, among other things, to depletion of forests. Thus, coal was coming into use and was being searched out. In an interesting twist, in Cheshire, England, the search for coal for fuel to boil brine led only to rock salt. However, the disappointed landowner declined to mine the salt itself and went bankrupt (Kurlansky, 2002, p. 194). Drilling, while not part of what normally might be considered part of the experimental tradition, was in a way, another method of "interfering with nature," and understanding of earth structure and related science sometimes came from it.

Mining: Ores and Coal

Of course, the search for metal ores and coal was also an activity from antiquity onward. This frequently involved the penetration of harder rocks than involved in the search for salt, although some of those deposits were also capped by hard rock. Vogel has credited the Egyptians as being the first in recorded history to tunnel through hard rock (Vogel, 1966, p. 206). Copper mining was pursued in Spain as early as 1600 B.C. (Vogel, 1966,

---

[38]There are two summary articles about drilling in Good (1998a), the first on a history of drilling, and the second on scientific drilling (v. 1, p. 173–178).

[39]There is an informative introduction to this history in Corapcioglu (1998, p. 835–836), with a bibliography for sources.

[40]I highly recommended two books from that literature: Multhauf (1978) and Kurlansky (2002).

p. 207). By the time of Biringuccio and Agricola, mining was well advanced. Rather than being a proponent of shafts descending from above to follow a vein from its surface expression, Biringuccio advocated using what we call a "drift mine," the tunnel situated at the bottom of the mountain, and driven nearly horizontally so as to intercept the vein(s) (Biringuccio, [1540] 1990, p. 18). As tunneling methods involved labor and danger, he admonished the reader to be sure of the direction of the main vein, and not be diverted by smaller offshoots.[41] The drift mine was easier to access than vertical shafts, and enabled easier removal of the ore and accumulated water. Biringuccio cautioned that the seeker after ore must not give up too easily and should employ enough men to do the labor.

However, drift mines were not always feasible, and Agricola gave instructions for arrangements of vertical shafts and connecting tunnels as well as drifts. The miners used picks to dig out the softest ores. For harder ores, hammers and iron tools were used. Biringuccio noted:

For excavating and breaking rocks strong and powerful tools are required, like large hammers and iron picks, long thick crowbars, mattocks and strong spades, picks both with and without handles, and similar iron tools, all of fine and well-tempered steel. It is not necessary to mention those used for excavating ores in softer stones, because ordinary tools are sufficient and necessity teaches which ones are to be used. (Biringuccio, [1540] 1990, p. 24)

In his Chapter VI, Agricola (1556) thoroughly described all the tools used in mines. More iron tools in combination were employed for increasingly hard ores. The final step, for intractable rock or ore, was fire, the fire being built against the working face, followed by water or perhaps vinegar to crack the heated stone. Miners were well aware of the effects in a poorly ventilated mine.[42] Agricola also mentioned a method to dislodge hard rock by driving iron plates into cracks, and then prying the rock apart with several kinds of tools.

Gunpowder, or blasting powder, which originated in China, was not easy to apply to blast rock faces in mines, and indeed, that difficulty led to the continued use of fire setting against the rock face into the twentieth century. The expense of both good (well-mixed) gunpowder and drilled shot holes, the danger from ignition of mine gases and possible collapse, the possibility of injuries, and the late (1831) introduction of the safety fuse, all delayed what would seem to have been a technological advance (Golas, 1999, p. 306–307).[43] Gunpowder was introduced in

mining in sixteenth-century Italy, and in Schemnitz in 1627. The British first began blasting coal faces with gunpowder in the eighteenth century (Wolf, [1952] 1961, v. 2, p. 631). The use of gunpowder continued until the end of the nineteenth century despite the introduction of dynamite (Agricola, [1556] 1950, p. 119 fn).

Mining and drilling technology did not change a great deal for centuries. Iron, and later steel, was substituted for bamboo. There were different designs for drill bits and coupling rods to extend the drill string. Metal ore mining still followed outward signs of the veins. The process of looking at material brought up from drilling in order to determine if economic deposits were below was followed to a certain extent. At the end of the seventeenth century, workers in England were boring small diameter holes in order to see if coal lay below. However, what they brought up was a paste of rock material that provided no clear indications as to the depth at which a certain rock type occurred. The material might be coal, lignite, or just carbonized wood. Moreover, as detailed by Torrens ([1997a, 1997b] 2002), drill holes could also be "salted" with a lump of coal, or the drilling rig sabotaged. Interestingly, the first device to provide a good indication of subsurface materials, invented by the Irishman James Ryan (1770–1847) in 1804, was not embraced by the mining community, despite its obvious utility. The device

consisted … of a T-bar to which any number of boring rods could be strung together. It could, as before, go down some distance, but it now carried a hollow cylinder with four cutting edges on a crown saw of hardened steel, which was rotated mechanically to cut through the strata. It left a core of rock in the centre of the cylinder. Recovering the cores from the borehole was achieved by a pair of Heath Robinson-type tongs. Ryan's invention now allowed cores to be recovered in sequence and oriented against a compass. (Torrens, [1997a, 1997b] 2002, p. 8)

The device could also be used to drill vents for air circulation in the mine. Despite the obvious utility of the device, its operation was more expensive than the older method of coring, and it was not adopted by the mining community for many years. Early on, geologists saw its utility in determining subsurface layers, and thus its value for determining where economic minerals would be found, as well as for building a useful stratigraphic column, but its use for this was also long delayed (Torren, [1997a, 1997b] 2002, Variorum I and VIII).

Drilling and tunneling technology steadily improved during the course of the nineteenth century, but their history lies outside the present account. As Torrens has pointed out, the practical world of the miner and the more theoretical world of the natural philosopher-scientist-geologist didn't intersect, especially in Britain. There is a kind of parallel here with the disjunct between the chemistry of glass making, tanning, metallurgy, textiles, and china/porcelain, and sometimes that of the alchemist and, later, that of the philosophical inquirers of the seventeenth and eighteenth centuries.

---

[41]The compass was used as an aid to direction-finding in mines, although I've not found reference to the alteration it would have shown near a large metallic ore body.

[42]The subject of ventilation was also well discussed, with ingenious methods to increase air flow. Few methods were very effective, so the use of fire to crack rock was discouraged. The Hoovers' footnotes remark that fire use was of ancient origin, and persisted through the nineteenth century (Agricola, [1556] 1950, p. 118). They also discussed the ancient use of such methods and the "myth" of the use of vinegar on p. 119.

[43]Golas reviewed the evidence for gunpowder use in mining from antiquity onward, and supplied copious references from mining technology literature.

OTHER METHODS

As would be expected from savants, miners, and others who had great interest in Earth, they used additional methods of trying to quantify the properties of Earth and its atmosphere. As just one example, in his explorations of the Alps, Saussure employed an array of instruments, a number of which he originated or substantially improved for use under field conditions. As mentioned previously, among these instruments were thermometers to measure the temperature of deep water, and of the subsurface earth. There were also an atmometer to measure evaporation, a cyanometer to estimate the color of the sky, a diaphanometer to measure atmospheric transparency, an actinometer to show the chemical effects of sunlight, and a clinometer to measure the inclination of beds (Saussure, 1796, v. 3, p. 536; Sigrist, 2001) as seen in Figure 8.6. It consisted of a half circle traced and graduated on a thin, rectangular copper plate. It had a plumb bob suspended in the center of the circle, and was intended to measure the inclination of beds, veins, and the slope of the terrain (Saussure, 1796, v. 4, p. 536). The clinometer, in particular, had probably been in use for a long while in one guise or another. There were few properties of Earth and its atmosphere that had not been investigated by the end of the eighteenth century, although the efficacy of the methods was spotty.

CONCLUSION

This chapter has discussed a number of other ways to learn about Earth and its materials, including gravity, magnetism, electricity, and early seismology. Two other methods that more or less increased knowledge about the way Earth was put together, namely, dowsing and boring or tunneling, were also considered. Unscientific as these latter methods were, they still aided conceptions of what might be found below Earth's surface, and aided the study of stratigraphy. A good bit of fervor, both for and against, was channeled into trying to explain ways in which there could or could not be a rational explanation for the apparent success at times of dowsing.

In all of this study, there was the strong influence of the practical man or woman, those who saw immediate relations with no need for mysterious interventions: Dowsing wasn't necessary to search out ores. It's probably safe to say that miners always related surface indicators to what might be found by digging. Yang cited a treatise from the fifth century B.C. that noted many useful minerals and ores. As aids for their discovery, he specified:

topographical surroundings of mines such as the top or foot, whether on the sunny or shady side of a mountain; the outward appearance and properties of a rock such as its hardness, colour, lustre, type of aggregation … etc. (Yang Wenheng, 1983, p. 258–259)

Associations of various rocks and minerals with each other were noted, with those of ores for different metals observed very early. The connection between specific plants and minerals had been

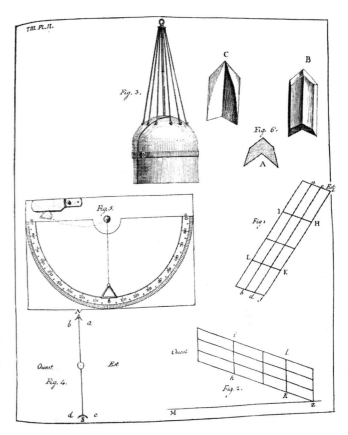

Figure 8.6. An illustration of Saussure's clinometer (*Voyages*, 1796, v. 3, Plate II, Fig. 5).

long noted, by the sixth century becoming a "new epoch of geobotanical and biogeochemical prospecting" (Yang Wenheng, 1983, p. 262), while still couched in rather fanciful language.

Biringuccio, Agricola, and other writers of mining treatises discussed ways to intercept veins with the least effort, while providing for ore and waste removal and the often necessary water drainage. At times, a water supply from above was also recommended to run machinery powered by water wheels. Tunneling and drilling for minerals contributed to later observations made in cutting into Earth to level beds for railroads and canals, and helped provide the additional understanding of layering needed for prospecting. One may regret the disruption, noise, and elimination of natural features caused by quarrying, mining, and construction, but it is still fascinating to be able to see concrete evidence of the way Earth is structured.

At the end of the eighteenth century, the gap between practical pursuits and science still showed. Just as glass-, soap-, and pottery-makers did not often contribute to the later development of chemical knowledge, observations made by canal diggers and miners were not often taken into account in geology, a well-known case being that of William Smith (1769–1839). Şengör (2001) has made the point that Werner's background in mining led to different skills among his students than might be ideal for science, although close observation of rocks and minerals in the field and of hand specimens was certainly useful, as was the mapping of field relations.[44]

For the eventually more "scientific" of this chapter's topics, there was a long period of observation before viable theories and mathematical treatments could apply. Newton and his contemporaries were the first to establish the influence that celestial bodies had on each other. The effects of gravity were slowly uncovered, and by the end of the eighteenth century, careful measurement enabled its use to describe the shape of our Earth. In the late eighteenth century, the variables that needed to be controlled for accurate measurement of gravity were familiar. Each material involved in the instrumentation might be subject to expansion and contraction of different metals under varying conditions of temperature, barometric pressure, moisture, and air currents, whether the method used pendulums or the spring or torsion balance. Accurate timekeeping was necessary as an aid to determining exact locations. The necessity of standards for length and mass was recognized. As with other instruments for other purposes, instrument makers were impelled to solve problems in reading very small increments of measurement. Examining anomalies, and the reasons for them, also proved fruitful.

The paired phenomena of electricity and magnetism were baffling through the eighteenth century. They both provided more facts about matter and its behavior, but were often applied in totally inappropriate ways. Eighteenth-century accounts of using the conductance of electricity to ignite alcoholic (or worse!) emanations from human subjects, and such investigations as Gray (1733, p. 39)'s "electrical boy" are appalling. Mesmerism and dowsing were both provided with hopeful electrical or magnetic explanations, or the new knowledge was used as an example that

novel things were being discovered constantly and thus later, more scientific, explanations would be forthcoming. Volta's battery provided proof of the electrical nature of matter, and led to Berzelius's explanation of affinities between elements, further clarifying the nature of an elemental substance. Much ingenuity was applied to instruments to measure and quantify electricity and magnetism and, like those for gravity, their stability was tested in both field and laboratory. While many principles of those instruments remain the same in the twenty-first century, many problems were solved in the nineteenth and twentieth centuries.

Earthquakes are a phenomenon separate from the universal qualities of electricity, magnetism, and gravity. Seismology is studied worldwide. Despite collection of much information about direction, depth, and strength of literally earth-shaking events, it is, practically, still basically a descriptive rather than a predictive science. By using strain meters, sometimes in volcanic regions, relatively short-term predictions can be made. The old folk methods of earthquake prediction are still in use, sometimes to good effect. But modern seismology provides a unique amount of information about, among other things, Earth's interior, and thus tectonic theory.

All of the properties and processes discussed in this chapter are intimately associated with Earth. There is a vast gap in conceptual scale between studying a hand specimen of a mineral, and determining the gravitational attraction at a specific locality. One inescapable fact that emerged from eighteenth-century studies was the necessity of considering the entire Earth. As geology became a recognized science, problems of scale from the microscopic to planetary were in evidence, as they are today. Some investigative methods discussed in this chapter aided the advance of our knowledge of Earth, and contributed to geological theory. The impact of the studies discussed in this and previous chapters is the subject of the following chapters, and is evaluated mainly from the primary literature. The comments of historians of science and how the secondary literature has treated the matter will be the subject of this book's final chapter.

---

[44]Şengör discussed this point in some detail, saying,

The deterministic and nonuniformitarian folklore, combined with an equally deterministic and nonuniformitarian scientific tradition, led to a tremendous development of the purely descriptive branches of geology such as descriptive mineralogy, but allowed very little latitude for interpreting geological history. (Şengör, 2001, p. 24–25)

CHAPTER 9

Rock origin: The investigations—Granite and limestone

INTRODUCTION

One of the most interesting things in the study of the history of science is the opportunity to investigate the thought processes of excellent minds from the past, and to attempt to determine their reasoning processes from their knowledge base. This has been called an "internalist history of science," particularly if the historian does not include the study of the social and institutional webs within which the scientists worked. The latter is difficult to avoid completely, when much of the work discussed here was done within a framework of friendship, correspondence, and occasionally rivalry, which framework sometimes shows the process of induction, observation, experimentation, and deduction, as they might be supported or questioned by the work of others. Nevertheless, emphasis here is on the work itself, and what its practitioners thought about it.

Much of what I've called "experimentation" was inconclusive. As noted earlier, even calling it "experimentation" is sometimes stretching a point, but our perception of the stretching is located in the twentieth and twenty-first centuries, and not in the eighteenth.[1] However, much of the work reported in these chapters was driven by a desire to understand theories. The theories themselves, especially those of rock origin as it fed into how Earth as a whole worked, were of such enormous breadth that any experiment on an eighteenth-century scale would necessarily be insufficient to explain the whole process, as the scientists themselves certainly recognized. Despite our twenty-first-century computers, lasers, sophisticated analytical methods, Global Positioning Systems, and satellites, Earth and its atmosphere still often prove to be of too large a scale for definitive answers.

But just as they do now, during the eighteenth century and the beginning of the nineteenth, the scale and the projected import of the investigations differed in the work of different people. At the extremes, some "philosophers" were content solely with determining a melting point, while others saw in several mineral fusions or a hand specimen a road guide to a theory of Earth. It appears, though, that over the eighteenth century, there was a general trend from more pragmatic instances of, say, isolated mineral analyses to greater and greater application of what was learned to larger and larger questions as data accumulated. That would not be precisely true. Theories of Earth abounded throughout the period. Perhaps the great bulk of them are dimmed in our consciousness due to the prominence of the Huttonian–Wernerian

debate. And despite their prominence in late-eighteenth- and early-nineteenth-century debates on how Earth was formed and functioned, both positions (origin by fire or water) had precursors, as was particularly noted in the basalt controversy.[2] Belief in the volcanic origin of basalt did not imply agreement with the Plutonist theory. Indeed, much of the pioneering work to connect basalt to volcanism was done by Neptunists.

After a short survey to see if there are trends with respect to experimentation in theories of Earth over more than a century, this chapter will survey how prominent reasoning from planned investigations and observations was, and will discuss specifics for several rock types. However, discussion of basalt, about which there was the greatest controversy, will be reserved for the next chapter. Chapters 9 through 11 will generally be limited to consideration of primary sources; Chapter 12 will examine what historians of geology have determined.

WHERE THEY STARTED: THEORIES OF EARTH

Certainly not the first or the best of the theories, but one that received a great deal of attention, was that of the English cosmogonist and earth-observer, Thomas Burnet (1635?–1715). Published in English in 1684, *The sacred theory of the Earth* engendered responses for at least the next 50 years, and it was reprinted in the twentieth century.[3] Not surprisingly for a book whose purpose was more theological than scientific, Burnet's explanations quickly made recourse to divine intervention. It is clear that Burnet closely observed the land and the sea as they could be viewed in his time. However, his full title, *The sacred theory of the Earth: Containing an account of the original of the Earth, and of all the general changes which it hath already undergone, or is to undergo till the consummation of all things*, indicated that he went considerably farther than direct observation. But Burnet was familiar with more earthly things. While his descriptions of volcanism ("Fiery Mountains") were predictably apocalyptic, he offered realistic comments about the empirical effect of heat on earth materials and said that great heat followed by cooling led to the formation of glassy materials. He described sedimentation quite clearly.[4] However, his density arguments

---

[1]"Experience" in the eighteenth century often involved repetition of the investigations of others.

[2]Young (2003) has drawn a particularly good distinction between the Plutonist and Vulcanist positions, as well as delineating them from Neptunist beliefs.

[3]Burnet's book was published first in Latin in 1681.

[4]For an excellent discussion of the content of theories of Earth in the sixteenth through eighteenth centuries, see P. Rossi's *The dark abyss of time* ([1984] 1987). Here I am concerned with looking for instances where the authors cited observed properties that influenced their theory. Oldroyd (1996), Rudwick (2005), and others also discuss aspects of theories of Earth.

regarding the natural place of soil, water, and oil as explanation for the formation of Earth and subsequent developments couldn't support his assumption of central fire, fire being the lightest of the elements.

Another clergyman, Erasmus Warren (c. 1643–1718), responded to Burnet in 1690.[5] There were a few allusions to things scientific, Warren's tract (Fig. 9.1) being mainly theological. He quarreled with the assumption of a fiery center of the earth. In discussing the constitution of the globe, he well knew that air was less dense than oil, and that oil floated on water. Fire being the "lightest" of the elements, it was difficult to see how it could be at the center. Warren was aware of earth's magnetism, but referred to "God's Divine Chymistry" (p. 242) when speaking of rain and clouds. No less than Burnet, he invoked the Almighty, but used that to refute some portions of Burnet's Theory.

English naturalist John Woodward offered a natural history of Earth that included minerals (Fig. 9.2), with a chapter titled

"Of the origin and formation of metalls and minerals" ([1695] 1978). He included several chapters dealing with the universal Deluge and alterations of the terraqueous Earth. He found it easier to account for terrestrial materials and strata than metals and minerals, because the latter were relatively rare. He described some minerals expertly, demonstrating his personal experience with them, and also noted the variable positions and conditions in which they might be found. His extensive mineral and fossil collections were renowned. In his view, minerals and metals owed their origin to the Deluge, just as rocks and strata did. Water carried them all. Some material was dissolved, and carried through cavities in the rocks until it was deposited. Woodward's comments about the action of heat, solubility, and gravity

---

[5]"Chymists by the force of fire" (Warren, [1690] 1978, p.15).

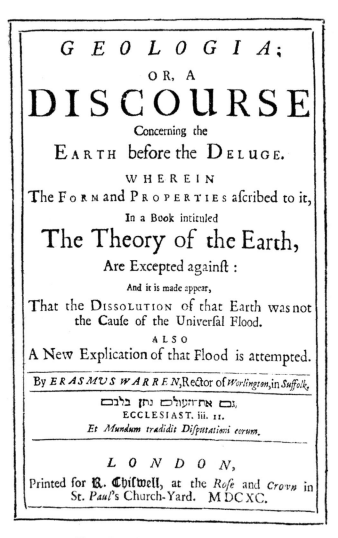

Figure 9.1. Title page of Warren (1690).

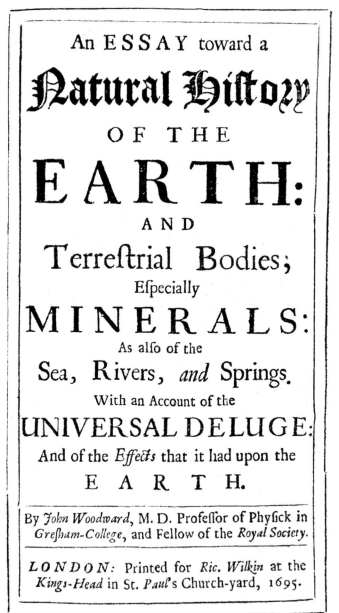

Figure 9.2. Title page of Woodward (1695).

reveal that he was familiar with the methods of philosophers who investigated earth materials. He also mentioned using a microscope. In general, Woodward applied his observations to comments about how Earth came to be as it was seen, and did not invoke the Almighty's power as a direct explanation. Perhaps this should have been anticipated by his titling his work as a natural history and not a "theory of the Earth." However, he was attempting to explain the same things earlier theorists had explored.[6]

William Whiston (1667–1752), the English cosmogonist and mathematician, published his *A new theory of the Earth* (Fig. 9.3) in 1696, just a year after Woodward's book. In the prolix style of the time, it was subtitled "From its original, to the consummation of all things wherein the creation of the world in six days, the universal deluge, and the general conflagration, as laid down in the holy scriptures are shewn to be perfectly agreeable to reason and philosophy" (Whiston, [1696] 1978, title page). Interestingly, Whiston did, as he promised, use reason and "philosophy" (science) to lay out scenarios for how the Deluge and the conflagration could have occurred. He invoked "Specifick Gravities" or relative densities to explain the sequence of solid layers. He suggested the Deluge could be understood by analogy to an experiment that consisted of a solid cylinder of marble with holes driven into it lengthwise and placed in a hollow cylindrical vessel that just allowed it to move. There was water in the vessel, and oil was poured into the holes of the cylinder. When the cylinder was pushed down, the denser water forced the oil from the holes at the top. He suggested that the weight of Earth acted likewise on the less dense water underground, which would be forced violently to the surface. He referred readers to the "Learned Dr. Woodward" for proof that "There is a constant and vigorous heat diffused from the Central towards the Superficiary part of our Earth" (Whiston, [1696] 1978, p. 163).

The Italian Antonio-Lazzaro Moro (1687–1764) is credited with showing the "empirical spirit characteristic of the enlightened intellectual climate of eighteenth-century Europe" (Thomasian, 1974, p. 532–533). His best known work, *Dei Crostacei e degli Altri Corpi Marini che si Trovano sui Monti*, which was concerned with the origin and development of fossil deposits, was published in 1740. Perhaps not surprisingly for an Italian who had personally observed active volcanoes and was aware of the emergence of new volcanic islands, Moro was a Plutonist. His observation of powerful igneous forces informed his thoughts on mountain formation and fossil placement, which to him required no divine intervention as the schemes of Thomas Burnet did. He wrote about Burnet's and Woodward's theories and refuted their Neptunist views on the basis of his observations.

In an appendix to his 1786 book, clock maker John Whitehurst (1713–1788) contended that even before Newton, thoughts about the formation of Earth had to be in agreement with the laws of gravity, fluidity, and centrifugal force (Whitehurst, [1786]

1978, p. 275). He relied on close description of minerals and earths, and his own observations, as well as the writings of others to explain Earth as it appeared to him in various localities. While he did not invoke the Deity for particular instances, and was "scientific" enough to provide numerical tables of some measurements (e.g., air temperatures at different locations), the Deluge was for him a real phenomenon, as he distinguished between "diluvians" and "postdiluvians." In this more scientific theory of Earth, Whitehurst still included tables that compared the longevity of particular peoples who lived both before and after the Flood, by means of references from scripture. He compared ages attained in different climates, and determined that people in

A NEW
THEORY
OF THE
EARTH,
From its ORIGINAL, to the CONSUMMATION of all Things.

WHEREIN

The CREATION of the World in Six Days,

The Univerſal DELUGE,

And the General CONFLAGRATION,

As laid down in the Holy Scriptures,

Are ſhewn to be perfectly agreeable to REASON and PHILOSOPHY.

With a large Introductory Diſcourſe concerning the Genuine Nature, Stile, and Extent of the *Moſaick* Hiſtory of the CREATION.

By *WILLIAM WHISTON*, M. A. Chaplain to the Right Reverend Father in God, *JOHN* Lord Biſhop of *NORWICH*, and Fellow of *Clare-Hall* in *Cambridge*.

LONDON:
Printed by *R. Roberts*, for *Benj. Tooke* at the *Middle-Temple-Gate* in *Fleet-ſtreet*. MDCXCVI.

Figure 9.3. Title page of Whiston (1696).

---

[6] John Hill's (c. 1716–1775) natural history, published in 1748, looks far more "modern" with mineral classification and properties, field descriptions of rocks, and no appeals to higher powers.

temperate climates lived longer. Despite use of the Deluge as a prime geological agent and our slight feeling of indigestion about the book as a whole, the plates that illustrated various strata were precise and detailed, with objective rock descriptions and presumably accurate depiction of strata contents, thicknesses, and dips. (See Fig. 9.4.) I could find no statement of rock origin, other than that of basalt, except in the most general terms of denudation and deposition. Among the strata of Derbyshire that he described so well, Whitehurst included toadstone:

A blackish substance, very hard; contains bladder-holes, like the *scoria* of metals, or Iceland *lava*, and has the same chymical property of resisting acids. ... It is likewise attended with other circumstances which leave no room to doubt of its being as much a *lava* as that which flows from Hecla, Vesuvius, or Aetna. (Whitehurst, [1786] 1978, p. 184–185)

What is now the best known theory of Earth (see Fig. 9.5), that of James Hutton, will be but little discussed, since he was well known for his belief that the laboratory could *not* provide

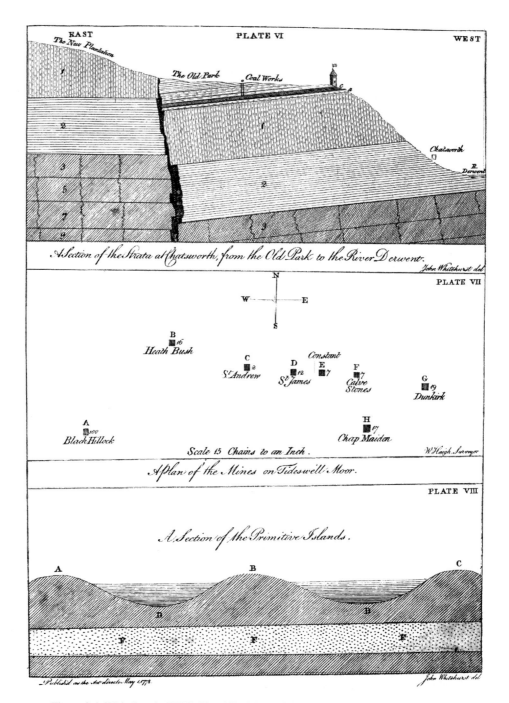

Figure 9.4. Whitehurst's (1778) Plate VI with carefully drawn map, strata, and a section.

answers to questions of rock origin, and he didn't advocate or support experimentation for that purpose (Hutton, [1795] 1972, p. 251). He did, however, conduct numerous chemical investigations, and he connected what he observed of zeolite in his analysis of it to what might be possible in the origin of basalts that included zeolite. His first paper on his theory was read in 1785 and published in 1788, while the more extensive statement was published in 1795.[7] Unlike many of his contemporaries, Hutton believed in the internal heat of Earth, and its ability to produce the rocks as he saw them. He included not only granite and basalt, or trap, but also limestone and sandstone among rocks produced

---

[7] The third intended volume was not published until it was found by A. Geikie in 1899. It is incomplete, according to Hutton's chapter numbering (Dean, 1997, p. 11).

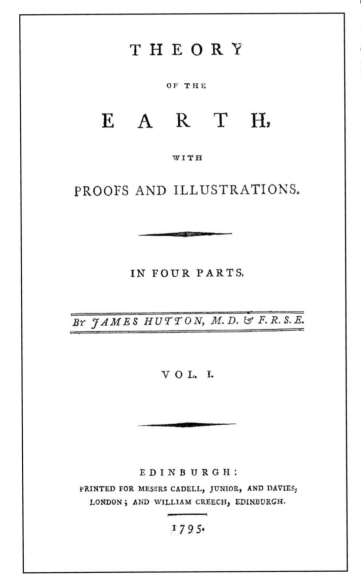

Figure 9.5. Title page of Hutton (1795, v. 1).

by heat. His conjectures about granite origin were partly based on crystallization characteristics of the constituent minerals, as in his discussion of Portsoy granite (Hutton, [1788] 1973, p. 256–257). He was known for being an accurate observer, and Playfair commented on his "cabinet," or collection of minerals, and his knowledge of them (Playfair, 1805, p. 89). Hutton used the results of an experiment that was not his own, in one of the prime tenets of his theory. Black's discovery that fixed air (carbon dioxide) is liberated from magnesia alba (magnesium carbonate) upon heating led Hutton to conjecture that pressure could prevent the release of fixed air if limestone (calcium carbonate) is consolidated by heat under great pressure. Playfair noted Hutton's awareness that water requires more heat to boil when it is under pressure (Playfair, [1802] 1964, p. 182). The pressure of the ocean or of rock masses supposedly could prevent the escape of volatiles and allow even coal to boil, as well as allowing limestone to retain its carbon dioxide.

In 1795, La Métherie published a lengthy three-volume *Theory of the Earth* (Fig. 9.6). The date of its publication is interesting because it is the same as that of Hutton's final exposition, yet the content is so different. Unlike Hutton's *Theory*, which is largely innocent of quantitative information, La Métherie incorporated virtually all the latest knowledge from chemistry/mineralogy and the physics of Earth, including electricity, magnetism, light, and observations about heat. He described rock and mineral characteristics, as well as reports of volcanism and field localities. He completed his *Theory* with an assessment of other theories of Earth. To demonstrate the breadth of La Métherie's work, as well as to see both the sheer number of the theories he felt it necessary to mention and the popularity of the genre, it is instructive to look at his list. He began with theories from the Orient, and those of the Egyptians, the first of which stated Earth was under the action of water, the second that of fire. He then discussed the systems of Burnet, Whiston, Woodward, Bernard Le Bouvier Fontenelle (1657–1757), Deluc, John Ray (1627–1705), "some Indians" (La Métherie, 1795, v. 3, p. 394), Phoenicians and Hebrews, Steno, Moro, Hutton, Pierre Simon Pallas (1741–1811), Saussure, Dolomieu, and Eugène Louis Melchior Patrin (1742–1815). After discussion of mountains and valleys, interwoven with comments on comets and submarine fires, he went back to the systems of Bélus, Sénèque (Lucius Annaeus Seneca) (b. ca 4–1 B.C., d. 65 A.D.), the Sages of the Orient, Descartes and Gottfried Wilhelm Leibniz (1646–1716), Benoît de Maillet (1656–1738), Buffon, Linnaeus, Bourguet, and François Bernier (1625–1688). Clearly, in the France of La Métherie, ideas about the formation of Earth and its landforms, minerals, and rocks were not just a debate between two dominant theories espoused by two dominant people. While it was still a question of fire or water, many more versions were included. In fact, Werner's was not among the theories discussed, though he is commonly (following Geikie) regarded as a key figure today.

Much of what La Métherie discussed is not germane to our interest in rock origin, being concerned with surface features. But basically, most of those theories considered rocks and minerals to be precipitated or physically deposited from the water that supposedly covered Earth. There were a number of suggestions about how the water could have diminished, and how inclined strata might have been formed. Of course, Hutton's reliance on heat to explain rock formation was noted. La Métherie felt that Hutton was able to support the igneous origin of lava and trap, but that the origin of the crystalline primitive rocks, including granite and gneiss, had been shown to be by aqueous precipitation (La Métherie, 1795, v. 3, p. 401–402). La Métherie questioned the source of internal heat sufficient to fuse all the claimed substances, but admitted that Hutton's question about how all the materials could be dissolved was difficult to answer.

As has been noted, Richard Kirwan was both a believer in the efficacy of the laboratory and an opponent of Hutton's theories. His laboratory experience with fusion temperatures and

knowledge of the ability to dissolve silicates led to disputes with Hutton's ideas. His reply to Hutton's first Theory, "On the supposed igneous origin of stony substances," is credited with galvanizing Hutton into his later, more extensive, theory. Kirwan's book of *Geological essays* of 1799 is perhaps not so well known as his earlier paper (1794b).[8] Kirwan showed himself to be well read, which again is in some contrast to Hutton. Even though Hutton was well informed, he generally cited fewer authors than did Kirwan, but included page after page verbatim from a relative few (Saussure prominently among them), and left some quotes anonymous (Hutton, 1795, v. 2, p. 389).[9] Much of Hutton's book consists of quotations from other authors, often in French. Many of those referred to by Kirwan, "Tilas, Gmelin, Cronsted, Ferber, Pallas, Charpentier, Born, Werner, Arduino, Deluc, Saussure, and Dolomieu" in one instance (Kirwan, [1799a] 1978, p. ix), can be recognized as active travelers, experimenters, and enthusiasts of instrument use. Kirwan's stated reason for writing his *Essays* was that he felt geology had not been noticed sufficiently, and thus felt it useful to present his findings and those of the authors he mentioned.

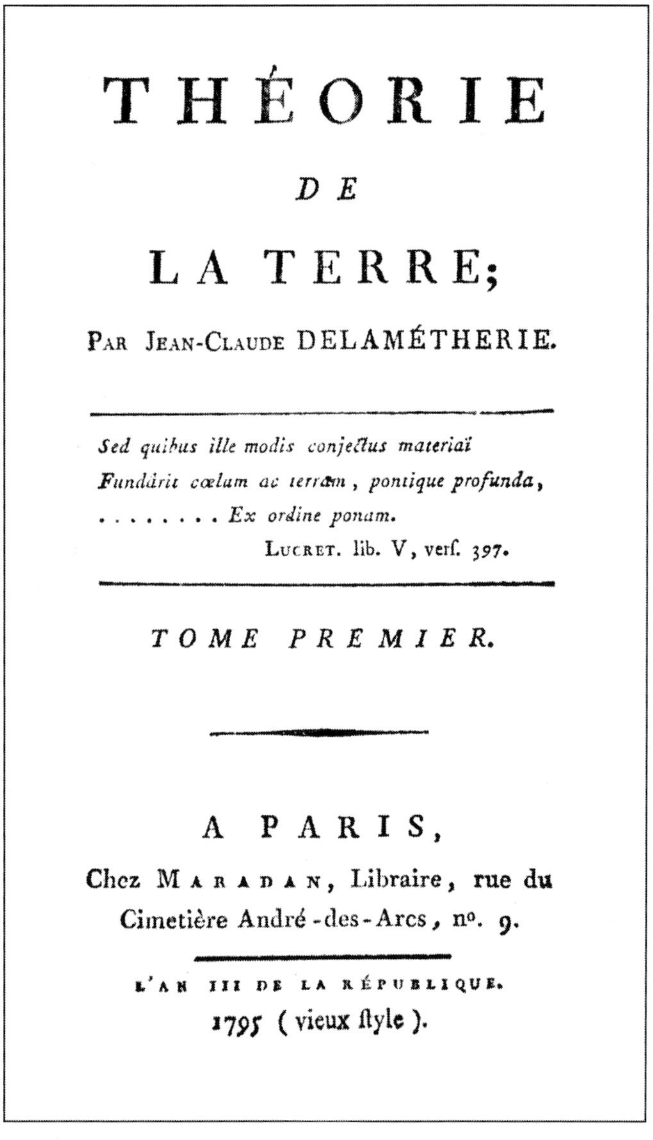

The *Essays* were different in plan from the theories of Earth previously mentioned. Like contemporary theorists, Kirwan spoke of the primitive state of the globe, the Deluge, and subsequent catastrophies. The last chapter consisted of his comments on Hutton's theory of Earth. But the bulk of the book consisted of clear descriptions of rocks and minerals and discussion of where they were to be found, as reported by numerous authors. Kirwan himself had done numerous analyses "in the humid way" for which he was often cited. In the *Essays*, he reviewed much of the work done on silica solution and deposition in the chapters "Of lapidification" (p. 107–142) and "Of the decomposition and disintegration of stony substances" (p. 143–153). He employed the Deluge to account for the occurrence in cold climates of the remains of animals known to live in warm regions, but also included notes on the appearance of minerals when fused. As in his earlier paper (1794b), he noted the lack of a heat source sufficient to fuse masses of rock and reminded readers of the role of solvents. He criticized Hutton, inquiring: "Does he think that from a view of Britain singly, a geology can be formed?" (Kirwan, [1799a] (1978), p. 482), although Hutton read and quoted (to excess) some work from the Continent. There was more in this vein, but by 1799, Hutton, two years in his grave, was unable to reply.

John Playfair's ([1802] 1964, p. 145) explication of Hutton's theory in 1802 was partly published to answer Kirwan. In terms of the behavior of matter, Playfair repeated observations on gran-

THÉORIE

DE

LA TERRE;

PAR JEAN-CLAUDE DELAMÉTHERIE.

Sed quibus ille modis conjectus materiaï
Fundârit cœlum ac terram, pontique profunda,
. . . . Ex ordine ponam.
LUCRET. lib. V, verf. 397.

TOME PREMIER.

A PARIS,

Chez MARADAN, Libraire, rue du
Cimetière André-des-Arcs, n°. 9.

L'AN III DE LA RÉPUBLIQUE.
1795 (vieux ftyle).

Figure 9.6. Title page of La Métherie (1795, v. 1).

---

[8]His *Essays* ([1799a] 1978) deserves to be better known for the excellence of its language, if not for its geological exactitude. One example, in discussing Hutton's words on quartz, is: "Hence the hypothesis grounded on the assumed identity of these substances and common glass, vanished like the unembodied visions of the night" (Kirwan, 1799a, p. viii).

[9]This comment is perhaps unjustified, as I have not done a quantitative study, which would be interesting. I was struck by the pages of verbatim quotes from the relatively few authors in Hutton's works, in contrast to Kirwan's crisp, brief, often multiple, citations backing specific facts.

ite, basalt, and limestone, but other than citing the experiments of Hall and Kennedy which supported Hutton's theory, I can find no mention of any new laboratory or quantitative evidence. Continuing the dialogue, it was again a Neptunist and follower of Werner, Robert Jameson, who built his case for the Wernerian system for the origin of rocks and Earth's form on the observed characteristics of minerals and rocks, many of which were repeated from other sources as was normal practice. This was included as the third volume of his *System of mineralogy*, published in 1808. The first two volumes consisted of detailed descriptions of minerals and rocks. The third volume discussed the occurrence and position of the various kinds of rocks, aided by Jameson's considerable travels in Britain and Europe, as well as by his more than a year of study with Werner. Jameson was well known for his experimental work. He found nothing to contradict his Wernerian interpretation of the way rocks were arranged or their sequence. He included three generations of granite, all primitive, the latter two crosscutting only other primitive rocks.[10] Cleaveland continued the discussion in 1816, and presented a summation of the arguments about basalt, and included origin by fusion, solution, and the possibility of both. Under "Geological systems" he presented both Wernerian and Huttonian views, and used basically Wernerian stratigraphy. When presenting Huttonian theory, he described it as "so ingeniously illustrated, but unsuccessfully supported, by Professor Playfair" (Cleaveland, [1816] 1978, p. 593).

Part of French naturalist Georges Cuvier's (1769–1832) larger work on evidence from fossils, particularly quadrupeds, was published in English, translated by Robert Kerr, and edited with notes by Robert Jameson. Rudwick (2005, p. 510) has pointed out that Jameson, by titling the translation as a theory of Earth, did not respect Cuvier's general suspicion of that genre. Cuvier's main arguments, impressive on close scrutiny, were from the fossil record, and thus we won't consider them at length here. With respect to the form of the surface of Earth, he quoted Burnet, Whiston, Woodward, and others. He noted the extravagant claims of many, but said:

Yet among those who have proceeded with more caution, and have not searched for geological causes beyond the established limits of physical and chemical science, there still remain much diversity and contradiction. (Cuvier, [1817] 1978, p. 45)

Cuvier said that many naturalists did not investigate facts before they assembled their systems, which he blamed on their either being "cabinet naturalists" who never looked outside or "mere mineralogists" who had not studied the variety and complexity of animals (Cuvier, [1817] 1978, p. 51–52). Cuvier praised the progress mineralogy had made under the influence of Saussure and Werner, and the connections made to the form of the layers of Earth. However, they had not investigated the "organized" fossils, which to Cuvier were what would tell of the revolutions, or

changes, in Earth. Jameson, who was a competent mineralogist, made few comments in his appendix to Cuvier ([1817] 1978) other than to expand the notes on where and how fossils were found. Jameson made no arguments there for rock origin based on mineral properties as demonstrated in his experimentation.[11]

Although much the same ground was covered, by the 1820s, books tended to be titled more along the lines of "Method of geology" or "An introduction to geology" than "Theory of the Earth." The word "geognosy" was still in use in its Wernerian meaning. By this time, there were copious, often quantitative, field observations.[12] The physical and chemical properties of many minerals and rocks were well known, and classification schemes for both were refined; investigation of temperatures continued, and mathematics was increasingly employed. The Deity was rarely invoked as the immediate cause of some geological action. Charles Lyell (1797–1875) realized that:

Many appearances, which for a long time were regarded as indicating mysterious and extraordinary agency, are finally recognized as the necessary result of the laws now governing the material world. … By degrees, many of the enigmas of the moral and physical world are explained, and, instead of being due to extrinsic and irregular causes, they are found to depend on fixed and invariable laws. (Lyell, [1830] 1990, p. 75, 76)

Lyell's opinion of former workers was not always kind, and he castigated those who had advocated explanations that were less than scientific, in his view.[13] Meanwhile, views about rock origin progressed, although the process was not smooth. Unlike the claims of some of his contemporaries, Hall's "proof" of basalt's igneous origin was not definitive, and both granite and limestone posed ongoing problems. Lyell felt that the alteration of existing rocks was more easily observed and documented than their origin, for alteration occurred at ambient temperatures, pressures, and oxygen concentrations that could easily be observed. By the middle of the nineteenth century, new methods of looking at rocks and minerals and new laboratory techniques made much of the earlier speculation no longer feasible.

GRANITE

Fusion

While the origins of basalt and limestone were disputed in the eighteenth century, and that of basalt engendered a long-standing controversy, the origin of granite proved to be the greatest conundrum of all, the source of controversy until at least the

---

[10]It has been pointed out that Jameson allowed his Neptunian views to skew his interpretation of field evidence on the island of Arran (Nicholas and Pearson, 2007, p. 34).

[11]Rudwick (2005, p. 510) commented on the 1813 edition. I have used the 1817 edition, which is substantially the same, differing in some page numbers. Rudwick noted that Jameson gave the work a Neptunian twist.

[12]There were many field observations in the third volume of Hutton's *Theory of the Earth*, published in 1899 (and reprinted in Dean, 1997), although they were not quantitative.

[13]Lyell described Kirwan as "a chemist and mineralogist of some merit, but who possessed much greater authority in the scientific world than he was entitled by his talents to enjoy" (Lyell, [1830] 1990, p. 68).

mid–twentieth century.[14] The very meaning of the word "granite" was unclear for centuries. Constants were its crystallinity and relatively coarse major mineral grains. The secondary minerals varied. Quartz and different feldspars were standard, but inclusion of accessory minerals such as various micas and small amounts of hornblende or other ferromagnesian minerals, and their amounts, as well as different ways to express mineral proportions, were the subject of lively debate throughout the nineteenth century and well into the twentieth.[15] The chemical similarity of gneiss and granite was noted, as were their spatial relations to each other.[16]

Arguments in the eighteenth century often focused on the crystal forms of quartz and feldspar, their perfection or lack of it indicating the crystallization sequence. Just as when a mixture of minerals is heated, the mineral with the lowest fusion point will liquefy first, so when a mixture of minerals is cooled, the mineral with the highest fusion point would be expected to crystallize out (solidify) first. In the eighteenth century, this fact had been observed with mixtures of salts and even with minerals. As always, we must be careful not to read our twenty-first-century ideas about complex solubilities and solid solutions into the logic of the eighteenth century. These were the precise uncertainties that led to the debate about granite origin throughout much of the twentieth century.

In 1748, John Hill (c. 1716–1775), the English botanist who expanded into mineralogy while writing his *Natural history*, published a description of granite that is easily recognizable today:

It is extreamly coarse, harsh and rough, of an uneven, but very firm and compact texture, considerable heavy, and of a mottled black and white colour: the greater quantity of its particles are of a pure marly white and opake, and these are generally form'd into the largest congeries, and emulate a coarse talcy or tabulated texture, among these there are a multitude perfectly transparent, and of a pure crystalline water and brightness; these are in general the smallest of all the particles, and among these are lodg'd in all directions very considerable quantities of differently colour'd, and constructed foliaceous Talcs; some of these are colourless and transparent, others of a fair white and semi-opake: others of the colour of brown Crystal, and a vast number perfectly black; these are not less different in their order and disposition, than in their colours; some of them intersect, and are divided by the other Granules; others lie parallel with, and some flatly and evenly upon them, and others are absolutely immers'd in them, like flies in amber. (Hill, 1748, v. 1, p. 498)

Hill's granite allowed water to get into its cracks and crevices, struck fire with steel, raised no effervescence with *aqua fortis*, and suffered little change in fire. Under the microscope, the rock showed clear crystals and "tabulated" crystals, some resembling those of spar or talc. Thus he pronounced his granite to be "a multitude of Mica, Crystal, dead Earth, and Talc." Evidentally, it was what we would call a "weathered" granite. The method of quarrying he discussed exploited the cracks in the rock by making a hole lined with clay into which water was poured. By the following day, a wedge could be driven in and large pieces broken off (Hill, 1748, v. 1, p. 498–499). This could be repeated.

On fusion characteristics, Hill had given this operational description of granite: "Stones compos'd of separate and very large concretions, rudely compacted together, giving fire with steel, not fermenting with acids, and slowly and imperfectly calcinable in a great fire" (Hill, 1748, v. 1, p. 498). Although the identification of some of the minerals in his examples raises questions, Hill gave a clear portrayal of granite's appearance, and noted that it showed little change in fire. He described several examples from several places and included reports on the microscopic examination of them.

The fusion points of feldspar, and feldspar combinations with other minerals, had been recorded. Let's see what the primary players themselves had to say about granite, using the known and empirically verified properties of the constituent minerals.[17] There were recurrent themes that drew on experiment: one was the fusibility characteristics of both quartz and granite rock; a second was alteration of fused components by the escape of volatiles; and a third was the solubility of the rock and the individual minerals that made it up. These arguments in various forms can be followed through the more than 50 years with which we are concerned. Some are part of the discussion of rock analysis in Chapter 7.

The behavior of quartz proved to be the stumbling block to finding an easy route to granite origin in terms of the two theories of crystallization—from solution or from cooling of fused melts. Silica was ubiquitous in the mineral world, as well-formed quartz crystals, amorphous quartz, pebbles and grains of sand, and flint. Siliceous earth was recognized as a component of many rocks and minerals, and was often called "vitrifiable earth." Its infusibility per se was well recognized.[18] Fusion experiments had demonstrated, as Pott said, that all vitrifiable earths and stones required the addition of alkaline substances in order to fuse them (Pott, [1746] 1753, p. 150). He noted that it was common knowledge among potters that a mixture of argillaceous earth (clay) with vitrifiable earth could be used to form vessels of a number of sorts (Pott, [1746] 1753, p. 139). In his experiments, he identified

---

[14]For a detailed presentation of the experimental basis of granite behavior in the eighteenth century and of the properties of the constituent minerals, see Newcomb (1990a). For an excellent discussion of the experimental work done on granite through the twentieth century, see Pitcher (1993), and for a thorough historical review, see Chapter 9, "Rocks and their formation," in Oldroyd (1996).

[15]Young noted that the "granite controversy" of the 1940s and 1950s was exacerbated precisely because of a lack of consensus on the meaning of the term "granite" (Young, 2003, p. 389).

[16]Hall stated, "Gneiss is found to pass by insensible degrees into granite; that is, specimens of every conceivable intermediate step have been found" (Hall, 1815a, p. 95).

[17]I was tempted to use a quote from Henkel (aka Henckel) in this section. It seemed to fit when he inquired, "Kiesel-stein, von dem möchte man wohl fragen, wo ist deine Mutter?" (Kieselstein … where is its mother?) (Henkel, 1744, p. 340). However, despite one translation of Kieselstein as being silica, or flint, it has a more common interpretation as a pebble, and the inquiry was more concerned with its placement than with its chemical origin.

[18]Calling a substance both vitrifiable and unfusible seems to be a contradiction. When silica does fuse in company with an alkali, it forms a glass. Pott noted the same contradiction when he said that in calling a substance "vitrifiable" he did not claim that it was fusible in fire by itself. On the contrary, when it was pure it could never be fused by itself, even in the most violent fire. But he called it "vitrifiable" because it could be used to make glass (Pott, [1746] 1753, p. 143–144).

siliceous earth as striking fire with steel, which clay would hardly do. His many experiments demonstrated the difficulty of fusing silica with even the strongest furnaces or burning glasses.

A few years later, Darcet recorded his fusion experiments with granite in his furnace, which he considered to be better than Pott's. A granite from Burgundy flowed and swelled a great deal in the heated crucible. It went somewhat past the stage of a frit,[19] and became mottled with areas of a brown melt. Another granite flowed under fusion after being powdered, but when larger pieces were heated they did not fuse, and the large grains of quartz didn't lose their transparency (Darcet, 1766, p. 66).

Saussure recounted Darcet's trials and pointed out that granite and basalt are different. He had read Wallerius and others on the properties of quartz and feldspar, and had attempted fusing quartz with an efficient furnace modeled on those of Antoine Baumé (1728–1804).[20] Saussure fused several isolated crystals of feldspars to a transparent glass, and he was familiar with the individual mineral properties. He stated that his fusion trials with granite had convinced him that it could not be reduced to homogeneous matter, even in a violent fire (Saussure, 1779, v. 1, p. 123). Not caring for the equivocal experiments others had done, he resolved to try his own. He chose a granite with well-characterized quartz, feldspar, and mica, describing each. After the rock was powdered and exposed to a strong fire, it was reduced to a gray-green glass that contained very small blebs, which under the magnifying glass proved to be quartz (Saussure, 1779, v. 1, p. 124). He then heated fragments of the granite, instead of powdering it. He broke the product that filled the base of the crucible, and could see the components: The mica had formed a brown and green glass; the feldspar produced a colorless transparent glass; and the quartz was preserved as a white mat, having lost its transparency (Saussure, 1779, v. 1, p. 125). This was quite unlike the behavior of basalt. He found similar results with other samples of different granites. Even a sample of what he called a granite, which consisted of a little quartz with schorl (probably tourmaline) that appeared to melt into a black glass, still contained the small pieces of quartz after fusion. His conclusion was that no rock of the granite class could give rise to basalt or homogeneous lava by melting (Saussure, 1779, v. 1, p. 127). Later, both Drée (1808) and Breislak (1812) approvingly cited Saussure's granite experiments.

Bayen (1779) included granite among the rocks he investigated chemically. He began with a complete description of samples from three different places. He treated each with vitriolic acid, and analyzed the efflorescences that formed. He also noted that each gave off a small amount of water when heated in a closed vessel. Bayen considered his work as extending the work of Pott, whom he admired. Close reading of his rock descriptions points out again how varied were the samples considered to be granite, and the number of accessory minerals they contained.

In these last decades of the eighteenth century, the fusion temperatures of minerals were recorded in many mineralogical studies, and after the 1780s, were given on the Wedgwood scale. It seemed logical that the mineral (quartz) that had the highest fusion temperature among the granite minerals should be the first to crystallize on cooling and thus have a perfect crystal form, itself well described. Feldspar, being next in the sequence, would be expected to have its shape somewhat altered by the physical presence of quartz. But this is often not the case with granite. To put it simply, feldspar crystals are often well formed, while the quartz rarely is.[21] This was the crux of many arguments in the fusion versus solution debate, in the course of which the two sides invoked various solution arguments, both understanding that it was also possible for one mineral to dissolve in the other. In describing the Portsoy granite (a "graphic" granite), Hutton wrote:

> It is evident from the inspection of this fossil [quartz] that the sparry and siliceous substances had been mixed together in a fluid state; and that the crystallization of the sparry substance, which is rhombic, had determined the regular structure of the quartz, at least in some directions.
>
> Thus, the siliceous substance is to be considered as included in the spar, and as figured, according to the laws of crystallization proper to the sparry ground; but the spar is also to be found included in the quartz. (Hutton, [1795] 1972, v. 1, p. 106–107)

As one can see, Hutton's reasoning did not and does not settle the question. In a sequence that might take an entire paper to summarize, many mineralogists discussed the question and presented arguments based on their observations of crystallization from both melts and solution. There were disagreements about both processes and results, which didn't interfere with the claims of a supposed "proof" of the true state of affairs.

Hutton made his arguments for granite origin from fusion on the basis of the crystalline structure of Portsoy granite, but made a scientifically suspect induction on the basis of a single hand specimen:

> We shall, therefore, only now consider one particular species of granite; and if this shall appear to have been in a fluid state of fusion, we may be allowed to extend this property to all the kind. (Hutton, [1788] 1973, p. 255; [1795] 1972, v. 1, p. 104)

He basically made his arguments on texture, not on the fusion points of the constituent minerals. Hall attempted to fuse granite and reproduce its crystalline structure without success, and in deference to Hutton, he did not continue fusion efforts until after Hutton's death (Newcomb, 1990a, p. 171–172). Thereafter, he confined his efforts to whinstone, lava, and marble. In the brief report of the earlier work, for which he did not supply a full paper, Hall made the analogy that the feldspar behaved like a solute to the quartz, as water did to salt. When a solution of salt and water is cooled, the lower melting point water excludes the

---

[19]A frit is a partially fused material.

[20]Saussure said that only furnaces made with mirrors or lenses of more than three or four feet in diameter would be more powerful than those of Baumé (Saussure, 1779, v. 1, p. 46).

[21]To enter eighteenth-century thinking about granite it is necessary to repress our present knowledge of Bowen's reaction series.

salt when it freezes, forming crystals of ice. So might feldspar exclude the quartz on cooling, which would account for the better feldspar crystals and the more irregularly shaped quartz (Hall, 1794, p. 10). Granite continued to be a riddle.

The English chemist and physician Thomas Beddoes (1760–1808), lecturer at Oxford for a short time, was sure, as evidenced by numerous field observations and by citations to other authors, that basalt and granite were both products of fusion, and could be seen grading into each other (Beddoes, 1791).[22] Beddoes referred to experimental work reported in 1787 by Karl Haidinger (1756–1797), the assistant at the Vienna cabinet of natural curiosities, who said granite formed a dark glass in the fire. The arguments about granite continued with each author "answering" the objections of those who espoused the other system. No matter what variations were made in attempts to fuse granite, the results were similar and unsatisfactory. In the absence of a strong flux, the quartz remained visible, and failed to melt completely. From our modern standpoint, we can also see that pegmatite veins with their very large crystals added a layer of confusion if it was presumed they formed from a melt, because the laboratory melts had not produced large crystals. Hall quickly realized that granite fusion provided no answers, and when he resumed fusion experiments after Hutton's death, he employed whinstone and lava. He, like Hutton, used solution (of quartz in feldspar) arguments to conjecture how the higher melting point quartz could be molded around the better-crystallized feldspar.[23]

Spallanzani included granite in his extensive experimental investigations. The granites used in public buildings in Milan, Pavia, and other towns of what was then Austrian Lombardy had both white and "deep flesh colored" feldspars (Spallanzani, 1798, v. 2, p. 156). Heating resulted in fused mica and feldspars that began fusion with many microscopic bubbles, but which didn't flux the quartz. The quartz became whiter, but didn't change other characteristics. He described a number of other granites of the Italian peninsula, which he also attempted to fuse, with generally the same results. Something he called a "granitous porphyry" imperfectly fused into "an ebullient scoriaceous enamel." Another granite contained "sulphur of iron, and red sulphurated oxyde of mercury (cinnabar)" (Spallanzani, 1798, v. 2, p. 160) and could be fused to produce mercury. After being in the furnace, the feldspar and included steatite blended into a porous scoria, but the quartz grains were not fused (Spallanzani, 1798, v. 2, p. 163). The quartz did not fuse in any of his attempts. Spallanzani used Wedgwood pyrometer pieces to gauge the heat of the furnaces, which he determined to be just below the fusion temperature of iron.

To understand the nature of granite, Kirwan drew on his extensive laboratory observations, and those of others.[24] In the 1790s, he reviewed the experimental evidence and found no way to produce the crystalline structure of granite under any conditions of fusion. To behave as a solvent (menstruum), and similarly to salt in water, the amount of feldspar should be greater than that of quartz, but this was not the case with some granites. He felt that if granite were truly fused and allowed to cool slowly, as Hall wished, the quartz, mica, and feldspar should crystallize in sequence and thus be present in layers; however, they are not found thus in the natural rock.[25] Also, Plutonic theory provided no explanation for what Kirwan and others had shown—that water is present in the minerals that make up granite. It should have long since evaporated if produced from the cooling of a melt. Quartz required so much heat to melt it that Kirwan could not conceive of its being done by natural (volcanic) agencies.

Kirwan (1802a, p. 6) again reviewed the efforts by several people to fuse pure quartz. Having performed his own fusion experiments, reported on in the several editions of his *Elements of mineralogy*, most at temperatures not exceeding 166 °WW, he felt that there was never enough feldspar to serve well as a flux to the quartz.[26] In 1784, he had cited the differences of opinion about the efficacy of calcareous or other earths as a flux for quartz (Kirwan, 1784, p. 9). Against the argument for slow cooling to account for granite texture, as Hall had drawn the analogy for basalt from the reports on cooling glass, Kirwan pointed out the difference between cooling glass and granite. Unlike granite, the glass lost volatiles in the cooling process when being manufactured. Slow cooling of granite should result in perfect quartz crystals forming first because of its high melting point, but most granites don't have well-crystallized quartz. Kirwan disputed the analogy made by Hutton and Hall between salt and water in accounting for the crystallization of feldspar rather than quartz, and also the analogy from slow cooling of glass drawn by Hall. Kirwan stated that the affinity between silica and alkali particles is increased with heat, supposedly demonstrated by the fusion. When cooled quickly, there was no chance for the silica particles to regain their affinity for each other because

the siliceous particles cannot move through it to reunite to each other, though their affinity to each other in a *low* temperature be greater than their affinity to an alkali, and thus they continue in that state which we call a *glass*. (Kirwan, 1802a, p. 8)

---

[22]Beddoes relied little on chemistry and said:

Should the matter of any given rock contain too little iron to be fusible by the blowpipe, and yet have other striking features of whinstone, would this be reason to conclude, that its formation has been different? Chemistry, if thus strictly followed, would perplex mineralogy instead of reducing it to order. Characters of minerals, purely chemical, would separate those whose natural history is alike, and bring together such as differ widely in their formation. (Beddoes, 1791, p. 53)

Dr. Beddoes has made a valid point—that is, of course, only part of the story. Chemistry is most useful.

[23]These arguments are followed in detail in Newcomb (1990a).

[24]His comments were generally plausible except in the dubious report of granite having formed "in the moist way" in the case of a mole in the river Oder filled with granite sand that became solid (Kirwan, 1794b, p. 81). The mole did begin with granite composition and one could suspect another concreting agent.

[25]Gravity settling was later noticed but was not so called.

[26]As stated in Chapter 3, Smith (1969, p. 334) views the Wedgwood scale as "horribly nonlinear." According to Wedgwood's own scale, 166 degrees Wedgwood would equal about 12,000 degrees Celsius (Wedgwood, [1784] 1809, p. 578; Newcomb, 1990a, p. 212) by converting Wedgwood's Fahrenheit to Celsius. Wedgwood used a conversion factor of 130 °F/1 °WW. However, the relation is not linear and is unlikely to be so neat.

In slow cooling, as in the slowly cooled glass that Hall had observed, there might be time for the affinity to be reestablished, and that could account for the observed crystallization. In another experiment, Kirwan noted that if a mixture of salt and water is cooled very quickly, only a congealed mass is formed, with no salt crystals evident (Kirwan, 1802a, p. 8).

Two books that followed the same arguments were also published in 1802, that of the mathematician Playfair, *Illustrations of the Huttonian theory of the Earth*, and that of the chemist and physicist Murray, *A comparative view of the Huttonian and Neptunian systems of geology: In answer to the Illustrations of the Huttonian theory of the Earth, by Professor Playfair*. It is interesting to see how two such erudite scholars could use the same data and observations and come to opposite conclusions. Playfair's book is the better known of the two, perhaps because he was arguing on the "correct" side, and thus historians of geology have cited it more often.[27] Playfair referred to Hutton on Portsoy granite texture:

Now, Dr. Hutton argued, that substances precipitated from a solution and crystallizing at liberty, cannot be supposed to impress one another in the manner here exemplified; and that they could do so only when the whole mass acquired solidity at the same time, or at the same time nearly … [page reference to Hutton's *Theory* here]. Such simultaneous consolidation can be produced in no way that we know of, but by the cooling of a mass that has been in fusion. (Playfair, [1802] 1964, p. 321)

In the following pages, Playfair expanded on this theme.

Murray agreed that the crystallization must have been "nearly simultaneous" (Murray, [1802] 1978, p. 13). He also agreed with Hutton that the texture of granite must result from the rock being fluid, but did not agree that if it were in solution, then all the minerals would have perfect crystal faces. Murray had previously discussed how simultaneous crystallization could occur from solution, and then said:

[T]he structure of granite can, from this circumstance, furnish no argument against its aqueous origin. But it affords the clearest demonstration, that it has not been formed by fusion. Felspar [*sic*] is a substance incomparably more fusible than quartz, the one varying from 120 to 150, the other being 4043 of Wedgewoods [*sic*] scale. (Murray, [1802] 1978, p. 241)

Clearly, with that objective evidence, quartz could not be fluid with feldspar crystallized, as the quartz molded on feldspar crystals in the Portsoy granite suggested. Murray came down as solidly on the side of the Neptunists as did Playfair for the Plutonists.

That evidence and those positions were reiterated over the next decades. Unlike basalt, whose connection to volcanoes could be observed or inferred—not to mention the "proof," debated as it was, of basalt's crystallization after slow cooling—granite experimentation didn't provide even clouded answers. Evidence of texture and crystallization sequence was equivocal. That

granite masses could be found at other than the base of the stratigraphic column was well documented, and more and more mapping demonstrated it. Nomenclature remained problematical. For example, Pinkerton, while discussing granite, quoted Dolomieu at length on the differences and similarities of granite and porphyry (Pinkerton, 1811, v. 2, p. 179–188). He included mention of Dolomieu's trials of porphyry with a blowpipe, and the contrast with granite under the same conditions. His arguments, however, included rocks that we would by no means categorize as granites now. The tendency of feldspar to weather into clay was noted, with the concomitant differences of hardness in the rock (Kidd, 1815, p. 67). The relation of granite to surrounding rocks was not clear, whether those rocks were chemically similar, as gneiss was, or whether they had an assemblage of minerals we now recognize as unrelated.

Breislak reported not only Saussure's work in the attempted fusion of granite, but also that of Dolomieu's brother-in-law, Étienne de Drée, who had repeated Saussure's fusion experiments "from a different point of view" in 1808 (Breislak, 1818, v. 1, p. 347). Breislak noted that when granite is fused, the parts don't mingle. Granite and porphyry could be fused, but didn't unite into a homogeneous mass. Breislak referred to Cordier's work with lava to insist that minerals melted together could recrystallize as separate minerals, although in that particular case, the resulting minerals were in very small grains and needed magnification to be seen. He followed other lava-related examples to emphasize that substances could be in fusion and yet crystallize separately. He found it necessary to refute the idea that fused minerals could only produce a glass. That might happen with small amounts of minerals, but Breislak noted that sheer mass might make a difference. We will return to Breislak's summary on fusion, and the work of Drée in later chapters.

Drée's work on granite was part of his extensive investigation of rock fusion, most being rocks he thought could be those that would be melted into various volcanics. He was interested in the amount of heat required for fusion, and, having read Hall's two papers on fusion in whinstone and lava and on heat modified by pressure, embarked on an extensive series of experiments in which he enclosed his samples so that pressure would be applied, no component would escape, and there would be no possibility of foreign matter reacting with the samples. His samples 19 and 39 (of 39) were granite, although he called no. 39 "amphibolitic." With both samples, he placed pieces of the rock in a crucible, filled all spaces with the same pulverized rock, and then covered them with mica and quartz that he thought would not melt at the temperatures he was using. He said little about no. 39, other than that the crucible was recovered intact. Having been heated for six-and-a-half hours, wherein the pyrometer registered 112–133 °WW, the crucible for no. 19 was broken; there was an effusion of glassy material, "mais sans que le feldspath et le mica de la roche aient quitté leur disposition granitique" (Drée, 1808, p. 46).[28]

[27]Both are available in twentieth-century reprints, which are mentioned in the reference list.

[28]"But without feldspar and mica the rocks wouldn't have their granitic arrangement" (my translation).

Solution

The solubility of quartz was another conundrum for those thinking about rock origin, as was the solubility of other minerals. Quartz was the subject of many investigations, because it was a component of granite, and also because people recognized that, as silica, it was a component in many other minerals, and had been found in mineral waters. Not surprisingly, solubility was rarely among the "characters" tested for mineral identification, because, with the exception of salts, most minerals are more or less insoluble, so it is not usually a property of interest. However, although minor components were not always detected, mineral and rock analysis that included detection of silica, as discussed in Chapter 7, proceeded and improved through the eighteenth century and into the nineteenth. As time went on, analytical procedures were refined, and it was common for specific analyses to be checked and rechecked by other workers.

As recounted earlier, the solubility of quartz via a series of reactions was demonstrated without doubt. Most of the people cited recognized, and stated, that there might be conditions of temperature or pressure, or so-far-unknown substances that could alter solution or fusion properties. Kirwan remarked on the discovery of what he called "sparry acid," or "hydrofluoric acid" (HF) as we would call it, by the Swedish chemist Carl Wilhelm Scheele (1742–1786).[29] This hitherto-unknown reagent could dissolve silica, and led Kirwan to suggest that there might be further unknown agents that could aid the solubility of minerals (Kirwan, 1794b, p. 63; Newcomb, 1990a, p. 173). In order for substances to crystallize, or achieve regular form, Kirwan said they "must be minutely divided, have liberty of motion, be placed at a due distance from each other, and be undisturbed by a force superior to that of their mutual attraction" (Kirwan, [1799a] 1978, p. 109).

Because crystals could be "minutely divided" by either fusion or solution, crystallization could occur by both means. However, far more perfect crystals resulted from sublimation (which would not occur easily with silica) and solution than from fusion, because in the latter case, the constituent parts were much closer together and would interfere with each other (Kirwan, [1799a] 1978, p. 110). Kirwan cited a number of instances wherein others had found silica in solution.

With this demonstrated ability to manipulate silica, as detailed in Chapter 7, it is not surprising that deposition from solution remained a viable option for the origin of granite well into the nineteenth century, given the inability to fuse quartz without temperatures that seemed impossible to reach, much less sustain, in normal terrestrial operations. The properties of silica were investigated throughout the nineteenth century. In the mineralogy books that appeared with increasing frequency as the nineteenth century progressed, the same arguments about fusion versus solution for granite origin kept being repeated. Pinkerton

stated that silica alone "is scarcely fusible; but when newly precipitated, [it] is soluble in 1000 parts of water" (Pinkerton, 1811, vol. 1, p. 144). In an earlier publication (1990a), I included fairly extensive tables that detail the fusion characteristics and solubilities, where available, of granite, its major constituents, and some accessory minerals as reported by a number of mineralogists and geologists from 1784 until 1816. There was essential agreement between the values, so laboratory investigation must have seemed a worthwhile activity.

LIMESTONE AND MARBLE

Although the conundrum of limestone and marble origin was more easily solved in the long run than that of granite, it did not lend itself conveniently to Plutonist theory. The properties of limestone and marble were well known because of their use as building stones and for sculpture, as well as in metal smelting. The use of lime, or calcined limestone, in agriculture and in making mortar went back to prehistoric time. Biringuccio spoke of limestone and its properties and uses with easy familiarity. The practical Henckel described it well, including its field occurrences. Pott had noted the various ways in which limestone and marble, as represented in his chapter on alkaline earths and stones, behaved in fusion with additions (Pott, [1746] 1753, p. 11–16).

The difficulties of the origin of limestone by fusion were discussed in Chapter 5. An obvious difficulty for Huttonians was that limestone was easily soluble in acid. But there were other serious problems. Animal remains (fossils) were frequently found in limestone. If one admitted suggested Huttonian temperatures and pressures, the fossils should have been destroyed. In modern notation, the chemistry in one of Hutton's cycles is:

$$CaCO_3 \text{ (limestone)} \xrightarrow{\text{heat}} CaO \text{ (quicklime)} + CO_2 \text{ (fixed air)}$$

Quicklime can then react with water:

$$CaO + H_2O \rightarrow Ca(OH)_2 \text{ (calcium hydroxide, limewater)} \\ \text{in solution}$$

And then, when exposed to the carbon dioxide in air, the following reaction occurs:

$$Ca(OH)_2 + CO_2 \rightarrow CaCO_3 + H_2O$$

The carbon dioxide normally present in air is present in water to a small extent, thus making surface water slightly acidic (a weak acid, H_2CO_3). Kidd noted that dissolved carbon dioxide (his carbonic acid) was the acid agent that enabled water to dissolve limestone (Kidd, 1815, p. 238), which accounted for the dissolution of limestone over the years. By doing tests, he also found that pure water would dissolve the carbonate, although we would say it is "sparingly soluble." Kidd agitated half an ounce of distilled

---

[29]Scheele found impure hydrofluoric acid in 1771, and a "fairly pure" version in 1786 (Partington, 1989, p. 151).

water with a few grains of pulverized limestone, and found that the solution would turn turbid if treated with oxalate of ammonia, calcium oxalate being almost insoluble. Natural systems display complex carbonate chemistry, the result of the interaction of the minerals dissolved in the water from the rocks through which the water flows, and the gases in the atmosphere.[30] In general, limestone and marble, both of which are calcium carbonate, were of concern for these experimenters. The concept of metamorphic rocks was some distance in the future. Sometimes rocks that were not calcium carbonate were referred to as marble, which problem persists until today with some ornamental stones. However, the chemical and physical properties allow us to distinguish which rocks are calcium carbonate.

In 1766, Darcet clearly described the properties of limestone: its conversion to lime; its solubility in acid with effervescence; and its employment as a flux in metallurgy, although it was infusible by itself. He attempted to melt various kinds of calcareous stones and saw a few fuse right at the edges of the crucible in particularly hot fires (Darcet, 1766, p. 21). He summed his findings by saying that only chalk, lime, and *spath calcaire* (calcite, quite pure calcium carbonate) were truly infusible. The other forms, less pure, could usually be at least partially fused (Darcet, 1766, p. 26). He attempted fusion of more examples of calcite, reported in 1771, and mentioned that many were completely soluble and effervesced more or less in acid. From the frequency of notes about fusion occurring only slightly at the edges where the sample touched the crucible, one may suspect that the clay of the crucible was acting as a flux.

In the third volume of his *Voyages* (1796), Saussure published an interesting observation. He discussed an example of beds of rock that he saw in his travels from Aix to Avignon and described a calcareous rock that effervesced in acid and had the expected properties of hardness, etc. A second rock he called just *pierre brune*, or brown rock, and thought it was the same that he had earlier identified as *silicicalce*. He again noted its properties, but stated that although it effervesced with acid it left many small blebs of a solid behind. He described those separately in some detail, and was sure that they were silica. From this occurrence, Saussure became convinced that this particular rock could not have resulted from fusion, because if it had, the fusing of those two components would have resulted in a homogeneous rock, because the calcareous material would serve as a flux for the silica. This reasoning is interesting, and was the result of Saussure's careful application of external character determination, the blowpipe, and chemical tests (Saussure, 1796, v. 3, p. 340–342).

Much of the experimental work that followed on fusion without increased pressure was repetitive. When investigating calcareous substances, the variables were the possible impurities and the material the crucible was made of, which led to a relatively small range of new results. As the latter decades of the

eighteenth century progressed, the same authors were quoted frequently, with consensus that pure calcium carbonate could not be fused.[31] As finally and fully reported in 1812, Hall's innovative calcium carbonate experiments were received in two ways: Huttonians thought they found proof of their theory; Wernerians did not. However, there was another strand in this web of ideas about limestone origin, again related to chemistry. Invoking pressure as a means of changing chalk to hard marble was not difficult conceptually (Hutton, [1795] 1972, v. 1, p. 112), but Hutton and his supporters supposed that all limestone and marble were formed from the remains of marine organisms. Hutton spent time verifying the idea that the remains of "organized bodies" could be found in limestone and some marble and that there was some specific explanation for when it was lacking. However, that lack in some instances seemed to leave open the possibility that the rocks' origin was crystallization from aqueous solution.[32]

As with granite, although it was not inconclusive for so long, the two sides had a running argument invoking field observations as well as laboratory evidence. Drée attempted the fusion of various rocks under pressure utilizing double crucibles to contain gases and increase pressure, which experiments will be discussed in the next chapter. However, carbonates were minor among his samples, and his conclusions not definitive (Drée, 1808). I have found no one else who attempted to duplicate Hall's experiments at that time, although certainly pressure work in other applications was carried on.[33] Friedrich Ludwig Ehrmann (1741–1800) did extensive work on fusion, including with carbonates, although to my knowledge he didn't attempt to raise pressures. But the pressure argument, tied to chemistry, continued. Hutton's assumption that limestone was composed of marine organisms was important at this time. Lavoisier had investigated the properties of phosphoric acid, and the indefatigable chemists of the era continued the work.[34] Part of Kirwan's response to Hutton was:

To these considerations [about fusion] I shall add a demonstration that pure limestone, such as Carrara marble, did not originate from shells; it is this, phosphoric acid is found in most shells, but none is found in pure limestone ... and its absence cannot be attributed to fusion, for phosphoric acid is indestructible by heat. (Kirwan, [1799a] 1978, p. 458)

The increasingly crystalline nature of shells subjected to the presumed pressure was noted. According to Playfair, the shells that had been the least mineralized would retain the most phosphoric acid. He faulted Kirwan for the above comment, thinking it was first necessary to see whether phosphoric acid had

---

[30]I have not discussed the deposition of limestone in the natural world, which subject is worthy of three or four volumes of dense prose. Suffice it to say that it may be virtually all a chemical reaction, it may be biologically aided, or it can be the physical deposition of the products of either method.

[31]The first clearly demonstrated fusion of calcium carbonate was reported in 1912, at 1289 °C and 110 atm, by H.E. Boeke (Eyles, 1961, p. 216).

[32]In following this reasoning today, we must yet again suspend our later knowledge.

[33]Although Hall's work was referred to frequently by British, French, Italian, and German sources of the time, Drée's is the only attempt at replication that I've found. In a personal communication, Bernhard Fritscher (2004) stated that as far as he knew there were no such experiments done in Germany before those of Gustav Rose (1798–1873) in the 1860s.

[34]Phosphorus had been discovered in 1675 (Brock, 1993, p. 85) or 1669 (Ihde, [1964] 1984, p. 747), and Lavoisier used phosphoric acid in his studies of air and other gases.

been present in those limestones to begin with, and that "nature has some process" that might remove it (Playfair, [1802] 1964, p. 195). Murray weighed in on the phosphoric acid question with the observation that Kirwan was correct about the resistance of phosphoric acid to heat. Thus if marble were made from the remains of marine animals, then the phosphoric acid should be there if they were formed by fusion. But if they were formed by deposition from solution, the absence would not be remarkable, since the phosphates in the shells are soluble (Murray, [1802] 1978, p. 202–203).[35]

Deluc criticized Hall's use of pressure in the experiments, saying that the pressures he could achieve were not of the same order of magnitude as those envisioned in Hutton's theory, and that water vapor pressure could not supply sufficient pressure. He cited experiments by himself, Lavoisier, and Pierre Simon Laplace (1749–1827) on the effect of water vapor on a column of mercury (Deluc, 1809, p. 363). He agreed that fixed air (carbon dioxide) would behave as Hall intended, because it was confined as in his fusion experiments. But he disagreed with the contention that it could be confined under the pressure of a column of water. The gas should rise rapidly due to its very low specific gravity, and it was also soluble in water. Therefore it could not provide the pressure claimed for it (Deluc, 1809, p. 366). However, Hall argued that his experimentation using water vapor for pressure inside his tubes had shown this method could provide the necessary pressure if a shell on the bottom of the sea encountered a hot lava flow, so that the shell could be calcined without losing its fixed air (Hall, 1815a, p. 90).

The two possible origins for limestone and marble were mentioned by John Kidd (1775–1851), who reaffirmed that limestone and marble were the same chemically (Kidd, 1809, p. 3). The formation of calcareous solids by evaporation from springs was a given. However, Kidd considered Hall's experiments to be conclusive in demonstrating that the crystalline nature (sometimes) of limestone and marble could also be the result of heating under great pressure. He noted that the presence of iron could alter their appearance (Kidd, 1809). In 1815, Kidd reiterated his observations about the solubility of carbon dioxide and carbonates, and used the turbidity test to illustrate the presence of dissolved calcium (Kidd, 1815, p. 238). He also noted two instances

---

[35]See also Newcomb (1990a).

of finding well-formed crystals of the carbonate deposited from solution: one was in a bottle of limewater; the second was in a tub of spring water that had been allowed to remain undisturbed for six or eight months (Kidd, 1815, p. 240). He continued by describing a number of instances in the natural world where calcium carbonate could clearly be said to have been deposited from solution. Elsewhere I have provided tables that summarize the findings about calcium carbonate in solution and fusion (Newcomb, 1990a, p. 188–189).

CONCLUSIONS

It is evident that knowledge of the chemical and physical properties of rocks and minerals was put to use in conjectures about rock origin, and all available equipment, increasingly sophisticated and standardized, was employed. By the beginning of the nineteenth century, there was better consensus about naming mineral varieties, and their characters were tested according to standardized methods. There were discussions about mineral content of rocks and their naming. Analysis of minerals and rocks improved, as did the equipment for analyses. New elements and minerals were discovered, and their properties delineated, while there was increasing agreement about crystal shapes and how they should be characterized. Many questions were taken to the field in an effort to corroborate modes of origin. Although basalt was the subject of the best known controversy, at least from the vantage point of our imperfect hindsight, the major theories involving fusion or solution were clearly insufficient for granite and limestone or marble as well.

Despite the claims for Sir James Hall's excellent work, those who thought he had proved limestone's origin by fusion were actually in the minority. As late as 1833, in the second American edition of the fourth London edition of Bakewell's *Introduction to geology*, many of the earlier arguments about rock origin continued, although new procedures had been developed to aid in finding the truth. Both procedures and equipment were tested for accuracy and reproducibility. While rock descriptions and their mineral contents were becoming more standardized, one can still read confusion between the lines, as in the difference between different kinds of rocks and what their definitions should include. The laboratory continued to pose more questions than it answered, but at the same time provided an anchor to reality.

CHAPTER 10

Rock origin: The investigations—Basalt and the central fire

INTRODUCTION

A great deal has been written about the basalt controversy, which was generally considered to be active from ~1770 to 1820. But as soon as such dates are set, exceptions are noted. In fact, the origin of basalt, and the related question of the origin of volcanic heat and action, had been of interest since antiquity, and the interest continued into the nineteenth and twentieth centuries. There is the same inconsistency with regard to dates in accounts of the defeat of the theory that basalt is of aqueous origin. It was not neatly and conclusively defeated in 1795, 1802, or 1816. Indeed, the basalt controversy has probably been given more importance in the history of geology than is warranted. The two possible modes of origin of basalt, through the agency of fire or water, were the same two agents that had been invoked for the origin of earth materials and landforms since the beginning of recorded history. The debate intensified in the eighteenth century, but if one pays less attention to theorists, and more to the natural philosophers who went into the field and back to the laboratory, making and checking observations as they went, their work shows a continuum that extended from well before to well after the 30 or 50 years of the controversy. Many of those workers were inspired not so much by the controversy as by a desire to answer much more than a single question about the origin of one sort of rock. Some of them related their findings to larger theoretical questions, and wrote or contributed to theories of Earth. Some did not. Entwined with the question of rock origins in general, and basalt in particular, was that of the source for volcanic heat.

The choice of either water or fire of a magnitude to render Earth liquid was necessitated from the earliest time when astronomers found Earth to be spherical.[1] That shape could best be explained by a prior liquidity. Thus, the earliest conjectures were about Earth as a whole, and not a single rock type.[2] The shape of Earth, and the association of heat with the depths of Earth as evinced by volcanoes, has been a constant throughout recorded history, delineated in images as well as in the written word. It is worth noting here that many observers held the view that volcanoes were a local phenomenon due to local conditions such as water infiltration. As described by experimentalists in the eighteenth century, the rocks associated with active volcanoes—lava, pumice, scoria, obsidian—were obvious. However, basalt, including columnar basalt, as well as trap, whinstone, and toadstone were often not associated with observable active centers of volcanism. The story of the gradual recognition of extinct volcanoes and their associated rocks in the latter part of the eighteenth century has often been told. Nonetheless, other than showing that volcanic heat could and did melt rocks, the ultimate origin of basalt was unclear. Many astute observers considered basalt and some other volcanic productions to be preexisting rocks simply altered by volcanic heat. A layer of rock precipitated from some solution that lay near a source of heat might be made to flow and resolidify, and thus basalt might originally have been deposited from solution.

The arguments about how much heat was necessary to fuse rocks were ongoing. A source of heat for volcanoes and volcanic action had long been a subject of speculation. Great winds rushing through voids underground, heated by compression, were suggested in antiquity, as was some sort of combustion. Volcanic eruptions were chronicled in detail, one of the first being the eruption of Vesuvius in 79 A.D., which event was recorded by an uncle and nephew, Pliny the Elder and Pliny the Younger. Supernatural forces were invoked, but even in Roman times writers and observers debunked such explanations and considered instead that water in contact with internal fires could lead to violent explosions. However, superstition still held sway for some.[3] Ideas about sources of material for combustion were refined as rational minds observed volcanoes more closely. Always, there was an effort to relate volcanism to smaller-scale observations.[4] Intertwined with questions of rock origin were those of just how much heat was necessary to melt rocks, which led, in circular fashion, back to sources of heat for volcanism. Rather than a "break" at some point in time, such as Hall's whinstone and lava experiment—begun in 1797, reported on in 1798, and published in 1805, delineating the "vectors" of an igneous origin—what one sees is a continuum of alternating arguments over the centuries. As instruments (e.g., the Wedgwood pyrometer) were invented and/or improved, and observations were tested in new ways and more rigorously, each instrument presented new data that necessitated more testing. To a greater or lesser extent, each conclusion was tested against previous knowledge of the results of fusion or solution processes.

Especially when directly observed, volcanoes are not an orderly occurrence. Casually viewed, some, like Kilauea in the

---

[1]Contrary to some of our history texts, there was never broad consensus for a flat Earth. The devil was in the details of the exact shape as exemplified by the latitude degree measurements in the seventeenth and early eighteenth centuries briefly mentioned in Chapter 8, when it was shown to be an oblate spheroid.

[2]For extensive background, see S.G. Brush (1979, 1983, 1996).

[3]Chester et al. (1985, p. 27) discuss early conjectures about the causes of volcanic action and include a useful list of literature from the sixteenth through the eighteenth centuries. Interestingly, they have their own criteria for the ending of the basalt controversy, with Leopold von Buch's (1774–1853) acceptance of basalt as a volcanic product in 1815.

[4]For an early example from Boccone (1674), see Newcomb (1998a, p. 443–444).

Hawaiian Islands, are relatively gentle, and produce flows that solidify into fairly uniform dark rock, sometimes with olivine crystals. But extruded products can include torrents of lava; rocks of varying colors, shapes, and sizes; pumice; obsidian; ash; mud; and combinations of all of the above. There can also be water or steam and gases, leading to explosive rather than gentle eruptions. Early in the seventeenth century, comparisons were drawn between volcanism and the effects of gunpowder or the "fermentation" of chemical reactions (Sigurdsson, 2000, p. 17). Robert Hooke was relatively free with references to subterranean heat (Drake, 1996, e.g., p. 191). In speaking of Etna, the Italian resident but German-born polymath Athanasius Kircher (1602–1680) wrote:

The reason of these Fires, is the abundance of Sulphur and Brimstone, contained in the Bosom of the H[e]ll, inkindled by Subterraneous Heats, Spirits, and Fires; with the free ventilation of the Sulphurous, and easily inflammable Air, and agitating Winds, through these open *Vulkanian* Vents and Funnels; with innumerable Chinks, Trunks, Pipes, and Caverns, with other conveyances through the Earth, &c. (Kircher, 1669, p. 43)

Kircher reported temperature increase with depth, based on miners' reports, into Earth (Buntebarth, 1998b, p. 409). It is important to remember that in the seventeenth century what fire actually did in either the analysis or the alteration of materials was not known.[5] In the eighteenth century, the idea of a molten earth interior gained favor via the hypotheses of Buffon, Immanuel Kant (1724–1804), and Laplace (Brush, 1982, p. 1185). Explanations of volcanism centered on deposits of sulfur and iron or bitumen or coal. For the practical experimenters we're concerned with here, there was much argument about two things: (1) the amount of heat that was necessary to fuse rocks as they were observed; and (2) whether volcanic heat was in itself sufficient to fuse rocks. As the nineteenth century progressed, evidence accrued that Earth's interior is very hot: Volcanoes extruded very hot material; the volcanic rocks on the surface of volcanoes showed signs of melting and recrystallization; and temperatures measured underground showed a progressive increase with depth (Brush, 1984, p. 26). In this chapter, we will discuss some of the data that supported views about how rocks originate. In the following chapters, opinions about rock origin by experimenters and those argued from experiment will be presented along with how some of the experimentation was subjected to further testing.

THE MEASURED HEAT OF EARTH

Temperatures

In the Field

The question of Earth's possible central heat was investigated in several ways. If Earth began as a molten mass, then there could be a residual central heat, so that the rate of cooling

was of interest, as was the total long-term impact of the Sun's radiation on earth temperature. Heat conduction was thus another facet of the problem. As will be seen, cooling experiments were done with model spheres.[6] Also, with the improvement of the thermometer, the investigative fervor of the eighteenth century extended to recording temperatures at the surface of Earth in sun and shade, in mines, caves, and volcanic fumaroles, as well as at the surface and at depth in bodies of water. Saussure was one of the great practitioners of this art. He and Deluc, among many others, designed a number of thermometers for specific purposes, and used them extensively in their travels. Among the instruments were thermometers that could record temperature at depth in water and not be crushed or change readings when brought to the surface. Another application was Saussure's "earth thermometer" that could register the solid Earth temperature if inserted into the earth in a tube and left in place for at least an hour. The rather extensive investigation of temperatures recorded by thermometers inserted in the solid Earth in the eighteenth and early nineteenth centuries rarely went beyond 24 feet (in solid Earth, not mines or crevices) and there were many difficulties (Middleton, 1966, p. 139–140). In another application, and to refute one of Deluc's theories, Saussure designed very slender thermometers that demonstrated that air temperatures are different in sunshine and in shade (Middleton, 1966, p. 211–212). Thermometers were taken into caves, and thrust into hot springs, as well as into gases and waters emitted in volcanic regions. A number of travelers had established that temperatures within caves were remarkably constant in summer and winter. All of this information figured into theories of rock origin.

Temperature Gradient

Thermometers were taken into mines, and as mines became deeper, temperature increases with depth were often more marked. The increase, postulated as heat from Earth's fiery beginning, became inextricably interwoven with the question of the age of Earth (Burchfield, [1975] 1990). At the time of Kircher's early reports about the temperature gradient, only the thermoscope had been invented. Records were begun, and a scale for the thermoscope was introduced early in the seventeenth century (Middleton, 1966, p. 11–12). As the long story of the thermometer's development continued, more and more records were kept and some were published, including those in 1781 by the crystallographer Romé de l'Isle. In one of the first reactions to the growing list of temperature investigations, Romé de l'Isle published his thoughts on the matter of central fire. In response to evidence cited by the French aristocrat Comte Georges-Louis Leclerc de Buffon, the astronomer Jean Sylvain Bailly (1736–1793), and Jean Jacques d'Ortous de Mairan (1678–1771), who studied ice and glaciers, Romé de l'Isle contended that their interpretations of measured temperatures were incorrect. In the pages of his 1781 *L'action du feu central démontrée nulle à la surface du globe*, he ques-

---

[5]Debus (1967) discussed this problem. Strong and gentle fires differed in their results. Boyle questioned whether the materials that separated from a substance when subjected to fire had been previously existent in it, or if they were new materials (Debus, 1967, p. 141).

[6]Buntebarth (1998b) has provided an excellent outline of the history of early efforts to understand the heat of Earth through the nineteenth century.

tioned both objective evidence of increasing temperatures with depth and the theorizing of those and other thinkers who might suppose a central heat left from the initial formation of Earth. The 1781 version was actually a second edition undertaken to answer doubts raised by the first, and in order to present a wealth of new evidence. Romé de l'Isle listed temperatures recorded in a number of mines, but simply thought that the measured temperature increases with depth were in error, and could be explained by local sources of heat in the mines (1781, p. 52). When challenged to explain how, if there was no central heat, volcanoes could exist, his reply was that water had penetrated to a pyrite bed that could begin "fermenting" with the aid of bituminous substances such as coal (p. 53). Romé de l'Isle's arguments were partly based on conjectures about how far the solar heat could penetrate Earth, and he included the effects of different degrees of insolation in summer and winter.[7] Later, Jean Baptiste Joseph Fourier also came to question whether temperatures continued to rise with greatly increased depth (Richet, 2007, p. 201).

Volcanoes

Volcanoes were a means of "accessing" internal heat, and as would be expected, it was common to measure the temperatures in and around volcanoes and of lava, ash, gases, and waters issuing in volcanic regions (Fig. 10.1). During a stay near Etna in 1787, Dolomieu reported that his thermometer rose by slightly over 4° Réaumur when held in air that contained fumes from an eruption (Dolomieu, 1788, p. 487).[8] In a direct approach, Breislak took his thermometers as well as other apparatus to the volcanoes themselves, as was recounted in the French translations (1792 and 1801) of his accounts of trying to tease out the secrets of volcanoes. Those publications received attention in the scientific world, as evidenced by reviews and abstracts that appeared in several journals. He collected and analyzed samples of gases, mineral waters, and solids, and kept temperature records of air and water at his collection sites, using thermometers scaled in Réaumur degrees. Breislak didn't hesitate to thrust thermometers into active fumaroles, and went perilously near molten lava. A small stream of water descended through the fumaroles of Solfatara near Vesuvius in Italy, which he measured at 32 °R in one place. In an excavation he made around a fumarole in order to collect samples, the thermometer registered 40 °R (Breislak, 1801, v. 2, p. 81). Inserted in another cavity, the thermometer rose from 12° to 70° in ten minutes, and after two hours it was steady at 77 °R (p. 82). In a tower that he built around a fumarole in order to have access for his instrumentation and sampling, the thermometer rarely went below 75 or above 78 °R.[9]

Breislak's colleague, Spallanzani, shared his interest in investigations in the field, and they agreed to work together. Grotta del Cane, a cave in the volcanic region between Naples and Pozzuoli, had roused much interest because gases that it contained could render a dog unconscious.[10] Spallanzani wrote that although he had hoped to join Breislak in his investigation of the cave, owing to a late return from Etna, the Lipari Islands, and

---

[7]It is interesting to see that, in the course of his discussion, Romé de l'Isle (1781) noted that "salt opposes the congelation of water," in other words, lowers the freezing point (p. 56), and recorded Black's observations that the level of a thermometer remains fixed as heat is added to change ice to water, until all the ice has melted (p. 104). At the beginning of the twenty-first century this is still a novel concept to some college freshman chemistry students.

[8]See Chapter 3 for the relations between thermometer scales.

[9]The relation between the Réaumur scale using alcohol and water and our accepted mercury thermometer with 100 °F is not a simple one, and depends on the thermometer maker and the country and date of its making. For a discussion of this complex subject, see Middleton (1966, p. 115–121).

[10]We recognize that the gas would have been carbon dioxide, which is heavier than oxygen.

View of the GREAT ERUPTION of VESUVIUS 1767 from Portici.

Figure 10.1. One of Hamilton's (1772, Plate I) depictions of Vesuvius in eruption.

Vesuvius, and the need to get back to Padua to begin his public lectures, he was unable to do so (Spallanzani, 1798, v. 1, p. 95). Spallanzani included a letter from Breislak about Grotta del Cane in his report of travels in the Kingdom of the Two Sicilies. In the letter, Breislak related that he had previously attempted to investigate the "mephitic" gas at another location, and thought he had discerned a slight increase in heat. However, he could not measure it because his thermometer had been broken on the road. He was later able to suspend a thermometer in Grotta del Cane. With it suspended at the aperture of the grotto, it

stood at between 13 and 14 of Reaumur's scale (62 and 64 of Fahrenheit's); and, placing the ball on the ground, so that it was immersed in the mephitic vapour, the mercury arose to between 21 and 22 of Reaumur (80 and 82 of Fahrenheit). (Spallanzani, 1798, v. 1, p. 103–104)

Breislak recognized that changes in humidity would alter the temperatures because the water evaporation rate would change. He repeated this measurement with different thermometers in order to see if the increase was real, because it had not been reported previously. While these investigations may seem mundane and repetitious to us, they still gave a quantitative aspect to the accumulating knowledge of Earth.

Some Results of Temperature Work

By the end of the eighteenth and beginning of the nineteenth centuries, the thermal gradient of the solid Earth had been determined at various places and levels by numerous investigators, which gradients ranged from 0.025 to 0.04 °C/m. Progressing through the nineteenth century, many geologists, notably Louis Cordier (1777–1861), took measurements both in the air of mines, and with thermometers in contact with the rock. Differences were found both in different types of mines, and at different levels below the surface (Buntebarth, 1998b, p. 410). Cordier noted that belief in central fire was ancient, and after what he called the discovery of the laws of the material world, scientists such as Halley, Leibniz, Mairan, and Buffon adopted that belief (Cordier, 1828, p. 3).[11] Their arguments were based on the figure of Earth; celestial appearances; what Cordier called the "instability of that subterranean principle," which causes magnetic action; surface temperatures compared with those at depth; or cooling of intensely heated bodies (Cordier, 1828, p. 3). He commented that their conclusions about temperature increase with depth "were supported by evidence too feeble to force conviction" and that heating by solar rays had come back into favor, supported by Neptunists such as Pallas, Saussure, and Werner (Cordier, 1828, p. 4). Later Cordier modified his position. He did claim that with so much learned lately by experiment about volcanoes and crystalliza-

tion of rocks, the passage of radiant heat and its conduction, and the latter confirmed by mathematics, Neptunism had been overthrown (Cordier, 1828, p. 5).

Cordier noted that a point of constant temperature below Earth's surface had been found, where solar radiation balanced the cooling expected at the surface. The depth of the constant temperature point would be different in different situations and materials. Cordier reported on three groups of investigations about the temperature gradient, the first done 150 yr previously, the next through the middle to the end of the eighteenth century, and the last those of the French mining engineer Jean-François Daubuisson de Voisins and several English investigators early in the nineteenth century. Temperatures of both air (two-thirds of readings) and water were noted, as were the types of thermometer (alcohol or mercury). Cordier discussed the effects of the ambient temperature of air introduced for ventilation, the size of the galleries, and the heat produced by different numbers of miners with their lamps. He then reported a range of depths that resulted in a temperature increase of one degree centigrade. Cordier followed this with a discussion of new experiments and described his thermometer—protected by what he called a leaf of silver paper, folded over seven times for insulation, the whole enclosed in a sheet tin case—that achieved the ambient temperature in 12 min. The supposedly mundane act of recording temperatures figured more and more in comments about rock origin.

Cooling Experiments

In another sort of investigation, it was conjectured that if Earth had formed from a once fused mass of material, and if it is now solid as we see it, some insight might be gained from observing the cooling times of spheres constructed for this purpose, in order to arrive at a time frame for the process. These investigations were not necessarily done to support the hypothesis of existing central fire, but instead were concerned with residual heat from the presumed initial state of fusion. That was Buffon's concern (Fig. 10.2). He considered volcanic heat to be local and limited, and thus separate from evidence of residual heat. Buffon was well acquainted with furnaces and fusion experiments. His experiments with the cooling rates of spheres of various materials are often briefly cited and unfairly discounted. His inspiration from Newton's experiments is less often mentioned by historians of geology (Roger, 1997; den Tex, 1996, p. 42). Buffon stated: "Un passage de Newton a donné naissance à ces expériences"[12] (Buffon, 1774, p. 152). However, Newton's comments about the relation of cooling time to the size of the sphere, as stated in two publications, were contradictory, one saying a larger globe would cool proportionally faster than a smaller one (in the *Principia*), while the other declared the opposite (in *Opticks*).[13] His evidence appeared to be mixed (Roger, 1997, p. 393). Buffon's extensive

---

[11]Note that Cordier does not mention Hutton here. He does later, after stating that Joseph Louis La Grange (1736–1813) and Dolomieu "revived the hypothesis of central fire," and then included Playfair and Hutton "even with their obscurities and the errors in physics into which they have fallen" by trying to apply their results to geology (Cordier, 1828, p. 6).

[12]A passage from Newton was the source of these experiments.

[13]Buffon (1774) has references to the appropriate passages in Newton on pages 152 and 154.

Figure 10.2. Comte de Buffon (1707–1788).

temperatures had been shown to be at nearly constant. There was no thermometer useful for his purpose, so after heating the balls to incandescence, he measured the time until he (or his assistant) could just hold each in his hand for half a second without being burned, and then measured the time until the ball reached room temperature. He determined the latter by keeping an identical ball unheated, and touching the two balls in turn until they felt the same. Buffon found a regular relation in the cooling time until he could touch the balls: an increase of about 24 min for each successively larger ball. He stated what the "perfect" relation would be, then gave his recorded times, which were remarkably similar. He did the same with the descent to room temperature. His results when he tried to repeat the experiment with the same spheres were not the same, and he suggested that they had lost weight during the heating. Further investigation showed that to be true (Buffon, 1774, p. 145–152).

From many historians' reports one might suppose that these cooling experiments were brief and inconclusive. However, Buffon continued this investigation over a period of years, and gave a full account of results in his *Supplément* (1774). As he continued the work of testing Newton's suppositions, Buffon found that the time for cooling increased with the diameter of his spheres (Buffon, 1774, p. 155). Since Earth was made of materials other than iron, he had spheres made of other materials—other metals as well as marble and sandstone—and measured the same two cooling intervals with them. For those interested in experimental methods, Buffon's methods and tables are admirable. And although inconclusive for some purposes, Buffon's meticulous tables did show proportional cooling times that allowed him to conjecture both about the structure of matter and the age of Earth. Using the iron spheres and the size of Earth, he calculated that nearly 43,000 (42,964 yr and 221 days) years would be required for Earth to reach the temperature where it would cease burning, and nearly 97,000 yr before it would reach its present temperature. After the second series of experiments using sandstone and materials other than iron, he arrived at a time of ~75,000 yr (Roger, 1997, p. 409–410). When he later considered rates for sedimentation and formation of calcareous rocks, he again extended the cooling time.[16] The important message from Buffon's work is not that his results were wrong, but that he ingeniously found a way to measure elapsed time since the beginning of Earth at all, rigorously controlled variables, figured out how to calculate the immeasurable, found regular relations in what he did, and applied them to observations of the real world. But like other models for Earth, then and now, the scale was too small to be meaningful. He had no idea of the internal composition of Earth, although density determinations for the whole Earth clearly showed an increase compared with the density of surface rocks.

investigations over a period of years showed clearly that larger globes cool proportionally more slowly than do smaller ones (Roger, 1997, p. 393–394).

Apart from their application to Buffon's extensive statements of how Earth worked and his conjectures about the age of Earth, the cooling experiments proved fruitful for their contributions to experimental methods, as well as to our understanding of how ideas were tested. We must remember that caloric was a mysterious fluid. Also, despite the care and efforts at standardization employed by Buffon, and sometimes the clarity of his results, he didn't attempt to apply a mathematical expression to what he had learned (Roger, 1997, p. 394), even though he employed a good deal of arithmetic. Although little was applicable to what was then known about Earth, Buffon found a reliable cooling relation and advanced knowledge of experimental methods and techniques using the tools he had available.[14]

Buffon first had ten iron spheres from one-half to five inches in diameter, each increasing by half an inch, or half a "pouce,"[15] made at a forge. Since all the balls were made of iron at the same forge, their weight was found to be proportional to their volume (Buffon, 1774, p. 145). He recorded their weights, as determined by a "very good" balance. The cooling experiments performed by Buffon himself or his assistants were done in cellars where

[14]Roger pointed out that Buffon's classification of metals by the speed of cooling corresponded to the product of density times specific heat, which relation Buffon did not know (Roger, 1997, p. 395).

[15]Or "thumb," the French measure of length, which is the equivalent of the current English inch.

[16]Rossi ([1984] 1987, p. 107–108) discussed Buffon's ambivalence about the time span, partly based on manuscript drafts. The best reference for Buffon's different chronologies is probably Roger's 1962 edition of Buffon's *Époques de la Nature*, p. lx–lxvii, especially the table on p. lxv (K. Taylor, 2008, personal commun.).

HEAT CONDUCTION, INSOLATION

It is safe to say that everyone who contributed to the discussion about rock origin recognized how complex was the question of heat in Earth. If one believed in the origin of Earth from an originally molten state, a fragment wrested from the sun, there was little disagreement about there being residual heat. In the middle and late eighteenth century, it was recognized that heat as an entity was not well understood, nor were there predictive ways of describing it mathematically. This was not accomplished until the work of the French mathematician Jean Baptiste Joseph Fourier (1768–1830), who originated a method for describing heat conduction in solids, along with a method for solving the resulting equations (Holton, 1973, p. 286). Fourier's book was published in 1822, and was a continuation of his earlier work on the mathematical theory of heat (Boyer, [1968] 1985, p. 599). Fourier's interest had been aroused by the question of temperatures of the solid Earth, the components of which were heat from the Sun, the primitive heat of Earth's interior, and "participation of the Earth in the temperature of the cosmos" (Brush, 1983, p. 82; Buntebarth, 1998b, p. 410).

Romé de l'Isle, who did not believe in central heat, discussed the contradictory indications of solar heat in summer and winter, and how it might be attenuated or changed in different localities. He noted that the "fluid of light" was not simple, as evidenced by a single light ray dividing into many different colors (Romé de l'Isle, 1781, p. 76). He quoted a number of people on the puzzles about whether light and heat were truly related, and about the source of the heat of the Sun. He failed to see how a central fire could be sustained, when it had been determined that the Sun's influence wasn't felt at a very great depth within Earth, and there was no "aliment," or food, for a central fire. He used as evidence temperature measurements taken by Genssane (c. 1780), Guettard, Deluc, Saussure, and others in summer, winter, on glaciers, and in water. Black's discoveries about latent heat were part of his general reasoning process. As noted previously, the lack of temperature rise while heating continues during phase change still puzzles students.

John Murray, who wrote the geology-informed response to Playfair's exposition of Hutton, some of which hinged on field observations of heat effects, wrote an article entitled "On the diffusion of heat at the surface of the Earth" which was read in Edinburgh in 1814 (Murray, 1815). He stated that his article was prompted by Playfair's reply to his previous arguments about central heat. It was difficult for Murray to argue for central heat when he could find no cause for its origin or continuation. It seemed highly implausible that it would be sustained for the duration of the repetitious worlds as postulated by Hutton. The transmission of heat through the "elastic fluid," or gas, was most inefficient. With respect to conduction of heat through the solid Earth, Murray raised the question about the enormous amount of heat that would have had to be transmitted through the solid Earth from the supposed hot center in order to melt all the rocks including mountains of quartz. To illustrate the impossibility of transmitting heat from the center to the surface, he used an iron bar 1000 inches long:

[I]f its temperature at the one extremity be 50° [he didn't state a scale], and if within five inches of this it is at white heat, then the heat increasing at the same rate, through every succeeding five inches, what must be its intensity at the other extremity? No effort of the imagination can form the most remote conception of it, nor can any argument be wanting to prove, that no such heat can exist in the interior of the earth. (Murray, 1815, p. 425)

In other words, there seemed to be no mechanism that could account for an intensity of heat within Earth that could be conducted to the surface in sufficient amount to fuse the great masses of rocks. Another part of the puzzle was how that enormous amount of heat could be dissipated into the atmosphere, leaving the moderate temperatures seen at the surface.

These are just two examples of the many people who were addressing the problems of heat and heat conduction as it applied to Earth. During this time, Fourier obtained his equation for conductance, which he conjectured could be applied to Earth. In so doing, he created a new branch of mathematics, which could include the geometry of the body (Richet, 2007, p. 184).[17] This is another instance where part of the solution, Fourier's mathematical advances, did little to explain mechanisms involving Earth. Fourier solved none of the geologists' problems, as the source of heat was still a mystery. Geologists continued to have many reasons for doubting that heat could be the chief agent of rock mobilization and consolidation. Opinions ranged from that of all terrestrial heat being due to the Sun to residual heat from formation to a combination of the two. Some thought furnace heat exceeded volcanic heat, while others advocated the reverse. It was in this context that thermal experimentation on rocks was done in the eighteenth century.

Fusion Experiments

Furnaces

By the eighteenth century, furnaces were ubiquitous in both scientific and technological procedures, as discussed in Chapter 3. The literature detailing their construction and use frequently included comments about properties specific to the particular furnace used, such as where it was the hottest, what sort of fuel was required, or how long heat could be controlled and/or sustained. With the advent of the pyrometer, some kind of temperature measurement or comparison was possible. Other important variables had been noted. Hall and others recognized the importance

---

[17]Richet wrote about Fourier's work:

But this was a new type of equation, and its solutions obviously depended upon the geometry of the body being considered: a bar would not conduct heat in the same manner as a sphere. In order to integrate such an equation under limiting conditions, Fourier created a new branch of mathematics. In particular, he demonstrated that any algebraic function could be expressed as the sum of trigonometric functions [Fourier series]. (Richet, 2007, p. 184)

of having not just enough heat to fuse samples, but also of the time during which the sample cooled. In addition, the duration of time during which heat was applied was increasingly seen to be important. As noted in Chapter 3, fusion experiments on rocks were sometimes carried out in what we would call industrial furnaces, those for glass, porcelain, or metal working. At other times, a laboratory furnace of one or another type might be used. The blowpipe was an invaluable tool, as observations on fusibility aided identification. At the other end of the scale, volcanoes provided directly observable terrestrial fusion and were available for some quantitative observations, either by direct temperature measurement or for providing analogies to forge or furnace processes. Dull red, glowing red, and white heat were recognized as a thermal sequence.

The Experiments

According to our current definition, basalt does not contain quartz as a mineral, but includes pyroxenes such as augite, olivine, some amphiboles, and calcic feldspars. The rock is dark colored due to its relatively high percentage of ferro-magnesian minerals, some of which also account for its attraction to a magnet. Local conditions of cooling determine a basalt's texture. If the cooling is quick, a mass of small crystals is formed, which can sometimes be discerned with a magnifying glass. Extremely fast cooling may produce glassy obsidian. Porphyry has large crystals in a fine ground mass, and results when the lava begins to cool slowly underground so that crystals of the higher melting point minerals begin to grow, and then the material is ejected so that the rest cools quickly. Some volcanic ejecta are lighter colored due to more siliceous rocks having been melted, but they are not called "basalt."

Darcet reported on a series of fusion investigations with volcanic rocks in both volumes of his *Mémoire* (1766, 1771). These were done at the time when the association between lava and basalt was becoming clearer, due to the observations of Desmarest and Guettard in the Auvergne. However, there was much that was still unclear, particularly about basalt. The magnitude of volcanic heat compared to that of furnaces was in question. In addition, data from different sources often did not agree.[18] Darcet's furnace, designed for making porcelain, could be heated slowly to a uniform heat that lasted for several days. The crucibles were very heat resistant, and often had covers to prevent contamination. Darcet fused a number of samples from the Auvergne, Italy, Ireland, and other places. Some he called *pierre ponce*, or pumice, and others he identified as basalt. Virtually all specimens melted easily into a brown, mostly transparent glass. He credited Henckel with being the first to have learned that pumice was fusible by itself, a fact confirmed by Cramer and Pott (Darcet, 1766, p. 72). Several of the samples supplied by Desmarest from

the Auvergne destroyed the crucibles in which they were fused, which, Darcet thought, indicated they had entered into a perfect vitrification (Darcet, 1766, p. 74).[19]

Many people worked on the puzzle of rocks found associated with volcanoes over the period from 1760 to 1820, and of course there were precursors and those who followed. In addition to his extensive travels in the Alps and elsewhere, often while using an extensive array of instrumentation and keeping meticulous records, Saussure performed experiments of many kinds in his laboratory. He stated that modern naturalists were unanimous in giving the name "basalt" to a matter that had (1) been fused by the heat of a volcano; (2) cooled; and (3) taken a regular form, in some cases prismatic columns, in other cases spheres, or sometimes plane and parallel (Saussure, 1779, p. 61). He reserved the name of basalt for a rock that had had a regular contraction (Saussure, 1779, p. 62). Saussure described various basalts in detail according to their external characteristics, and also reported exposing some fragments of basalt to an intense fire, which resulted in a completely vitrified slag that was dark and sprinkled with very small grains. There were areas where there appeared to be no grains, but when viewed with a magnifying glass, extremely small grains could be seen (Saussure, 1779, p. 130). This easy fusibility helped to support the idea that volcanoes had lower temperatures than furnaces, because in general they didn't supply enough heat to vitrify basalt as completely as was the result in trials with a furnace.

Dolomieu had been writing about volcanic rocks in a number of papers and books published since 1783, and had had extensive field experience, having described many colors and sorts of volcanic products. In his 1788 "Mémoire," he included a detailed catalog of Etna's lavas. In a footnote to an article in 1794 (p. 117), he mentioned that he and others had tried in vain to replicate the formation of lava. No matter what sort of fire was employed, fusion resulted only in glass or scoria. In a trial with a porcelain furnace, melted and cooled dark lava, collected from the base of a trap, resulted in a very compact mass with a glassy surface. Internally, it was less vitreous, and resembled a paste of dark porcelain in its grain, fracture, and hardness. None of it resembled the starting material, and the crystals of schorl and hornblende could no longer be seen (Dolomieu, 1794a, p. 117). Dolomieu then quoted Saussure about an experiment he had done hoping to see the crystals reappear after fusion, but they had not done so (Dolomieu, 1794a, p. 117, quoting Saussure, 1779, v. 1, p. 121–122). Dolomieu stated that the results of volcanic fire were quite different from those of experimenters' trials. The latter denatured the molecules, and destroyed the relations between constituent

---

[18]As just one example of many, Darcet cited asbestos as a rock that for a long while had been said to be refractory, but mentioned Henckel had found a kind that would melt without addition, while Pott mentioned several varieties that would melt not only in a fire, but also with a burning mirror (Darcet, 1766, p. 69).

[19]Buffon remarked on the furnaces used by Pott and Darcet, and thought their experiments were equally good. However, Buffon felt that it was incorrect for them to say that they had fused everything fusible in nature, since they had not employed a burning mirror in conjunction with their furnaces. They could have classified things as refractory that would yield to the correct conditions (Buffon, 1774, p. 60, 63). Pott did mention that very refractory argil (clay) with an addition of iron would fuse not only in a large burning mirror like that of Tschirnhaus, but also in a violent fire (Pott, [1746] 1753, p. 122).

molecules. Volcanic fires appeared to respect the composition and aggregation of the primitive bodies (Dolomieu, 1794a, p. 118). He thought experimentalists' fires couldn't be controlled sufficiently to do that now.

Kirwan frequently weighed in with his thoughts on the properties and origin of basalt, and contributed a running commentary for at least 30 years. We will put his thoughts into the web of reasoning about basalt origin in Chapter 11, but for now we will just mention his experimental work. Kirwan used Wedgwood's pyrometer pieces extensively. He determined that "by the help of a large bellows, loaded with 130 lb. weight, I raised in my forge a heat of 168° in less than half an hour" (Kirwan, 1810, p. 40).[20] He described successive states of fusion as "emollescence," porcelain, "slaggs" or enamel, semitransparent, and transparent. He noted that "All lavas are more or less magnetic, give fire with steel, are of a granular texture, and fusible *per se*" (Kirwan, 1784, p. 135). About the Rowley Rag stone (a variety of basalt) he remarked:

Heated in an open fire it becomes magnetic, and loses about 3 per ct. of its weight. It does not redden; but at 98° [WW] melts into a somewhat porous black mass, partly porcelain, and partly an enamel. (Kirwan, 1794a, v. 1, p. 230)

Kirwan found that a similar rock from Scotland had melted at 130 °WW. A basalt from Saxony melted into a black glass at 100 °WW. He also determined specific gravity for his samples before and after melting, and found it to be similar across samples (Kirwan, 1794a, v. 1, p. 233). Kirwan found that what he called "obsidian" melted into a gray, opaque mass. However, he also said that he had no reason to think that his sample of obsidian was of volcanic origin (Kirwan, 1794a, v. 1, p. 265). Later, Kirwan found most stones called "basalt" lacked what he considered marks of fusion, the porcelain grain and glazing, and some air bubbles (Kirwan, 1794a, v. 2, p. 437). From this, he concluded that most volcanoes rarely exceed 100 °WW (Kirwan, 1794a, v. 2, p. 407).

Spallanzani and Breislak did not need to travel far to see the effects of volcanism. Both did extensive investigations of volcanic phenomena in Italy, and their work was well known to interested scholars throughout Europe. (See Fig. 10.3.) They also did extensive laboratory work and made observations and investigations out in the field. Spallanzani is perhaps better known for his work on physiology than for his geology, but historians of geology have also recognized his major contributions (Morello, 1981; Vaccari, 1996). The careful fusion work of Kirwan, Hall, and others was somewhat eclipsed by that of Spallanzani, although the latter is less frequently cited in English-language texts.[21] Spallanzani stated that he had fused literally thousands of samples from volcanoes. He used the results of this work to

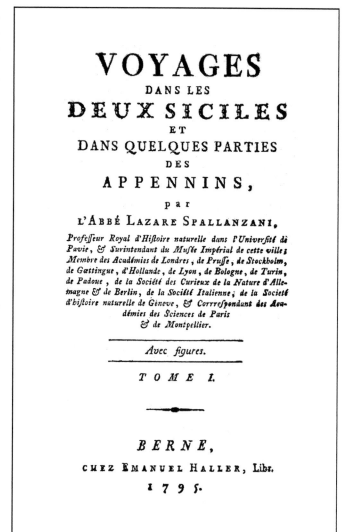

Figure 10.3. The title page of Spallanzani's *Voyages dans les deux Sicilies* (1795).

investigate rock origin and to conjecture about the causes of volcanism. He used the results of the analysis of various "volcanic productions" in the same quest.

It is difficult to choose just a few of Spallanzani's fusion investigations for examples, in part because some of his convictions arose from the breadth of his experience and because there were so many trials. He rarely used the word "basalt," generally favoring "compact lava," "prismatic lava," etc., perhaps not surprisingly for someone so concerned with active volcanoes. His questions about the various meanings and usages of the word "basalt" will be discussed in Chapter 11. Spallanzani employed a glass furnace, which he felt was completely capable of fusing the volcanics. In order to be more accurate, he used Wedgwood pyrometer pieces, placing them in a case of the same clay in the furnace next to his samples. He found the furnace attained a temperature of 87 1/8 °WW (Pinkerton's 1811 translation says 87½), which is ~12,777 °F

---

[20]The 1810 edition of Kirwan's *Elements of mineralogy* was said to be an edition made without his consent. I have had brief access to it, but prefer the second edition (1794a).

[21]I worked chiefly from the 1798 English translation of the four volumes of his 1792 *Viaggi alle due Sicilie*.

(Spallanzani, 1798, v. 1, p. xxii).[22] Some short experimental inquiries were included in his accounts of his travels; the more extensive trials were in the later volumes, which discussed the experiments and his conclusions about volcanic heat and rock origin.

Spallanzani first described his samples by location and characters, as well as texture. Any large porphyritic crystals were also described. In some, the accessory crystals fused, in others they did not. He carefully described any bubbles that might be present in the melt, and as we shall see, found ways to determine their composition. Some large schorls, both alone and along with the lava they were combined with, fused into a "black, compact enamel, giving sparks copiously with steel" (Spallanzani, 1798, v. 3, p. 228). A sample of lava from a prismatic column was converted by the furnace "into an enamel inclining to a black colour, with a few but large bubbles; a fusion takes place in the shoerls [*sic*], but not completely in the feltspars [*sic*]" (Spallanzani, 1798, v. 3, p. 230–231). A trial with another sample, which he called a "pure pitch-stone lava," showed the thoroughness of Spallanzani's work:

A few hours in the furnace deprive [*sic*] this stone of its colours, rendering it of an ash grey: it likewise loses its friability and softness, and becomes capable of giving sparks with steel; resembling then a paste of porcelain. Continuing the heat longer, the cinereous [*sic*] color remains, with the appearance of very numerous bubbles; and the lava passes into a homogeneous vesicular enamel, with a fusion of the feltspars [*sic*]. (Spallanzani, 1798, v. 3, p. 252)

Multiply these observations many times and one cannot help but be more impressed with Spallanzani's conclusions about rock origin than with those of theorists who stayed home or at best saw a limited number of field situations. Also, Spallanzani was familiar with most of the European literature that was related to his interests.[23]

In his eagerness to understand more about an active eruption, Spallanzani took risks that many of us today would consider unacceptable. He suggested putting both flammable and fusible substances in a cavity in the lava and observing closely how long it took them to begin burning or fusing. Then he could time the same effect on the same substances in a furnace whose temperature he could track, and thus have an idea of the intensity of volcanic fires. He also realized he could use Wedgwood's pyrometer directly on the lava:

[T]o ascertain the absolute heat of the superficies of the lava, one or more of the cylinders of clay should be let down upon it, inclosed in the box of the same earth adapted to them, fastened to an iron chain that it may not be carried away by the current, and the experiment prevented. This being taken up after having been suffered to remain there some hours, the shortening of the cylinders would shew the quantity of absolute heat they had suffered, and, consequently, that of the lava on which they had rested. (Spallanzani, 1798, v. 1, p. 17–18)

He suggested trying to find the internal heat of the lava by putting a clay cylinder in a hollow sphere of iron held by an iron chain. Spallanzani thought that the temperature of the lava probably wouldn't exceed that of iron fusion, so he could retrieve the cylinder, but if it were exceeded, he would still have a reading for volcanic heat. He realized that all lava flows might not have the same temperature, and also that there might be practical difficulties with the iron sphere, fearing it was not heavy enough to penetrate the lava. It is uncertain whether he ever actually performed these experiments, as he recorded no results. After those suggestions, he continued with describing the eruption that he did see.

A colleague and sometimes field companion of Spallanzani, Breislak was equally inventive in devising means to learn about volcanoes, in the field and in the laboratory. He read extensively, and later became well aware of Hall's results with volcanoes. Breislak's publications on volcanic observations and experiments began in the last decade of the eighteenth century and extended to the 1830s. After hearing of Hall's work, he observed a number of volcanic samples that had been exposed to the heat of a limekiln in Palermo. Some of them had been completely vitrified, others showed some crystallinity, while others had neither transparency nor polish, nor any of the other characters of glass. He listed several things he had observed that influenced theories of volcanic heat. First, all parts of the rocks would not receive heat to the same degree, so the degree of fusion was not the same, and it was thus not surprising to find some remnants of the primitive rocks and their original minerals. Second, some minerals might be present both from the primitive rocks and as residue (alteration products?) from the original rock. Third, to explain some mineral formation, he argued that minerals originally present could form by sublimation in the cavities of lava, as evidenced by such formation in furnaces. He stated as further evidence of formation by sublimation the pyroxenes found on the walls of a church after it had been enveloped by lava (Breislak, 1801, v. 1, p. 287–289). Breislak's thoughts on fusion continued, and we shall see later how he reacted to the continuation of fusion speculation and experiment in relation to basalt.

Hall had begun his fusion investigations before the last decade of the eighteenth century, but had the misfortune to begin with granite, with inconclusive results. He dropped those investigations, but continued to think about the problem and pursued the slow cooling of fused glass, finding that he could induce a "stony" texture by slow cooling. The glass could then be re-fused.[24] He began fusion experiments with whinstone (basalt) and lava after Hutton's death. At that time, Hall did not yet try to alter pressure in his work, although he knew of Hutton's ideas about rock consolidation under the pressure of incumbent masses of rock or water. Hutton had also assumed that laboratory methods could never supply heat intensities similar to Earth's vast resources. Hall, however, had noted that whinstones and lavas could be melted rather easily, and that "it is impossible that a substance should congeal at a higher point

---

[22]This rather fearsome temperature is based on the admittedly imperfect Wedgwood scale.

[23]See Vaccari (1996) for a discussion of Spallanzani's many interests, and when his geological work was done. He has also described the general content of each volume of the *Viaggi*.

[24]Hall mentioned the commonly known work of Keir (1776).

than that at which it may afterwards be melted" (Hall, 1805b, p. 45, which paper he read in 1798). Hall believed that he could extend what he called an "imitation" of a natural process to an analogy of laboratory efforts to those of the natural world. He likened the process to that of astronomers who could extend observations on gravity made with a pendulum to predictions about heavenly bodies (Hall, 1805b, p. 45).

Hall determined to use the group of rocks called "whinstones," also called "basaltes, trap, wacken, grünstein, and porphyry," which he considered varieties of the same class (Hall, 1805b, p. 45). To those he added lavas that he himself had collected when on Vesuvius and Etna, and in the Lipari Islands in 1785. He began by fusing a sample of a whinstone contained in a black lead crucible in an iron foundry furnace, and found that it fused in 15 min. Allowed to cool quickly, it formed a black glass. Other samples, cooled more slowly, produced a vitreous product embedded with small spheres with an earthy fracture. Hall then induced still slower cooling by transferring the sample from the furnace to an open fire, surrounding it with coals. The fire was allowed to go out after several hours and the coals cooled. The result was what Hall called "a substance, differing in all respects from glass, and in texture completely resembling whinstone" (Hall, 1805b, p. 48). After a colleague's objection that perhaps the basalt had not completely lost its initial crystallinity, Hall adopted the procedure of first fusing his samples completely and letting them cool, then fusing them again before the slow cooling. He found that those samples had to be heated more intensely in order to fuse them the second time. A sample from that procedure was fused, the temperature reduced to 28 °WW, and held there for 6 h.[25]

Having determined that he could reproduce something other than a vitreous texture, Hall proceeded with samples of seven whinstones from localities in Scotland, the lavas he had collected, and one that had been supplied to him from Iceland. He used a muffle furnace, and for the slow cooling surrounded the sample with burning coals in a long and narrow muffle. The temperatures of fusion and re-fusion were recorded by pyrometer pieces placed next to the crucibles containing the samples in the muffle. He found that the temperature of re-fusion was not the same as the original temperature of fusion, but it occurred after a period in which the sample softened but did not fuse. Holding the temperature at a more moderate level seemed to induce a crystalline texture. For one of those re-fused samples he wrote:

Numerous and varied experiments have since proved, in the clearest manner, that, in any temperature, from 21 to 28 [°WW] inclusive, the glass of this whin passes from a soft, or liquid state, to a solid one, in consequence of crystallization; which is differently performed at different points of this range. In the lower points, as at 23, it is rapid and imperfect; in higher points, slower and more complete, every intermediate temperature affording an intermediate result. I likewise found, that crystallization takes place, not only when the heat is stationary,

but likewise when rising or sinking, provided its progress through the range just mentioned is not too rapid. Thus, if the heat of the substance, after fusion, exceeds one minute in passing from 21 to 23, or from 23 to 21, the mass will infallibly crystallize, and lose its vitreous character. (Hall, 1805b, p. 51–52)

This range for crystallization answered for Hall the reason why a higher temperature was required for the second fusion, namely, that the forces of crystallization had to be overcome. While Hall's work is well known, his specific reasoning is not often stated. In Chapter 11, we will see how that work influenced interpretations and those of others on the origin of basalt.

As one translator noted, Klaproth did not have knowledge of the fusion work of Hall, as Klaproth first reported in volume 14 of the *Bibliothèque Britannique* (Klaproth, 1802, p. 126 fn). Klaproth based his opinions about basalt on observations he had made in basaltic mountains, which convinced him that it was formed in the humid way. His fusion and analytical work was reported in the first volume of his *Beiträge zur chemischen Kenntniss der Mineralkörper* in 1795. The six volumes of this work were published between 1795 and 1815. There was an English translation of part of that work in 1801, entitled *Analytical essays towards promoting the chemical knowledge of mineral substances*. His basalt analyses were translated into French in 1802 (*Journal des Mines*, v. 13, An. 11, p. 123–134). That record illustrates several things: One is the persistent market for mineralogical information over a period of years. It also points out how often, and how quickly, information traveled between countries. And it also illustrates how commonplace was the theory of humid origin for basalt, well after Hutton advanced his theory.

The first 38 pages of the English edition of Klaproth (1801, v. 1) were titled "Experiments concerning the habitudes of various species of stones and earth in the heat of a porcelain-furnace," and included 111 samples. Klaproth briefly discussed earlier work with fusion, and noted that some investigators, such as Pott and Gellert, had been more interested in combinations of substances than the substances themselves, while others, such as Darcet, made the error of using clay crucibles. He referred to the German mineralogist Carl Abraham Gerhard as one of those who "has communicated true results, because in his operations he employed crucibles of charcoal, besides those manufactured of clay and chalk" (Klaproth, 1801, v. 1, p. 3).

Although a charcoal crucible did not combine with ingredients to alter fusion as might crucibles of clay, it was recognized that charcoal could reduce iron compounds in the sample to a metallic regulus,[26] thus not producing a true fusion product (Klaproth, 1801, v. 1, p. 36). For such crucibles, Klaproth made a hollow in a piece of charcoal into which he inserted the sample, then closed it with another piece of charcoal, after which the whole was enclosed in a clay crucible which was then sealed with lute. He recorded the crucible type that held each sample.

---

[25]According to Wedgwood's table of his scale compared to Fahrenheit's, 28° of Wedgwood is equivalent to 4717 °F, or the temperature at which fine silver melts (Wedgwood, [1784] 1809b, p. 578).

[26]This is an impure intermediate product obtained when smelting metal ores.

He used the furnaces of the royal porcelain manufactory at Berlin, and his samples were included along with the porcelain in the furnace runs (Klaproth, 1801, v. 1, p. 3). The samples were weighed before and after exposure to the fire and any change of weight was recorded, along with a careful description of the fusion product. Because of its often relatively high iron content, Klaproth felt he could not obtain a true fusion product for basalt if he used a charcoal crucible. In that case, after fusion, he found a somewhat unvitrified mass with the appearance of scoria rather than a black glass. The iron content had implications for both origin and classification, since it was possible that instead of being in the class of earths and stones, it should instead be regarded as an iron ore (Klaproth, 1801, v. 1, p. 37). He recorded that one sample of a basalt from Bohemia contained in a clay crucible fused into a black-brown glass, losing 2% of its weight in the process (Klaproth, 1801, v. 2, p. 198).

Drée had been impressed with the debate about basalt and lava, and how difficult it was to decide the origin of those rocks. He was charged with publishing his brother-in-law's (Dolomieu's) works, and with putting his collection in order, and thus wanted an unambiguous way to classify volcanic rocks. To see if they were the result of fusion, Drée put small pieces of the rocks into a crucible or porcelain cylinder, filled all spaces with the same rock pulverized, completely covered the sample with a layer of an infusible substance such as quartz or mica, and then inserted a safety plug, luted in place with an easily vitrifiable substance. Most containers were enclosed within another container, also with quartz powder, sometimes covered with a sheet of mica or platinum. They were luted with argil, and finally fastened with an iron wire. Drée employed several variations of these methods. He next subjected the containers to what he called a "median heat" in a furnace for a long while to concentrate the heat, with pyrometer pieces placed in the crucible or porcelain container next to the sample. The sample would thus be protected from any substance that might decompose it. He hoped that his procedure would closely approximate the situation of a rock within a volcano (Drée, 1808, p. 36).

Compression had been suggested by Hall's account of his experiments with marble, in which he closed his crucibles tightly to increase the pressure during heating. Drée thought that would assist in producing "lava lithoides" or stony as opposed to vitreous lava. He noted that Hall had said earlier that for basalt this process was only the result of devitrification, whereas Drée intended to add confinement. Drée summarized other experimental arguments intended to support or refute Hall, and finished the discussion by referring to Gregory Watt's support of Hall. Arguments supporting Hall were in opposition to those of Dolomieu, who had supposed that vitreous lava was produced from "stony" lava that rose to the surface of the volcano and became vitrified by combustion in the atmosphere (Drée, 1808, p. 38). Drée continued his experiments to try to resolve the question, and showed he could produce what he called "nonvitreous liquefaction." The more detailed results from Hall led him to refine the questions being asked, which were:

Les laves sont-elles le produit d'une liquéfaction ignée particulière, et différente de la fusion vitreuse, ou sont-elles le résultat de la dévitrification?
Les cristaux inclus dans les laves sont-ils préexistans à la fusion, ou sont-ils de formation potérieure à cet acte? (Drée, 1808, p. 39)[27]

Those questions had figured in lava and basalt problems for more than 30 years.

Some of Hall's and Drée's results are shown in Table 10.1 for easy comparison. The first few samples are Hall's, and they follow his numbering; the remainder are Drée's. Hall's experiments were simpler in conception. Drée included 39 samples, but I have omitted nos. 19 and 39, which were granites. Most of the rest were rocks that he felt were representative of the country rocks that could have been fused by volcanic heat into volcanic rocks. Generally, his samples were not of volcanic origin. He didn't report on the results of some of the samples, so they are omitted (e.g., nos. 7 and 10). As can be seen, Drée seems to have had no problem in fusing country rocks and producing a substance that he felt was similar to a "stony" or partially crystalline volcanic product. He noted that when he began, he had no idea of the degree of heat that would be necessary to liquefy the diverse samples, the best means to apply it, how to contain the gases, how much samples would expand, or how to contain them (Drée, 1808, p. 49). Hall's samples, which were volcanics rather than country rocks, were much smaller and of a different composition.

Drée noted that his results were variable, so he grouped similar samples from his eight separate trials in order that he could report on them. His first group consisted of nos. 6, 12, 17, 38, and 39. These were two of porphyry *c*, one of *d*, amygdaloid *i*, and what he called an "amphibolitic granite." Despite different results from the different regimes of heat and cooling to which they were subjected, the samples from this first group all seemed to show indications of feldspar crystals. He compared the results to five lava samples that he labeled A–E. He said they demonstrated the same texture, grain, appearance, color, and hardness. Their specific gravities were similar, as were their magnetic properties (Drée, 1808, p. 52). They were not as compact, but he found that unsurprising, considering that their natural products were far more compressed. He was convinced that a true devitrification and then cooling resulted in a product composed of new materials that obeyed the laws of attraction to take a crystalline form. By contrast, his products included original crystals. The products, especially for nos. 6 and 17, were not the result of the compression, but were there because nothing could escape from the double crucible, resulting in a new form of liquefaction (Drée, 1808, p. 56). He noted that he didn't subscribe to the views of Dolomieu, who believed that sulfur aided liquefaction, or Breislak, who attributed the liquefaction to water "charged" with muriate of soda (NaCl).

---

[27]"Are the lavas the product of an individual fiery liquefaction (or igneous fusion) and different from vitreous fusion, or are they the result of devitrification?
"Are the crystals included in the lavas preexistent to the fusion, or is the formation subsequent to that action?" (my translation).

TABLE 10.1. HALL'S AND DRÉE'S FUSION RESULTS

Sample	Container product	Additional material	Covered?	Furnace type	Duration heat/cooling
Hall					
1. whin (basalt)	Clay / Somewhat stony	None	No?	Reverberating	21–31 °WW
3. whin	Glass / Feldspar not melted, crystallite	No		Muffle	100 °WW / 60 °WW
2, 4–7. whins	All formed glass, then crystallite	None	No	"	See immed. below
1–6. lavas	All formed crystallites when held at softening, given temperature / Orig. softened: 18–55 °WW / Glass softened: 14.5–24 °WW / Crystallite softened: 32–43 °WW / (Hall, 1805, p. 75)	None	No	"	Table of original fusion temperatures, of glass and crystallite softening
Drée					
1. porphyry *m*	Porcelain cylinder / Removed quickly, powdered rock formed black scoria, pieces not fused	No	Yes	Domed	24 hours, 40 °WW, or perhaps 25 °WW
2. porphyry *b*	" / Long cooled, powder agglutinated, rock pieces not changed	No	Yes	"	"
3 & 4. porphyry serpentine *a* & porphyry *b*	Glass, except for feldspar and quartz	No	With platinum sheet and luted	"	18 hours at 43–46 °WW
5. & 6. serpentine *a* & porphyry *c*	Hessian crucible 5, black glass with feldspar crystals / 6, NOT fused	No	Enclosed and luted	"	"
8. porphyry *a*	Porcelain tube / Not broken, powder fused to a non-vitreous scoria, pieces softened to porcelain aspect	No	Yes	"	23 hours, 40–42 °WW, cooled
9. porphyry *b*	Hessian crucible / Powder liquefied, but was 'stony', pieces had only softened	11 gr. NaCl, 69 of S	Yes	"	Same as no. 8, removed before cooling
11. porphyry *c*	Crucible / All powdered, powder became red and scarcely agglutinated, was black on bottom	No	Yes	"	Same as no. 8, but removed quickly
12. porphyry *c*	Crucible liquified, then appeared perfectly like a lava, as sample 6 did	>3 oz powdered rock, 3 gr. NaCl	Yes	"	Same as no. 8, but removed after cooling
14. porphyry *d*, a kind of *a*	Crucibles, due to lack of cylinder: only report 14, the crucible split, powder fused, part like scoria	No	In another crucible	"	36 hours, 50 °WW
17. porphyry *d*	Crucible / Crucible split before cooling; some material oozed out. But pieces fused and formed true feldspar crystals.	No	Covered with quartz and mica	Forge fire	6.5 hours, 112 and 133 °WW
21. porphyry *d*	Square porcelain crucible / Crucible split, removed after cooling, all a glass, part escaped / Like no. 17	No	Yes	"	"
23. porphyry *c*	Removed before cooling, split, some of glassy material foamed. Feldspar did not resist heat, melted like a paste.	No	Yes	"	"

(continued)

TABLE 10.1. HALL'S AND DRÉE'S FUSION RESULTS (*continued*)

Sample	Container product	Additional material	Covered?	Furnace type	Duration heat/cooling
Drée (*continued*)					
24. porphyry *b*	Like no. 21 Same as no. 21, but removed before cooling	No	Yes	"	"
25. porphyry *d*	Porcelain cylinder Removed intact, powder liquefied, then stony, pieces softened to porcelain aspect	No	Yes, with quartz and mica	Brongniart's porcelain furnace	Nearly 6 hours, 80 °WW
26. hornblend, rocks *f & g*	Crucible Crucible broken, the two pieces liquefied, some material flowed outside. The remainder lithoid, thin pieces like scoria.	No	Yes, with quartz and mica	"	"
33. porphyry *b*	Double porcelain crucible 27–33 all similar. 33 formed obsidian glass without fusion of quartz crystals	No	Well closed, without compression	"	Time not recorded, 110 °WW
37. porphyry	Crucible placed Crucible intact inside large cast iron crucible	No	Yes, space between filled with quartz powder, then plugged	"??	69 hours, 45–50 °WW
38. amygdaloid	Crucible intact Perfect liquefaction, same crystals as no. 6	No	"	"	"

Note: Where the category is blank, it wasn't reported. "gr." refers to grains, not grams.

Drée's second group of products included samples 1, 8, 9, 24, 25, 26, and 37. For each, the powder had liquefied and resolidified into what he called *lave porcelanite* (Drée, 1808, p. 57). The solid pieces had softened to the same state on their surfaces. Only 26 had melted completely and then reconsolidated into a state intermediate between a stone and a glass. All were in a state between glass and lava, and thus not remarkable for their similarity to the lava. However, this raised the question of whether that state was due to very strong, or very prolonged heat, which issue he felt his data were insufficient to resolve.

The third series contained nos. 3, 4, 5, 19, and 21, all products of vitreous fusion. Both powder and pieces had undergone vitrification, but the included crystals had not. The crystals were the same as in porphyritic lavas. He said no. 19 was distinguished by having formed a granite. Drée felt the "porphyry paste" had melted into a glass without the feldspar crystals having been sensibly changed; they resembled his lavas F and G, from Ischia and Ténériffe. The feldspar crystals in those, however, had not been there prior to their fusion. But his trials showed that the feldspars he saw were evidence for Dolomieu's position that they had been preexistent. He would later withdraw his support for that position for no apparent reason (Drée, 1808, p. 59). He didn't intend to say that they could not form after liquefaction; there appeared to be some liquefaction in nos. 6 and 8. However, close examination showed him the difference between the two modes of formation.

Drée's fourth series, samples 21, 23, and 24, showed that higher temperatures could overcome the resistance of feldspar crystals to fusion, and they completely disappeared, while leaving the small white quartz grains. Sample 23, in which all crystals fused, demonstrated to Drée's satisfaction that porphyry could be the base of obsidian (Drée, 1808, p. 60).

The trials on the samples in his fifth series, nos. 2 and 11, were designed to show their state immediately preceding liquefaction. Sample 2, in the same position as sample 1, compared those two different porphyries. Drée remained unhappy with the crucibles he used, because they were not strong enough to contain the reactions. It is interesting that Drée incorporated two new modifications in his fusion experiments. First, he didn't begin with rocks that had either directly or apparently resulted from volcanic action. Rather, he used those that he thought might be the initial country rock and might be fused by volcanic heat. Second, he thought that pressure could prevent the escape of volatiles that might influence the result.

In 1801, Breislak was already speaking of Hall's fusion experimentation with approval. He thought it notable that Hall's and Kennedy's experimentation supported the ideas of the British experimentalist William Thompson (1760–1806) on volcanic origin so well. Breislak found it interesting that Hall and Kennedy were corroborating Thompson's conjectures by experiment virtually simultaneously (Breislak, 1801, v. 1, p. 285).[28] Later, Breislak reminded his readers that in his investigations he had

---

[28]This is William Thompson, or Thomson. Discredited in England, he had gone to Italy, and continued to work in science (H. Torrens, 2006, personal commun., from his entry for the forthcoming *Dictionary of national biography*).

access to the vast laboratories of chemistry in nature, namely, the volcanoes of Italy (Breislak, 1812, p. ix). He spoke admiringly of Hall's experiments under pressure, but said they did not allow him to support the appropriate parts of Hutton's theory (Breislak, 1812, p. 115). While noting that small experiments were no match for the immensity of volcanic action, he still subscribed to Drée's descriptions of fusion that took pressure and length of time into consideration. His observations led him to support Drée's suggestion that there are two kinds of fusion:

[D]ans la liquefaction ignée, le calorique détruit la cohésion fixe des substances, sans changer leur nature; dans la fusion vitreuse, au contraire, toutes les substances composantes sont dissoutes pour former le verre, matière homogène qui n'a plus aucun rapport avec les substances primitives. (Breislak, 1812, p. 138)[29]

Small samples of lava taken into a laboratory would be cooled quickly in the air. But in the much larger operations of nature, vitreous fusion would give way to *liquefaction ignée*, in which some mineral features are preserved (Breislak, 1812, p. 138–139).[30] Breislak repeated virtually the same arguments in his *Institutions géologiques* (1818). He used the term *fluidité pierreuse*, which corresponded to Drée's *liquefaction ignée*, and explained what the terms meant to him (Breislak, 1818, v. 1, p. 352, 354). Breislak questioned whether fusion temperatures of minerals would be the same in a volcano where the softening rock would be under enormous pressure as they would be in a much smaller sample not under pressure and in contact with the atmosphere (Breislak, 1818, v. 1, p. 358). As we shall see in Chapter 11, Breislak went on to review theories of basalt origin in the light of experimental evidence.

One can see from this rather extensive sampling of fusion experiments that the cause of the origin of basalt was anything but obvious. Clearly, the rocks eventually associated with volcanoes could be fused, and at easily attainable temperatures. The frequent blowpipe fusions had underscored the fact that vast heat sources were not necessary. Hall began his fusion work with both lavas and rocks such as whinstone, which immediate association of with volcanoes had not been documented. Drée began with country rocks. Both ended with products not unlike the basalt, lava, scoria, and obsidian that could be seen in the vicinity of active volcanoes. But neither provided definitive evidence of original igneous origin. Similar rocks could be produced by temperatures similar to those recorded in volcanoes, so the argument that basalt could be a rock originally deposited from solution and then altered by volcanic fire was still possible.

Analysis

Chemical

As we have seen in Chapter 7, mineral and rock analysis was well developed by the end of the eighteenth century, and that of basalt had been a matter of great interest. A good deal had also been learned about mineral waters and the gases that came from volcano openings. As noted earlier, Breislak and others had had no problems associating water content with salts found on a volcano. Those salts that accumulated from fumaroles had been identified, at least to the limits of the chemistry of the time. All of that knowledge contributed to theories of volcanic origin. Some of this work was done at the volcano itself, utilizing the traveling chests of analysis materials and instruments, such as those mentioned earlier, while samples also were collected to be taken back to the laboratory. Our main "characters"—Saussure, Dolomieu, Spallanzani, Breislak, and Hall—all included descriptions of how and where they collected, and sometimes analyzed, samples while in the field. The samples taken back to the laboratory were sometimes broken up and shared.

Examples of the latter are many, but three discussed earlier are the mineral water analyses that Joseph Black did of the Icelandic mineral waters sent to him by Sir Joseph Banks (1743–1820); Breislak's reports on the mineral water analyses done by chemists from the hospital for incurables; and the lava, whin, and basalt samples that Kennedy analyzed for Hall, and which were in part supplied by Hall. Hutton is not now associated with being an analyst, but his work with zeolite, in which he identified soda as a component, was cited by Kennedy as an illustration that it could be found in minerals (Kennedy, 1805, p. 87). Kennedy's discovery of soda (sodium carbonate, Na_2CO_3) in basalt both explained some anomalies in basalt analysis, where there had often been totals of <100%, and gave support to some theories of basalt origin. Perhaps unfortunately for the relation to Hutton's theories, soda was used by some Wernerians to postulate that the different characters of fused basalts resulting after heating were due to the escape of volatile elements such as the soda, and not to slow cooling after an igneous origin.

Blowpipe

The blowpipe demonstrated not only fusibility at easily attained temperatures, but also provided a clue to composition. Spallanzani remarked, "Though the blowing pipe did not in general greatly conduce to the success of my experiments, I sometimes found it useful" (Spallanzani, 1798, v. 1, p. xxiii). He did find it useful when augmented with oxygen. Kirwan estimated that the heat of a blowpipe rarely got as high as 125 °WW, and never exceeded 130 °WW (Kirwan, 1794a, v. 1, p. 43).[31] In 1819, Bakewell used the figure of 125 °WW, which he identified as "the strongest heat that can be produced by a blacksmith's forge"

---

[29]"[I]n igneous fusion, caloric [heat] destroys the regular relationship of the substances, without changing their nature; on the contrary, in vitreous fusion all the component substances are dissolved to form glass, a homogenous material that has no relation to the original substances" (my translation).

[30]In Breislak's 1818 book, *Institutions géologiques*, some passages are word-for-word those in the 1812 work, such as the discussion of Drée's two kinds of fusion (p. 352+ in 1818, v. 1). Others are different, either by only word substitution, sometimes "heat" for "calorique" although he continued to use the latter, or with new material. I have not had an opportunity to compare the two books page by page. Breislak's ideas about fusion didn't appear to change that much over that time.

[31]Cleaveland noted that although that was what Kirwan reported, Brongniart extended it to 150 °WW. However, the compound blowpipe utilizing hydrogen and oxygen exceeded it (Cleaveland, [1816] 1978, p. 65).

(Bakewell, 1819, p. 59). In 1784, Kirwan had reported that several substances such as argillaceous earth and "iron in a more or less dephlogisticated state" served as a menstruum in making earths and stones more fusible (Kirwan, 1784, p. 14). It was not a great leap to associate the well-known iron content of basalt with its fusibility. Kirwan also recorded the reactions of a good many specimens of lava as observed under the blowpipe by a number of people, and defined that rock as:

the immediate product of liquefaction or vitrification by volcanic fire, which should carefully be distinguished from the subsequent productions affected by the water either in a liquid or fluid state, which is generally ejected at the same time. (Kirwan, 1784, p. 135)

He designated three varieties of lava—cellular or frothy, compact, and vitreous—and cited both fusion results and analyses of a number of observers. He noted that Saussure had "ingeniously imitated all these species of lava" by starting with a compound argillaceous species that was mainly hornstone (Kirwan, 1784, p. 138). In 1794, Kirwan described lavas in the same way.

Saussure subjected many rocks to the blowpipe, including some lavas and the minerals that filled voids in them. A material on the surface of a somewhat decomposed basalt melted at the first touch of the blowpipe. The basalt itself melted easily in the blowpipe flame (Saussure, 1796, v. 3, p. 321). Saussure felt obligated to explain why he identified this particular rock (from Beaulieu) as a basalt. When he described the lava associated with it, he noted the external characteristics, and then stated that the blowpipe rendered it into a black and brilliant enamel, from which the standard amount produced a globule of 0.8 (no units given), which corresponded to 71 °WW (Saussure, 1796, v. 3, p. 323). In 1794, Saussure had published his table of fusibilities for a great many minerals,[32] in which he noted the inverse relation between the size of the globule formed under the blowpipe and the Wedgwood temperature required, which he was able to quantify (Saussure, 1794, p. 33–44). To measure the size of his globules, Saussure used a micrometer (Ellenberger, 1984, p. 47).

By 1816, the still-nascent field of American geology was well informed. About basalt, Cleaveland noted:

Before the blowpipe it melts into an opaque, black or grayish black glass, which is often attracted by the magnet. Its melting point is not far from 100° W. and, when *very slowly cooled*, melted Basalt resumes its former *stony* aspect. (Cleaveland, [1816] 1978, p. 279)

He continued by citing the values Kennedy got for its analysis, and mentioned that it was subject to decomposition. He described columnar, tabular, globular, and amorphous basalt. When large

enough to be separated from the matrix, the minerals found in basalt and lava were separately subjected to the blowpipe. The expected olivine, pyroxenes, hornblendes, and calcium-rich feldspars were investigated by the blowpipe in this manner. Additionally, both Breislak and Cleaveland reported on the characteristics of idocrase, or vesuvianite, an interesting mineral that results from contact metamorphism near limestone. On Vesuvius it is formed as the result of calcareous blocks ejected from the volcano.[33] Breislak called it the most beautiful of the crystals on Monte Somma (part of the Vesuvian system). La Métherie called it *hyacintines*, Werner dubbed it *vésuviennes*, while Haüy called it *idocrases* (Breislak, 1801, v. 1, p. 157). Cleaveland reported what a number of mineralogists had said about it, including that under the blowpipe it easily melted to a yellowish translucent glass, which afterward turned black. He noted that it was "found abundantly in the vicinity of Vesuvius in the cavities of a rock, composed chiefly of quartz, feldspar, mica, talc, and carbonate of lime" (Cleaveland, 1816, p. 302).

Cordier's Methods

Chemical analyses of basalts and lavas continued into the nineteenth century and beyond. Cordier extended the methods of mineral identification available for fine-grained rocks, such as basalt and lava, in the early nineteenth century. For this, he was another candidate for the honor of doing the definitive work to establish that basalt is igneous (Ellenberger, 1984, p. 44). While some of his contemporaries recognized the importance of his work, there were still doubters about basalt origin.

Ellenberger credited Cordier with being the first to utilize the microscope in geology in a methodical way. It had not been unusual for the magnifying glass or microscope to be employed to see more detail in crystal fragments or small crystals, or to observe the results of fusion or blowpipe analysis. As one example, Saussure had discussed a yellow powdery substance he had found, of a sulfur color, but incombustible, which he observed with a lens to consist of transparent white and yellow semitransparent grains, neither of a regular form. He identified the white grains as "spath calcaire," or calcium sulfate, which was insoluble in nitric acid. The yellow grains were a ferruginous mineral that at first was not magnetic, but after exposure to the blowpipe was attracted to a magnet from farther than a line of distance. He also noted that those grains lost their transparency after heating, and were covered with dark enamel (Saussure, 1796, v. 3, p. 320). Saussure was applying these techniques to samples from near the extinct volcano of Beaulieu in Provence, and he noted their correspondence to those from other volcanoes.

Many samples of basalt and lava don't have visible crystals. On close inspection, particularly with magnification, shiny facets of miniscule crystals may be visible. But the actual mineral content, whether of one or several kinds, cannot be determined by simple inspection. For this reason, it was difficult to fit any

---

[32]In the table, he first found the diameter of a globule of a standard sample fused in the blowpipe with the Wedgwood temperature required to fuse a cube of the same mineral. Then he could determine the WW fusion temperature of another mineral by comparing its globule size to that of the first mineral. He first discussed using vitrified glass as the standard, then compared it to feldspar to get the fusion temperature of feldspar (Saussure, 1794, p. 10).

[33]Idocrase is a sorosilicate, $Ca_{10}Mg_2Al_4(SiO_4)_5(Si_2O_7)_2(OH)_4$.

fine-grained rock into any of the rock classification schemes.[34] As mentioned in Chapter 6, to solve the problem Cordier combined several procedures for learning about small fragments. Saussure had found a way to keep very fine particles (the size advantageous for fusion) of rock in the flame of a blowpipe, rather than being blown away, of which he said:

Cette facilité d'examiner les fragmens d'une extreme petitesse m'a été souvent avantageuse pour la connoissance des pierres composées; j'en ai trouvé qui résultoient de l'assemblage de différens grains, tous si petits que je n'aurois pas pu les assujetir séparément sur le charbon. (Saussure, 1785, p. 410)[35]

He commented on the usefulness of the technique for examining small samples of rare rocks, and also stated that he used a strong lens to observe the results carefully. Interestingly, he remarked on the color of the flame at the blowpipe in some instances (Saussure, 1785, p. 412). These techniques were refined and adapted, and Saussure, who was in correspondence with a number of other practitioners, made improvements as published in 1794. One can see the germ of Cordier's method in this earlier work.

Cordier published two "Mémoires" on his investigations of volcanic products in *Journal des Mines* (the first in 1807; the second in 1808). His samples were those he had collected himself, or were from the collections of a number of other workers (Haüy, La Métherie, Dolomieu, whose samples were then in the possession of Drée), all of which had well-recorded localities (Cordier, 1808, p. 56). His stated purpose was to investigate whether the rocks were a product of subterranean fire. In the process of investigating the *fer titané* minerals, one can see the beginnings of his better-known efforts to determine all the minerals in basalt (Cordier, 1808, p. 58–62).[36] He couldn't always see particles of the mineral in his samples, and they did not always respond to a magnet as might be expected. However, after powdering and sieving the samples, the magnetic grains could be separated from the earthy portions. He could then also see metallic grains by means of a microscope, aided by a micrometer, so in this way he could identify the mineral. The second "Mémoire" provided much detail about his methods and results. He commended Kennedy's analysis of a basalt that Cordier also analyzed, which included soda and muriatic acid, but had missed the *fer titané* and a little manganese oxide (Cordier, 1808, p. 66).

Cordier read his paper on the determination of minerals in volcanic rocks in 1815, and it was published in *Journal de Physique* in 1816. Brongniart gave a detailed accounting of it in *Journal des Mines* in 1815 (Cordier, 1815). In it, he noted that the origin of many rocks was still contested. Cordier discussed the difficulty of separating the minerals in fine-grained volcanic rocks so that they could be identified. His method was:

1. Grinding solid rocks until they had a "thinness" (particle size) of 1/10th to 1/100th millimeters
2. Washing, to separate the particles by density
3. Looking at the parts with a microscope to see their form and to see their fracture
4. Using acids, the magnet, and the blowpipe by Saussure's methods, to determine their nature
5. To relate the separated minerals to those found most often in volcanoes, and compare them to known minerals of those kinds. (Cordier [Brongniart], 1815, p. 383–384)

Here, knowledge of methods of technology, never far removed from blowpipe studies, came to Cordier's assistance. It was a normal act in mining to separate heavy ore minerals from gangue, or matrix, by flotation. As discussed in the chapter on characters, specific gravity was another known and useful quantity. So, in addition to magnetic separation, Cordier used specific gravity. He put the rock powder on one side of a glass plate, inclined it slightly, and then gently tapped the plate. The heaviest particles fell to the bottom, while the others spread out. He cautioned that the particles must be nearly equal in volume to each other. Portions of the weight-segregated particles could then be examined under the microscope. They clearly showed cleavage faces. Cordier put much effort into the use of his microscope and lighting arrangements, settling on a background of white paper. He followed the methods of Saussure in identifying the powders under his blowpipe, which, with a large bulb, was said to be able to produce a temperature of 500 °WW (Cordier [d'Orbigny edition], 1868, p. xiii). This is in contrast to the figure used by Robert Bakewell in 1819: that the strongest heat of a blowpipe used with a candle was at the tip of the internal cone of blue flame within the outer orange flame, and corresponded to "125° of Wedgewood's [*sic*] pyrometer" (Bakewell, 1819, p. 59). In 1833, Bakewell cited Cordier's work in an extensive footnote. He listed feldspar, leucite, chrysolite, augite, hornblende, "iron sand," and olivine—along with their characteristics—which Cordier had been able to identify with his methods (Bakewell, [1833] 1978, p. 280).

Fluid Contents

Water. Fluid inclusions had been noted in rocks and minerals for a long time. The first instances were in Roman times, and by the eleventh century there was a complete description. In later times, we find the names of Boyle and Steno, followed by Dolomieu, Davy, Brewster, and others associated with studies of fluid inclusions (Wiesheu and Hein, 1998, p. 312–313). Many of them included water solution inclusions in quartz, where they were very noticeable. The solutes included in the

---

[34]Nicol's 1815 initiation of the thin section was discussed in Chapter 6. Thin sections were subsequently employed in the polarizing microscope, which he invented in 1829.

[35]"This facility to examine extremely small fragments has often been advantageous to me for knowledge of compound rocks; I have found in them an assemblage of different particles, all so small that I was not able to fix them separately on carbon [to subject them to the blowpipe]" (my translation).

[36]Titanium was identified as the oxide in 1791 by William Gregor (1761–1817), and was rediscovered and named by Klaproth; the metal was isolated by Berzelius in 1825 (Ihde, [1964] 1984, p. 747).

water—often halite (mineral NaCl), but sometimes the gases carbon dioxide or hydrogen sulfide—were soon identified. Dolomieu "may have been" the first to find petroleum inclusions in quartz, and Breislak definitely reported this phenomenon (Wiesheu and Hein, 1998, p. 313). Cavities within basalt had been seen to contain water, and could serve as arguments in favor of Neptunism, as they appeared to be the residuum of the solutions from which the rock or mineral crystallized, as well as proving that they could not have been subjected to the heat invoked by the Plutonists (Cleaveland, [1816] 1978, p. 279, 286). The fluid inclusions seen in basalt were cited by most Neptunists as being inimical to origin by heat, as were those in quartz for granite. Basalt analyses routinely included water as a constituent, determined by weight loss in preliminary heating, while lava did not contain it.[37]

Gases. The presence of gas within rock or mineral voids was not visually so obvious, although a little later vapor bubbles within the liquid were noted by an English student of optics, David Brewster (1781–1868), in his investigations of fluid inclusions in a number of minerals (1826a). Clearly, volcanic rocks, lava and pumice, contained numerous cavities, while basalt, especially that found away from known volcanoes, rarely showed visible cavities. Spallanzani, Breislak, and numerous other workers identified the gases emanating from volcanoes, Breislak being aided by his ingenious apparatus, as detailed in Chapter 6. Spallanzani stated:

[I]n the present age, in which naturalists and chemists are so earnestly employed in analytical researches relative to the nature of aëriform fluids, it is not sufficient to assert and prove the presence of gases in liquefied volcanic products; it is likewise incumbent on us to discover their peculiar nature, and thus prepare the means to ascertain what part they may take in the eruptions of volcanoes. (Spallanzani, 1798, v. 3, p. 324)

He had noted that gases are the apparent agent that raises lava in the crater. Samples subjected to the heat of the furnace acquire vacuities as well, sometimes foaming over the edges of their crucibles. Spallanzani wanted to identify the gases responsible. He put samples to be fused in a "matrass," a round flask with a long, narrow neck used in distillation, to the neck of which he attached a glass tube which extended under the mercury of a pneumatic apparatus. The matrasses were made of clay, six lines thick,[38] and he tested their tightness to gas by blowing into them, reporting that not one bubble of air escaped. He coated the flasks, and then tried them with the air pump, and determined that no gas was escaping. He also checked after the experiment to be

sure that no gas escaped, and that none entered from the outside (Spallanzani, 1798, v. 3, p. 326).

The first samples Spallanzani examined were pulverized volcanic glasses. Small amounts of what he showed to be atmospheric air appeared above the mercury. However, to summarize his detailed descriptions, he found that when cooled, his samples had bubbles of various sizes. By means of close scrutiny of the flasks and the sample, it appeared that the material itself must have vaporized, and redeposited (by sublimation) on the upper, cooler, parts of the flask. He found the same effect when he used covered crucibles. He performed similar experiments on six samples of lava, which he had found less likely to inflate with bubbles on heating. With only a little air appearing over the mercury, and the other appearances the same after strong heating, he declared:

I conceive myself sufficiently warranted to conclude, that those bubbles and inflations of various sizes, which we so frequently observe in the products of volcanoes, are by no means caused by the action of any permanent gas, but by that of an aëriform fluid produced by the excessive attenuation of those same products in consequence of heat. (Spallanzani, 1798, v. 3, p. 345)

However, he noted Joseph Priestley's experiments that seemed to indicate that some carbon dioxide could be present, and decided that some volcanic products might include the gases. Spallanzani did suggest that most volcanic action attributable to gases was due to gasification of the lava itself (Spallanzani, 1798, v. 3, p. 348).

However, Spallanzani felt that this was probably not enough to explain the force with which stones were thrown from volcanoes. That great force seemed to require different gases, which he determined to investigate. He noted that several workers had observed and identified gases rising from quiescent volcanoes, for which he listed "hydrogenous gas, sulphurated hydrogenous gas, carbonic acid gas, sulphureous acid gas, azotic gas, etc." (Spallanzani, 1798, v. 3, p. 353). But the most powerful aid to the explosive power that he could imagine was water. He expressed the old idea that volcanoes are either surrounded by the sea or not distant from it, and were supposedly connected thereto by secret passages whose existence he felt was proved by the evident presence of water expanding into vapor as a result of powerful subterranean explosions. He used the vivid example of the violent explosion that resulted when a cannon, hot from being fired, was cleared by a wet rammer. The resultant force of water changed to vapor "will force the rammer out with such violence as sometimes to carry away the arm of the gunner" (Spallanzani, 1798, v. 3, p. 356). He followed this with an example of molten metal being poured into a damp mold, resulting in death for some bystanders. Spallanzani also repeated experiments that involved pouring water onto molten glass, which he observed to result only in the rapid disappearance of the water. From his perspective, this showed that water falling into the mouths of volcanoes could not result in violent eruptions. To test this idea further he melted various metals in a furnace in open crucibles and dropped water onto them. Melted tin produced the most violent explosions, but he could find no consistent explanation for the various results.

---

[37]We can easily second-guess this to be moisture acquired from hydration or simply from atmospheric conditions, but the samples used were visibly dry. Moreover, minerals occupying the cavities such as zeolites, retained their water of crystallization. This was objective fact at the time, prior to the knowledge of secondary mineralization.

[38]The line, or "ligne," was 2.256 mm.

Spallanzani resolved to try the same experiments using lava as the sample. He found the same lack of "eruption" when he dropped water onto hot lava that was not porous, as had occurred with respect to glass, iron, and copper. A violent reaction occurred, however, when he dropped water onto a basalt sample with a large bubble in it. This didn't seem to be related to the lava itself, but to the accidental inclusion of the bubble. He then resolved to fuse lava of the kind that swelled and overran the crucible when heated strongly, and then pierced it so that water could be dropped into the resultant hole. He was concerned that the reaction would be violent, so he removed the apparatus to an outside courtyard and poured the water in through a long tube that

reached the vessel through a hole in the door that shut in the courtyard. As soon as the water entered the cavity, the pieces of the receiver and of the lava were forced violently to the distance of many feet, with a detonation equal to the report of a musket. (Spallanzani, 1798, v. 3, p. 373)

He had used fresh water in those investigations, but as he thought volcanoes were probably affected by seawater, he tried the same experiments with it and found the same results. This proved to him that water falling into a crater would not cause an explosion, but water entering from below, where it could not easily expand, would, and that water vapor was capable of causing volcanic eruptions (Spallanzani, 1798, v. 3, p. 374). These experiments convinced Spallanzani that water vapor was a likelier explanation for volcanic explosions than Dolomieu's supposition that "hydrogenous" gas was the cause (Spallanzani, 1798, v. 3, p. 378).

CONCLUSION

The experimenters of the eighteenth and early nineteenth centuries used all of their ingenuity to investigate the problems associated with volcanism. In so doing, they built on the legacy of both alchemy and technology, with associated instruments and techniques. We see them rather as the blind men with the elephant, trying to grasp a problem of enormous scale by methods that could only be used on very small samples. In addition to the workers who've been discussed, there were many others interested in the questions they raised, and who considered them to be important. How else can one explain the interest in the copious accounts of travels and observations of volcanism? The work of Sir William Hamilton (1730–1803), George III's ambassador to the court of Naples, is particularly noteworthy. Though not trained as a natural philosopher, he became enamored of volcanoes, climbed Vesuvius many times, often during eruptions, and traveled to other volcanic regions. His careful reports to the Royal Society gained him membership, although they were basically purely descriptive.[39] Trips to see volcanoes were part of the standard Continental tour for travelers of many nationalities, and Hamilton frequently played host to British travelers.

The simple observation of real, erupting, volcanoes translated to interpretation of those that were extinct, although the parallels were not always obvious. The German geologist Leopold von Buch (1774–1853) remarked that seeing real volcanoes didn't resolve questions about basalt (Nieuwenkamp, 1970, p. 554). Each new observation brought new possibilities for interpretation, rather than being confirmation of either Neptunian or Vulcanist origin. Throughout reports of the experimental work, there runs the frequent thread of exasperation with methods that didn't work, that were inexact, and/or equipment that failed in function or reproducibility. Great inventiveness was often displayed in experimental setups, instruments, and interpretation of results. We will continue with the ideas the experimenters had on the question of rock origin, as informed by their work.

---

[39]It has been pointed out that Hamilton did suggest that volcanoes could be generative instead of merely destructive (K. Taylor, 2008, personal commun.).

CHAPTER 11

Rock origin from the experimenters

L'eau n'a eu aucune part à la génération des crystaux que je viens de décrire.
La division primordiale de la matière dont ils sont composés, a été opérée
par le feu qui a réduit la substance métallique dans une fusion parfaite.
—Grignon (1775, p. 477)[1]

INTRODUCTION

The above quote, although about metal crystals, demonstrates the ongoing interplay of thought about the roles of fire and water in rock formations in the latter half of the eighteenth century. At this point, we can generalize about the importance of experiment in the geology of the late eighteenth and early nineteenth centuries. In earlier chapters, we looked at the methods and equipment that were used, what parameters were considered, and how the information thus obtained was applied in experiments on the rocks themselves. This chapter sets out the opinions of many workers who experimented on earth materials during that period, and some who may not have experimented themselves but argued from the experimental data of others.

Surveying the writings about Earth during the eighteenth century shows a gradual decrease in the proportion of earth science literature occupied by "theories of Earth," and the increasing presence of technical/quantitative articles, aided by the appearance and proliferation of scientific journals. There is often an unclear distinction between technology and science. As mentioned in Chapter 1, science and technology have always informed and complemented each other. Perhaps because of the nature of the materials, this is more the case with geoscience than with other sciences, although certainly chemistry runs a close second. Because the tools of technology were used extensively with materials such as ores, glass, and clay, with the appropriate additions, it is not surprising that those tools, for example, furnaces and the containers used within them, would be employed to try to investigate the complexities of rock origin. The processes of metallurgy, pottery and ceramics in general, and glass making, and even soap and salt production and dyeing, supplied analogies or information that could advance conjectures about rock origin. What we now call "physics" and "chemistry" also had a large role.

In Chapter 1, I suggested that experimentation aided by solution or fusion was more influential than theoretical systems or

field observations in determining opinions about rock origin. We see that those questions remained unresolved for much longer than is often suggested, and that there was no consensus about the formation of "igneous" or some sedimentary rocks even after Hall's experiments at the turn of the eighteenth century to the nineteenth century. Experimentation began sooner, lasted longer, and was more widely practiced than is often realized. It did not provide positive proof of either sort of origin, aqueous or igneous, although experimenters had begun to delineate the necessary conditions of composition, heat, and pressure for rock formation. General understanding of rock-forming processes thus increased enormously. No less importantly, essential variables were identified and myths discredited. Although unsolved, the nature of the puzzle was recognized in a way that the combination of theory and the field could not resolve by itself. So what were the arguments from experiment?

LIMESTONE ORIGIN

Marine fossils were rather familiar objects in the eighteenth century. Martin Rudwick ([1972] 1976) has provided an excellent history of fossil studies from the sixteenth through the nineteenth centuries, after noting the early use of the word "fossil" as meaning most things taken from Earth, including minerals. The meaning of the word gradually assumed the more limited sense in which we use it today. At least from the time of the Swiss naturalist Conrad Gesner's (1516–1565) extensive work, the relation of petrified or otherwise preserved remains to living organisms was recognized, although not by everyone (Rudwick, [1972] 1976, p. 30). It is a great oversimplification to assume otherwise. In 1667, Steno convincingly demonstrated that so-called tongue-stones were the remains of shark teeth (Rudwick, [1972] 1976, p. 50). The remains of marine animals found high in mountains did pose a conundrum, but to the increasingly observant natural philosophers of the eighteenth century, there was not as much mystery as popular literature might have us believe. Conjectures about a mechanism to raise mountains or lower sea level were many. James Hutton used the fact that seashells were found on mountains as evidence that land was consolidated from

---

[1]"Water has had no part in the generation of crystals that I have described. The primordial division of the matter of which they are composed had been carried out by fire, which reduced metallic substances in a perfect fusion" (my translation).

sedimentation in the sea. None of this was incompatible with religion, as the Deluge could always be invoked as an explanation.

Limestone was a familiar substance, having been used as a fertilizer, a flux in smelting, a component of glass, and in mortar for untold years. Marble was recognized as essentially the same substance, although its increased crystallinity, hardness, and ability to be polished were noted, and it was employed for statuary and other ornamental uses. Roasting to change limestone from the carbonate to the oxide, and with the addition of water to form the reagent "limewater" $(Ca(OH)_2)$, was a common operation. Among the conjectures about rock origin, perhaps Hutton's ideas about limestone were the most plausible. That it might collect at the bottom of the sea as a sediment, which included marine animal detritus, was credible. That those deposits could be found at high elevations was perhaps not too unlikely, given the evidence of earthquakes and contorted layers, especially if one had doubts about sediments accumulating on sloping surfaces. But the reasons for the greatest doubts about this process were drawn from experience—the knowledge that limestone gave up fixed air (carbon dioxide) when heated, and Hutton's scheme required heat. It was difficult to imagine that marine shells, many of them fragile, could remain, sometimes intact, in the process envisaged by Hutton. He suggested, however, that the enormous pressure of superincumbent rock masses and/or the weight of water in the ocean would constrain the loss of carbon dioxide. Hall's famous and much admired experiments on limestone, first published briefly and then read in 1805, and published in full in 1812, seemed to corroborate that view. But apparently that evidence wasn't conclusive for everyone. And what gave rise to the questions was not necessarily theory or fieldwork, but was more considered thought on the part of the experimenters.

Chemistry contributed more than just analysis. Solubility characteristics of both carbon dioxide and phosphates entered the arguments of heat versus solution. The arguments about phosphoric acid, to which numerous workers contributed, were discussed in Chapter 9. Other arguments about limestone and marble origin based on knowledge of their chemical and physical characteristics had been frequent. In 1742, when Argenville discussed the origin of marble, he gave a brief reference to an earlier work (p. 59). He concurred with previous suggestions that marble must have been formed from a solidifying juice, variously colored by dissolved substances from nearby mines, such as copper or iron. He said:

Les Congellations qui croissent en longueur formant des cylindres s'apellent *Stalactites*, elles sont transparentes comme l'eau & de diverses figures souvent pyramidales, différentes en cela des *Stalagmites* qui sont opaques & toujours rondes. (Argenville, 1742, p. 80)[2]

And he illustrated those congellations. He was clear about their mode of formation. Argenville also distinguished petrifactions from stones. He felt that stones could hardly be formed in the same way that petrifactions were, because there was no space in their solid matrix for lapidifying juices to penetrate.

Buffon observed the reactions of the "pierres calcaires" used for the exterior walls of his furnace. When exposed to that very moderate heat for a long time, the hardness increased proportionally to the duration of the time heat was applied and the walls gained weight (Buffon, 1774, p. 69). However, when the same stones were placed in a more intense heat for a long time, they lost nearly half their weight as well as the cohesion between the constituent particles. He found the two things difficult to explain. On the other hand, Werner described limestone in detail, and discussed the differences between those he called "primitive" and those that belonged to the later floetz formations. As always, he listed their properties carefully, and included discussion of stratigraphy, albeit of a rather local variety. Werner made many comments about rock origin, based on observation. He said that a rock had to be of aqueous origin if it contained remains of organic bodies, or if it included water of crystallization (Werner, [1786] 1971, p. 103).

Using his multiple field and laboratory observations, Saussure thought deeply about rock origin, ever ready to see convergences that hopefully led to understanding. In a kind of summary comment, having reported detail about the appearance of precipitated calcite crystals in a solution from a year-long investigation into mineral water, he wrote, "Ce fait, petit en apparence, me paroît être d'une grande consequence pour la théorie de la formation des montagnes au milieu des eaux" (Saussure, 1779, p. 213).[3]

In a lengthy letter to La Métherie, Deluc cited Saussure as an authority who supported his views of the aqueous origin of limestone. He began by saying that the ideas of several naturalists that calcareous beds were formed by the remains of marine animals acquired some plausibility from chemical analysis, but couldn't stand up to examination by geology (Deluc, 1790b, p. 441–442). In opposition to the idea that limestone beds (*couches calcaires*) were formed from marine animals, he pointed out that similar beds might be formed with very different marine creatures contained therein, whereas quite different kinds of beds might have the same animals in them. Chemical analysis couldn't account for that. He noted the difficulty of progressing to general causes from specific examples. As he went on to say at some length, there were limited cases where calcareous concretions were clearly formed from marine animals (Deluc, 1790b, p. 444). To explain what he saw, he postulated a number of epochs on Earth where conditions and operations differed from those today.

Dolomieu favored precipitation from an aqueous solution for primitive (crystalline) calcareous beds, the result of limestone's solubility in water.[4] Those he identified as marble, a rock

___

[2]"The congellations (concretions) that grow lengthwise forming cylinders called *stalactites* are transparent like water and have different, often pyramidal faces, differing from those of *stalagmites* that are opaque and always round" (my translation).

We recognize that the calcium carbonate of those formations is distinguished from marble, which in our definition is subjected to some metamorphism.

[3]"This fact, appearing small, seems to me to be of great consequence for the theory of the formation of mountains by means of (in the midst of) water" (my translation).

[4]Formation of stalactites, stalagmites, and tufa, as well as the result from mineral water analysis, gave evidence of slight solubility.

that could be polished to a luster. Marbles retained some of their water of crystallization. However, he rejected that origin for Secondary and Tertiary limestones, and all those that contained shells, because he couldn't explain how there could be enough water to dissolve those great masses (Dolomieu, 1791b, p. 387). He admitted the difficulty of suggesting a plausible mode of origin for them, stating it was one of the problems in geology that he found the most difficult to resolve. Obviously those with shells had been deposited on the bottom of the sea, and the sea had aided in the deposition. He asked three questions: Where had the material come from? How had it been taken up? And how had it been transported and deposited? Calcareous earth was formed along with the other earths, and was included in or with many of them. For the rest, in which he included salt beds and sand movements, he invoked actions during the extensive epochs of Earth, and actions and deposits presently observed on Earth, to end with a plausible explanation. To do this he suggested three questions to be answered: Where did the sea get the material from which the beds were formed? How did it take it up? And how did it transport and deposit it? (Dolomieu, 1791b, p. 390). The answer to the first was the easiest, because the calcareous material had been there since primitive times, and it could be dissolved and re-precipitated. There are no animal remains to be found in primitive limestones. Subsequently there was a great catastrophe, disrupting the beds and causing them to be mixed with fragments of animal and vegetable life. Finally, water rose, some dissolved, and the secondary limestone was deposited or re-precipitated with its accompanying material, sometimes horizontally, sometimes on uneven ground. Dolomieu followed with a number of examples that illustrated his suggestions (Dolomieu, 1791b, p. 389–394).

However, La Métherie reviewed the opinions of many workers about limestone origin in secondary deposits, and noted several places in nature where there was what he identified as *chaux* (caustic lime, CaO) as opposed to *crai* (calcium carbonate, $CaCO_3$). He found an explanation for this in the presence of volcanic heat or heated water (La Métherie, 1795, v. 1, p. 309–310). He noted that caustic lime was not found in primitive terrain.[5]

Kirwan didn't alter his views about limestone too much in the 1790s, although he used increasing evidence from experiment. Many arguments were concerned with observations about solubility and crystallization, and conditions for both. He cited Werner, Johann Carl Wilhelm Voight (1752–1821), Charpentier, Saussure, the Swede Johann Jakob Ferber (1743–1790), Herman, etc., when discussing that granular-textured limestone is primeval, or primitive and "no petrefactions [*sic*] are ever found in it" (Kirwan, [1799a] 1978, p. 213). Limestone in Secondary formations might include twigs, leaves, or petrified shells. In lengthy discussions of solution, concentration, and precipitation, Kirwan showed his familiarity with the solution behavior of calcium carbonate and lime, as well as of silica. He reviewed the inability of furnaces and burning mirrors to fuse limestone, and wondered how granular limestone, supposedly subjected to great heat, could be bordered by the lower-fusing argillite without the latter showing evidence of heat (Kirwan, [1799a] 1978, p. 457).

Lining up arguments for and against the Neptunian or Huttonian systems was a frequent activity for experimenters at the beginning of the nineteenth century, as a good bit of literature attests. Often members of opposite camps would fasten on the same piece of evidence and claim that it supported their thesis. Playfair maintained that Saussure's and others' lack of success in fusing marble didn't disprove Hutton's theory (Playfair, [1802] 1964, p. 184).

Hall (1812) believed that he had fused marble under pressure, which would allow it to retain its fixed air, and, in a long explanation that brought in the pressure of superincumbent basalt, he explained how limestone could remain uncalcined even when exposed to volcanic heat. He had clearly shown that the heat required for fusion under pressure was far in excess of that needed for basalt and lava fusion. The resulting pressure where lava originated could easily retain carbon dioxide in limestone, which a lava had passed through. His calculations, summarized in a table, showed that

the carbonic acid of limestone cannot be constrained in heat by a pressure less than that of 1708 feet of sea, which corresponds nearly to 600 feet of liquid lava. As soon, then, as our calcareous mass rose to within 600 feet of the surface, its carbonic acid would quit the lime, and, assuming a gaseous form, would add to the eruptive effervescence. (Hall, 1812, p. 161)

The arguments by no means stopped there. During this time the Irish experimenter William Richardson (1740–1820) discovered fossil marine shells, ammonites, in what he thought was basalt on the north Ireland coast (Wyse Jackson, 2000, p. 43). He also mentioned the anomaly of the presence of silica particles in limestone. Supposedly, they should have fluxed together if that limestone had been subjected to heat. Bakewell disastrously argued that "lime and flint are convertible into each other by natural processes," citing instances where shells known to be made of calcium carbonate were found as fossils of flint (Bakewell, 1813, p. 171).[6]

In 1819, Bakewell had been influenced by Hall's discoveries. Knowing that shells and limestone had the same chemical composition, he thought that

their [the shells'] transition into the finest varieties of lime-stone or statuary marble is not so extraordinary as many changes observable in the mineral kingdom. It would not be difficult to conceive, that any cause which could effect a disintegration of the shell or coral, would be sufficient to form the earthy varieties of lime-stone; for from the partial solubility of carbonate of lime in water, the particles would soon be united into one compact mass. (Bakewell, 1819, p. 410)

---

[5]We wonder if he saw caustic lime anywhere in nature, considering its rapid reaction with water.

[6]Gillet-Laumont gave an excellent description of the process that would replace calcium carbonate with silica dioxide in *Journal des Mines* (1796–1797, An. 5, p. 491).

But on the next page he noted that Hall's fusion experiments could explain the crystalline character of marble. Observation and experiment seemed to show that some naturally occurring varieties of calcium carbonate could survive and become more crystalline under pressure and the influence of heat. But that did not erase the evidence of origin under aqueous conditions. This duality is seen repeatedly, with Neptunian terminology applied to both formation and location long after the beginning of the nineteenth century.[7]

GRANITE ORIGIN

Like those about limestone, many theories about the origin of granite were hindered rather than helped by experimental evidence. That evidence seemed mainly to show that granite in its usual crystalline form should not exist. Yet, indubitably, it did. Field evidence was similarly confusing. John Hill described all the mineral constituents of granite, and the tests he performed on the separate minerals and the whole rock, including the fact that it was not much changed in fire. He then wrote:

[I]ts particles seem not to have all coalesced at the same time, since the black Talc bedded in the pure crystalline particles speak: that's having been fluid when the former was already coalesced; their concretions must however have fallen nearly at the same time, since their firm, mutual coalescence into this solid mass, can be owing only to their having subsided together while soft, and in a condition to cohere like the particles of melted metals, which will refuse to do so when the absence of what kept them soft or fluid has suffer'd them to acquire any degrees of their natural hardness. (Hill, 1748, v. 1, p. 499)[8]

Fusion as a means of rock origin was thus rejected. Werner and others postulated that granite was one of the primitive rocks, always at the base of the stratigraphic column. But granite clearly existed in other situations, perhaps most famously as the intruded veins in Glen Tilt seen by Hutton. Some workers had pronounced that basalt and granite graded into each other (Beddoes, 1791, p. 58; Murray, [1802] 1978, p. 243). Metal workers had described the crystals of metals and/or fluxes and slag that formed on the walls of furnaces. Like other questions of rock origin, it was a problem seemingly beyond the comprehension of humans and their laboratories. But that did not stop the speculation.

Saussure referred to Desmarest's statement that as one approached a volcano, granite could be seen to be more and more altered by fusion until there was basalt. However, Saussure pointed out that composition of the rocks under and surrounding volcanoes was variable, and the products of increased fusion of various rocks closer to volcanoes could give products unlike basalt (Saussure, 1779, p. 129–130). He raised the question of whether quartz could infiltrate other minerals and unite them, but said he was forced to abandon that idea because in most granites, quartz was not only the "glue" but often the main component. Other minerals didn't necessarily seem to have preexisted. Saussure discussed granite texture and noted that sometimes there were masses of mica in one place, quartz in another, and heaps of feldspar elsewhere. He could think of no mechanism that would explain a sequential deposition, even assuming that voids might have existed, but as had not been observed, in sand or gravel (Saussure, 1779, p. 102). He wrote:

Je crois donc que les parties du Granit sont toutes contemporaines; qu'elles ont toutes été formées dans le même élément & par la même cause; & que le principe de cette formation a été la crystallisation. (Saussure, 1779, p. 102)[9]

If the three minerals could be dissolved in water, he saw an analogy to the situation where different salts are likewise dissolved and then crystallized, and where they may be deposited more or less mixed with each other. He enumerated many instances of granite in the field that showed the mixed crystal structure. He did not believe that the materials of granite could ever be rendered homogeneous (Saussure, 1779, p. 123).

Dolomieu was convinced from his many observations of the Ischian volcanic materials, including pumice of different densities, that they were always accompanied by lava that was nearly granitic with a feldspar base (Dolomieu, 1788, p. 30). Chiefly due to its relation to basalt, which he thought was contiguous with granite and which he considered definitely fused by volcanic heat, Beddoes also considered the granite to be formed from a melt. Although some volcanic heat wasn't very considerable, he saw no reason why it couldn't vary enough to fuse the minerals in granite. Other than a nod to the common knowledge of basalt fusion during his time, Beddoes' arguments were nearly all from field observation of the relative positions of basalt and granite. He did note the very large crystals in rocks of granite composition found in veins.[10] To refute the possibility of aqueous deposition he suggested that such large crystals couldn't have been formed from the small amount of solution that could be held in the space of the vein. He used the example of what would happen if one added more of the same solution to an assemblage of niter crystals already precipitated. Instead of causing the original crystals to grow, the addition resulted in a new crop of small crystals (Beddoes, 1791, p. 70). Beddoes was respected enough in the scientific community that he was often quoted on the relation of granite to basalt, namely, that they graded into each other.

Kirwan wrote in several publications that granite had been observed to have formed "from the agglutination of its own sand" (Kirwan, 1794a, v. 1, p. 340). One instance was said to have occurred in the river Oder in 1722, where a mole made of granite

---

[7]Neptunian stratigraphic terminology was both useful and entrenched in common use. It is more difficult to understand why Hall's experiments with marble were not considered more definitive. Perhaps they seemed out of place in what was viewed as a field science. It is also tempting to postulate that at least some of the later accounts of the history of geology were written by people not conversant with experimental practice, but it would take a great deal more investigation to support that contention.

[8]This slightly ungrammatical statement is a direct quote.

[9]"Thus I believe that the parts of granite are all contemporaneous; that they have all been formed in the same element [water], and by the same cause; and that the source of this formation has been crystallization" (my translation).

[10]We would recognize these as part of a pegmatite, which was not identified as such in Beddoes' time.

with chinks filled with what he called "granite sand" was seen to have solidified completely. He also noted an instance from the literature where granite was seen to pass into sandstone (Kirwan, 1794a, v. 1, p. 341). This suggestion of an origin for granite was ridiculed by several of his contemporaries.

La Métherie had an interesting comment about granite formation that certainly arose from the numerous analyses done in his time. He first listed the 18 known metals: gold, silver, platinum, mercury, copper, lead, tin, iron, zinc, bismuth, cobalt, antimony, arsenic, nickel, manganese, tungsten, molybdenum, and uranium.[11] He considered that those metals must be primitive substances, formed before the crystallization of Earth, because they, especially iron, are found in all of the parts of granite (La Métherie, 1795, v. 1, p. 324). In other words, the metals had to exist prior to the formation of the granite to be included in it. That was a novel addition to thoughts about granite formation.

In his continuing discussion of rock origin through the 1790s, Kirwan again made the observation that the minerals of granite could not have been subjected to heat because when the minerals that compose it were heated strongly, they were considerably altered in appearance from those in the rock. He repeated the evidence that the quartz in granite didn't fuse (Kirwan, 1799a). In an infrequently cited paper from 1802(a), although versions were printed in several places, Kirwan sought to rebut the arguments of Hall and a number of other experimenters on the origin of both granite and basalt by fusion in meticulous detail. For each of their "proofs," he had a counterstatement based on specific appearances, and which suggested the possibilities of origin from solution. Murray reviewed most of the evidence about granite's origin from both sides of the argument. He argued with Hutton's theories from the standpoint of fusion temperatures and texture, and found more to object to in igneous than in aqueous origin. In 1806, Murray took notice of the action of fluoric acid on silex (silicon dioxide), as well as its preparation. The existence of a substance that could produce a menstruum, fluoric acid, which could dissolve intractable minerals, made the Neptunist position more plausible (Murray, 1806, v. 3, p. 545, 552). After a short discussion of the colors imparted to glass by very small amounts of metals, he noted the different composition of several feldspar samples. But despite his recognition of the importance of chemistry to geology, it was not sufficient in itself to identify minerals. Additionally, while Davy gave credit to Hall for his "masterly experiments" on basalt and limestone, he found no strong analogies with the case of granite or well-formed gems (Siegfried and Dott, 1980, p. 57).

Meanwhile Deluc continued to be convinced that granite was "evidently produced by chemical precipitation from a liquid," and implied that it was the most ancient of formations (Deluc, 1809, p. 49). In 1815, Kidd pointed out some reasons for the confusion about the origin of granite. He extensively reviewed Saussure's comments about granite and came to realize that, although

Saussure had defined granite as a rock consisting of quartz, feldspar, and mica, in fact he had identified as granite some rocks that contained a great many other minerals, which included tourmaline, hornblende, chlorite, steatite, and epidote (Kidd, 1815, p. 66). Kidd was concerned not only with identifying granite properly, but also with pointing out its relation to other rocks, specifically those we now call "stratified." He described various granites, both well consolidated and those that were friable, as well as noting places where the granite appeared to grade into schists. He saw that the relation of granite to gneiss could not always be clearly observed. Sometimes it appeared older, sometimes younger, than granite (Kidd, 1815, p. 69). Kidd applied those observations to musings about the origin of rocks from the view of the Huttonian or Wernerian schemes. Virtually the same comments were voiced by the capable chemist Robert Bakewell in 1813 and 1819.

In 1818, Breislak noted the extreme differences in fusibility of the minerals that make up granite as a reason to doubt its origin by fusion (Breislak, 1818, v. 1, p. 356). The gas (hydrogen or hydrogen and oxygen) blowpipe made possible the fusion of many minerals considered infusible. As would be expected, the method was applied to a great range of minerals. It was also conjectured that volcanoes, particularly Vesuvius, could be likened to a great oxygen-hydrogen blowpipe, the gases supplied by the dissociation of water by volcanic heat (Otter, 1825, v. 2, p. 418).

BASALT ORIGIN

Today basalt's origin seems so obvious that we have difficulty understanding eighteenth- and nineteenth-century debates about it. As did many of his contemporaries, Hill likened the regular columns seen at Giants Causeway to large crystals formed by precipitation (Hill, 1748, v. 1, p. 468). The columnar shapes had been discussed for several centuries, although the interpretation of their formation had undergone many changes. In 1565, Gesner mentioned them as an interesting natural "fossil." (See Figs. 11.1 and 11.2.) Grignon noted the similarity of slag and scoria in his furnaces to the products of volcanism and questioned a system that invoked currents of hot water (Grignon, 1775, p. xviii). After his observations of the *fritte* (literally, "fried material") adhering to his furnaces, he called it lava, because of its similarity to the lava of volcanoes (Grignon, 1775, p. 297). He found many examples of volcanic rock types among the furnace products.

John de Magellan, the editor and translator of the 1788 version of Cronstedt (1776), brought readers up to date on the thinking about volcanic occurrences. In extensive footnotes, he discussed the latest findings and listed Bergman, Guettard, Kirwan, and others as those who believed that basalt was of aqueous origin. But Desmarest, Ferber, a certain Baron de Dietrich, Faujas de Saint-Fond, and others said it was produced in the dry way, being cooled from a melt. As evidence Magellan stated that basalt is not in the vitreous state in which volcanic ejecta are found (Cronstedt [Magellan note], 1788, p. 912). Basalt columns were not the result of crystallization, but the result of cooling from a soft state (Cronstedt [Magellan note], 1788, p. 914). He noted that basalt and lava were similar in composition, but perhaps basalt had

---

[11]In 1795, uranium and the not-mentioned titanium and zirconium were known only as oxides, which were not reduced to the respective metals until 1841, 1825, and 1824 (Ihde, [1964] 1984, p. 747). At this time, tellurium was not commonly identified as a metal.

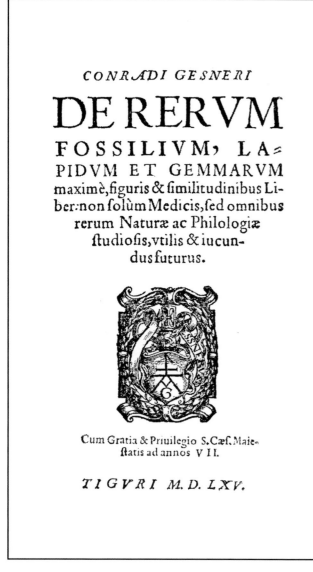

Figure 11.1. The title page of Gesner (1565).

Figure 11.2. Gesner's illustration of columnar basalt (1565).

been boiled, baked, or roasted in the bottom of the volcanic crater. As evidence he listed analyses done by Bergman and Faujas.

Dolomieu stated that he did not intend that the columns formed by lava near a volcano should be considered as crystallization from a water solution. He suggested that it was problematic to use the word "basalt." Crediting Pliny the Younger with the first use of the word, he listed a number of different instances in which it had been applied over the ages. Some groups of naturalists had applied it to stones formed by fusion or by aqueous deposition, while other restricted it to a product of fire. Ornamental uses of "basalt" encompassed stones of different colors, not just those that were dark and compact. Dolomieu himself was unable to decide how to use the word, and felt that until there was better agreement there would be great confusion about it in the works of naturalists (Dolomieu, 1788, p. 95–99 fn). Later, he also stated that analyses of basalt and lava, even if done on the same sample, could be different.

Beddoes was not as unsure. He commented: "Notwithstanding the recent objections of Mr. Werner, I shall assume the origin of basalts from subterraneous fusion as thoroughly established by various authors" (Beddoes, 1791, p. 49). He then listed 11 leading authors who are familiar to us today. In a comment on Beddoes' article by an anonymous author, that was echoed (Anonymous, 1791, p. 139). But the author of the comment took some exception to Beddoes' belief that crystalline rocks were all formed by fusion and sometimes forced up between other rocks. His counter-example was the correct observation—that pipes that carry water are sometimes filled with a mass of what he called "compound stone," clearly deposited by or from the water.

In 1794, Kirwan declared that various opinions about the origin of basalt formed an interesting part of the mineralogical history of his century, because of the many investigations which were triggered by the question, and the discovery of both truths and errors (Kirwan, 1794a, v. 1, p. 431). He reviewed the extensive observations that linked basalt rock to extinct volcanoes in some instances, as well as noting Desmarest's contention that fused granite was the "mother-stone" of basalt. He compared the appearance of basalt with that of fused substances and found that basalt lacked the glazed porcelain grain or the conchoidal fracture of the fused material. In 1799, Kirwan still agreed with Sir William Hamilton that the heat of volcanoes was not excessive (Kirwan, 1799a, p. 270–271). In 1802, Kirwan used analytical findings to make a strong argument that Hall's results were not due to slow cooling, but to the loss of volatiles (Kirwan, 1802a).

At about the same time, La Métherie noted that basalt must be a product of fire because it had the correct characteristics of hardness, as seen in glass, scoria, and frit. Lava that was dark and compact, he thought, was made of basalt (La Métherie, 1795, v. 2, p. 46, 47).

But also in 1795, the accomplished mineralogist Schmeisser sat firmly on the fence with respect to the origin of basalt:

From a variety of characters, as well as from its appearance and component parts, it may be supposed to be produced by subterraneous fire; or it may be considered as of volcanic origin ... but it appears upon the whole, that some basaltes are formed in the moist way, or, at least, water has contributed something to their formation; and that others are of volcanic origin. (Schmeisser, 1795, v. 1, p. 186)

In his detailed discussion of the properties of volcanic products, the dichotomy remained, some being true volcanic products (the glasses) while others were of aqueous origin, more or less altered by heat.

Spallanzani made the sensible comment that, before one inquired about the origin of basalt, whether by fire or water, "it would be proper to decide what we mean by the term" (Spallanzani, 1798, v. 3, p. 192). He began with Pliny the Younger's and Strabo's (ca. 63 B.C.–ca. 25 A.D.) description, but remarked that no one to his knowledge had gone to the stated points of origin to determine whether the rock they described was volcanic. However, he wrote that Dolomieu had investigated some of the monuments referred to in Rome, and had decided the material was all of humid origin (Spallanzani, 1798, v. 3, p. 194). This agreed with the conclusions of Bergman and Werner about basalt. Nonetheless, in some of Spallanzani's investigations of columnar basalt, he found it to be indisputably volcanic. However, "Observation, likewise, teaches us that the same combination of earths, according to different circumstances, forms prismatic basaltes, sometimes in the humid, and sometimes in the dry way" (Spallanzani, 1798, v. 3, p. 197). Spallanzani's extended experiments on volcanic gases and on the effects of putting water on fused glass or rocks also led him to believe that if water falls on a burning crater, it produces no explosion, but if it is introduced to the conflagration below the erupting volcano, it results in explosive eruptions (Spallanzani, 1798, v. 3, p. 374). The power of water changed to vapor was enormous. In his quest to understand volcanoes, Spallanzani went on to question whether oxygen rather than water could cause the explosions, and tried to figure out how muriatic acid (hydrochloric acid) could be produced in a volcano and mixed with its products. He also thought about whether volcanic or furnace fire was stronger. He noted that he had more than once inclined to the idea that furnace fires were hotter. However, after numerous experiments fusing lava in different furnaces, he decided that even powerful furnaces were not enough to completely liquefy lava as he had seen it pouring from volcanoes. He differed in this belief from Dolomieu, who believed volcanic fires to be relatively weak and that they did their work by long-sustained heat (Spallanzani, 1798, v. 4, p. 57).

Spallanzani's colleague, Breislak, asked some of the same questions. In *Voyages physiques et lythologiques dans la Campanie,* with respect to oxygen entering into the heat of volcanism, he asked whether the same substance could be at the same time both cause and effect. It was perhaps unreasonable to say that oxygen provided the fuel for volcanic fires (cause of fire), but also, as was conjectured, that it could be released from carbonic acid (effect of fire) (Breislak, 1801, v. 1, p. 244). The review of this work mentioned that of the more than 200 authors who had written about Vesuvian volcanism "very few have been guided in their investigations by scientific views" (Watt, 1804b, p. 28). However, Watt had a high opinion of Breislak's ability in observing and reporting. As to the origin of basalt, he believed that if there had been a central fire, the molten material extruded in volcanoes should have melted to a homogeneous mass. Therefore, the observed variability of basalt, which Watt called "a lava," and other volcanic products would argue against such an origin (Watt, 1804b, p. 33).

Another possibility for the cause of volcanic heat was aided by Volta's discovery of the storage battery. Davy utilized it, actually Cruikshank's improvement on it, to isolate certain alkaline and alkaline earth metals that had previously been known only in their compounds or, as then termed for some, "earths." The pragmatic Davy (see Fig. 11.3), who worked from experimental evidence, found it difficult to espouse either an aqueous or an igneous origin for basalt. Davy found Hall's experiments on basalt interesting, but not conclusive. He was informed about the chief conjectures about sources for volcanic heat, coal layers or the action of pyrites, metal, and water, as Lemery had suggested. But Davy rejected the suggestion of central heat, considering that if it been there it would have been dissipated evenly to the surface rather than being concentrated at volcanic orifices. When he discovered the properties of the very active metals, he suggested that their presence deep in Earth, when combined with water, could produce explosive eruptions (Siegfried and Dott, 1980, p. xxxviii).

Conjectures about the origin of basalt continued to proliferate. Murray had noted that because a substance could be formed either by fusion or in the "humid way," it stood to reason that if the composition of the parts was the same, a similar substance would be formed in both cases (Murray, [1802] 1978, p. 250). Not everyone was so even-handed. Hall found no reason to question whether heat for fusion originated from a great central reservoir or a limited, local chemical process. Erupting volcanoes provided all the necessary proof that sufficient heat existed (Hall, 1812, p. 157). Addressing the question of basalt's origin in 1816, Cleaveland listed three possibilities and their adherents, as well as arguments in favor of each position. In support of the Neptunian position, he listed Bergman, Werner, Kirwan, Jameson, most of the German mineralogists, and a few of the French. Those in favor of a volcanic origin were mainly the French and Italians. And those who believed it was sometimes one and sometimes the other were Spallanzani, Dolomieu, Fortis, and a few others (Cleaveland, [1816] 1978, p. 284). Cleaveland listed nine reasons in favor of aqueous origin and four for volcanic, most from the standpoint of field observation, although he mentioned Hall's experiments. Cleaveland

Figure 11.3. A caricature of Humphry Davy in *Vanity Fair*, 20 February 1875.

Whether there exist a mass of heated matter under the whole surface of the globe may be uncertain but that there is subterranean fire, under a considerable extent of the surface, can scarcely be doubted. (Bakewell, [1833] 1978, p. 361)

There were plausible modes to explain most forms of basalt consonant with igneous formation during this time. Theory was certainly discussed, but specific arguments about rock origin were more likely to be the result of observation and experiment.

MATURING THOUGHT

The discussions above were not purely for the purpose of enumerating instances, but were rather to show that the debate about rock origin was constant and ongoing, both well before and after the time of Hutton, Werner, and other theorists with whom we are now familiar. The water/fire controversy also was evident in cosmology, but here we are concerned with a more immediate origin of the rocks seen on Earth. It is more difficult than might have seemed apparent in earlier historical studies to label a worker as Huttonian, Wernerian, Neptunist, Vulcanist, or Plutonist. In the following section, we will let some of them speak for themselves, showing that was more complicated than two or two-plus modified camps, and that each actor brought his own experience and background into questions of rock origin. Even after Hall's experiments that showed fixed air (carbon dioxide) could be retained if heat was applied under pressure, there were still questions about the preservation or disappearance of marine and other fossils. Additionally, after the realization that volcanic rocks could end solidified with basalt characteristics, there was still doubt about whether all instances could be ascribed to that cause. There was still much doubt about the existence, intensity, and characteristics of central heat. Although data in its favor were accumulating, counterarguments were also supported. Vulcanists had less trouble invoking volcanic fires than did Plutonists in maintaining documented "hot spots" supported by a pervasive central source of heat. Cooling experiments such as Buffon's, and heat conduction studies such as Murray's, indicated that residual heat from a molten beginning should have dissipated or at least be distributed evenly.[12]

The state of experiment was not primitive. Although procedures were in flux, they were generally more rigorous than we sometimes credit our eighteenth- and early nineteenth-century counterparts. It was increasingly accepted to run duplicates or triplicates of samples. A great deal of thought was given to controlling variables, and reagents were purified and guarded from contamination. The interaction of vessels and samples was noted and guarded against. Instruments were calibrated and checked. If heat was involved, the interaction of the temperature and the measuring instruments was investigated. Some work was what we would call observational; for example, Grignon's depended

suggested that arguments in favor of aqueous origin of basalt appeared to predominate, although basalt often resembled real lava. If there were two modes of origin, the basalts in Saxony, Ireland, and Scotland were of aqueous origin, and those of Auvergne, Italy, and Sicily would be volcanic (Cleaveland, [1816] 1978, p. 286).

By 1833, the volcanic origin of basalt was more or less, if not unanimously, accepted. Bakewell noted that inflammable substances produced by volcanoes were sulfur, carbon, and hydrogen (Bakewell, [1833] 1978, p. 276). Water and atmospheric oxygen admitted to volcanic chambers resulted in eruption. He suggested that Davy's theory was no longer accepted due to the difficulty posed by the existence of such active metals under Earth for years (Bakewell, [1833] 1978, p. 277). The question of a central fire was still addressed:

---

[12]In many ways, the origin of granite still provides conundrums not easily answered even in the twenty-first century, as evidenced by continuing international conferences on the subject.

simply on what he could see in the slags of his iron furnaces, although his interpretations were well informed. At the other extreme were the precise investigations of the behavior of thermometers, investigations which compared the physical characteristics of their glass, metal or wood backing, and the included liquid.[13] Treatment of numbers and calculations was becoming more rigorous. The multiplicity of standards for length and volume was yielding to a generally accepted few.[14]

Experimentalists were increasingly speaking the same language, not necessarily a national language, but the language of science and geology. Definitions within geology and other sciences were being refined, if not perfected. And overall, in geology, the science was more and more seen to be amenable to experiment—not that the tension between the laboratory and the field was resolved. In the first decades of the twentieth century, that tension was still apparent in debates about the origin of granite and other instances, just as the tools of the laboratory became more powerful and versatile, as evidenced in high temperature–high pressure apparatus.[15] As in some other seemingly intractable problems, the contributions of both field and laboratory were invaluable.

By the early nineteenth century, there was little consensus as to an overarching theory of rock origin. It is inappropriate to assign most of these thinkers to one or another theoretical camp, although it was sometimes attempted then, and some historians have done so since. But in reading the primary literature it can be seen that support of an aqueous origin of one sort of rocks didn't imply acceptance of an aqueous origin for other rocks. Geologists might be dubbed Neptunists or Plutonists, but that was often just with respect to basalt origin. The workers of that time can only be presented, and judged as they were—individuals who thought for themselves, and who used the evolving tools of experiment to uncover some small portion of the enormous conundrum of rock origin. Over the time period investigated here, some sort of consensus eventually emerged, or at least some possible alternatives disappeared from rational assumptions. As has been suggested for a later time, perhaps some of the early theories met their demise more by the death of their adherents than by compelling evidence for the new theories. The rest of this chapter contains four snapshots in time, letting the experimental protagonists speak for themselves on the question of theories of rock origin.

Theory of Rock Origin before 1750

The great mining treatises of the sixteenth century laid the groundwork for understanding the origin of rocks and minerals. There was still some magic in tales of helpful or hurtful gnomes or goblins that inhabited the depths of dark mines, but

essentially these treatises were eminently practical (Agricola, [1556] 1950, p. 217). Pervasive infiltrating water in the neighborhood of mineral deposits led to belief in their necessary association, as did minerals colored by traces of neighboring metallic ores. The observation of crystallization from aqueous solutions seemed to be a connection to the formation of crystalline rocks such as granite. Angular basalt columns appeared to be large crystals, and stalactites and stalagmites were observed in the process of formation. Fire often seemed more an agent of dissolution than of formation in liberating parts of a substance. Sherley's opinion about the origin of solids was an example of seventeenth-century thought, basically shared by Boyle, Hooke, and most other experimental scientists of the time. In his treatise on the formation of stones, he quoted authors from antiquity through the Middle Ages on the petrifying power of various springs and lakes. However, he credited "Chymical Philosopher's opinion that the Art of Pyrometry is the only true means of informing the mind with Truth" (Sherley, [1671] 1978, p. 15). He showed some doubt that salt was the only proper agent of solidification. For the formation of stones, he wrote:

The Hypothesis is this, *viz.* That stones, and all other sublunary bodies, are made of water, condensed by the power of seeds, which with the assistance of their fermentive Odours, perform these Transmutations upon Matter. (Sherley, [1671] 1978, p. 24)

He discussed petrifactions at length, and cited many observations and experiments including those of Jan Baptiste von Helmont (1579–1644) and Boyle, in which they planted a willow tree and squash in earth in pots, and added only water. Both plants grew and increased in weight, showing to their satisfaction that the water could indeed produce solids (Sherley, [1671] 1978, p. 59). In the seventeenth century, many investigations of solid bodies by way of strong heating in a closed vessel appeared to show that water, or at least a liquid, was a constituent.

For Erasmus Warren (fl. 1700), explanation of Earth was inextricably interwoven with religion and the biblical account of Creation. Fire and water were the two chief agents of change, but the hand of God was never absent. By 1695, Steno's work had been disseminated enough that Woodward could declare that the shells of marine creatures had indeed been formed in the sea despite their being found even in the mountains, as a result of the Deluge (Woodward, [1695] 1978, p. 20). In trying to justify their current distribution, he indulged the notion of specific gravity as well as the hand of God. Unconsolidated stony materials were "digested" into strata. Besides the heat of the Sun, volcanic heat resulted from great stores of sulfur and niter. Metals and minerals owed their positions in nooks and crannies to the Deluge. Metallic and mineral matter no longer grew, but could be transported by fire or water (Woodward, [1695] 1978, p. 196). At about the same time William Whiston commented:

The *Strata* of *Marble*, of *Stone*, and of all other solid Bodies, attained their solidity as soon as the *Sand*, or other matter whereof they consist, was arriv'd at the bottom, and well setled [*sic*] there. And all those *Strata* which are solid at this day, have been so ever since that time. (Whiston, 1696, p. 203)

---

[13]For detailed accounts of this work, see Middleton (1966) and other books about historical instruments.

[14]Good accounts of the complex story of the transition from multiple, locally derived systems to international standards can be read in Frängsmyr et al. (1990), Heilbron (1993), and Wise (1995).

[15]A few useful references are Bowen ([1928] 1956), Read (1957), and Tuttle and Bowen (1958).

He believed that irregularities could be accounted for by the many things that had happened since, which included conflagrations and the Deluge.

In his commentary on Burnet's and Whiston's theories of Earth, the English mathematician and natural philosopher John Keill (1671–1721) couldn't accept that Earth was ever a completely liquid mass because "there is not such a quantity of water in the earth, as would be requisite to soften and liquefy them all" (Keill, 1698, p. 48). Many things could not be liquefied by water. The presumption that before the Chaos the surface of Earth would be even was also in doubt, and Keill said that according to what he called "*Hydrostatical Principles*," the lightest fluid would rise the highest, and thus the surface would not be uniform (Keill, 1698, p. 49). In 1699, Hooke postulated four possible origins for rock consolidated from sediments: heat and fusion, salt from an aqueous solution, sand agglutinated by drying bituminous matter, and long and cold compression (Drake, 1996, p. 125). All were possibilities suggested by observation and experiment. Even in an age accustomed to accepting the biblical account of Creation and the testimony of Moses, the realities of observation and testing were beginning to be felt.

The Midcentury

By the middle of the eighteenth century, the literature that considered geological materials was varied, to put it mildly. A tract on the origin of pebbles read to the Royal Society invoked the Creator who supposedly knew the proper place for each kind of substance, so that they could perform their correct function in the world (Arderon, 1748, p. 467–470). It shared space with articles about electricity, magnetism, and burning mirrors, which emphasized practical, sometimes numerical functions. In his *A general natural history* (1748), Hill mentioned that the origin of basalt had puzzled many, as had that of granite. By the time Henkel (aka Henckel) wrote his 1744 *Kleine mineralogische und chymische Schrifften*, great descriptive detail had been added to books for those interested in mines and metal working, but his volume was still basically a practical book for miners. Fire and water got their due, but in conjunction with realistic, useful description, and reports of many trials. Pott's *Lithogéognosie ou examen chymique des pierres et des terres en général* (French edition of 1753) was similar. Those authors were more concerned with the behavior of rocks and minerals and earthy substances and the uses to which they could be put than with their origin. There were numerous books of this sort with investigations of the properties of useful minerals and the beginning and development of both wet and blowpipe analysis, as discussed earlier for these were the properties that influenced theory.

From about 1750 to 1770, new field discoveries, augmented by fusion investigations, were informing some assumptions about rock origin, particularly basalt. Improved and improving wet analysis was showing that rocks of supposedly different origin, with different names such as "trap," "whinstone," "toadstone," "lava," "scoria," or "basalt," might have very similar compositions. As Carozzi pointed out, the definition of "basalt" was vague (Carozzi, 2000, p. 10). Guettard had identified volcanic products associated with the extinct volcanoes in the Auvergne region of France. Later fieldwork by Desmarest in the same area showed that prismatic basalts were indeed connected to extinct centers of eruption, which opened the possibility that all prismatic columns could be so caused. Grignon and others had seen, or would soon demonstrate, that well-formed crystals could be found in iron furnace slag, as well as in metals. But the relation of different rocks such as granite and basalt to each other was not at all clear. At this early time there were no overall theories of rock origin, although water was probably the preferred agent in the absence of direct evidence of volcanism.

1770 to 1799

During this period, the two best-known theories of rock origin from the eighteenth century were suggested, by Werner in 1786 and in Hutton's public statement of his views in 1788. Neither experimenter was remarkable for his laboratory work on rocks or minerals, although both men were very competent mineralogists and chemists. Neither was the first to suggest the respective origins, although both fleshed out their systems in more detail and over a broader range than had previously been the case. Both admitted only a very minor role for the opposing theory. As stated previously, not everyone felt compelled to support one or the other theory. However, the tension between the two certainly led to a great deal of investigation and innovation, in the laboratory and the field as well as in instrument and equipment design. Increased use of mathematics was also a feature as it was in all the sciences, along with more sophisticated use of numerical data.

As a practical person, Grignon saw nature and volcanoes in the light of his knowledge of furnaces and forges. To account for the placement of minerals near volcanoes he suggested that water could perhaps operate for a succession of centuries, whereas fire would operate for a much shorter time (Grignon, 1775, p. vii). Furnaces could most closely approach the working of nature. After describing the well-formed crystals found in his furnaces, he suggested that chemists' furnaces couldn't have the same effect as his very large furnaces. In a passage that may have influenced later fusion work with basalt, he mentioned that the smaller furnaces heated and cooled very small amounts of material quickly, while his large furnace could heat a great mass of matter and take as long as three years to cool. Therefore, there were possibilities for the combination of materials into different substances that were more similar to those of the immense forges of volcanoes, which chemists' furnaces could not mimic satisfactorily (Grignon, 1775, p. xxxj–xxxij). He argued with Romé de l'Isle, contending that crystals could be produced by fire, or the dry way, and not just in the humid way (Grignon, 1775, p. 475).

Publishing over a number of years, Buffon, not surprisingly, modified his thoughts about the origin of Earth during his long career, and he considered that surface rocks solidified from fusion during a long period of cooling. Volatiles such as air and water separated to form the atmosphere, while vitrified matter crystallized into interior rocks and those that formed the kernels of mountains. At first Earth was still too warm to enable water to exist on the surface. Calcareous rocks were formed later, after the sea had formed, because water was needed for their materials (Buffon, [1778] 1971, p. 64). He discussed many things that only existed and were modified because of the remaining residual heat of Earth. In his view, water was responsible for much that happened on Earth. He suggested that clay was the decomposed remains of vitrifiable sand (silex), which was the origin of acid. This acid found in argil he regarded as the combination of vitrifiable earth with fire, air, and water (Buffon, [1778] 1971, p. 90).[16] Heat, with water incursions and the aid of pyrites, explained the immediate violence of volcanoes, but, interestingly, Buffon was convinced that electricity had a role in earthquakes and volcanism. Informed by his experiments, he found that "le fonds de la matière électrique est la chaleur propre du globe terrestre" (Buffon, [1778] 1971, p. 121). He believed that electricity worked on the fire, water, and air in subterranean cavities to produce volcanism and earthquakes. That interpretation by the eloquent Buffon serves to remind us of the background of reasoning about the ways Earth worked.

In 1779, Saussure began the publication of his accounts of his travels, the instruments he used, and the laboratory work on collected specimens. From his extensive observations, and drawing on his experimentation with basalt and granite, he thought that what he called "Pierre de Corne" and soft or brittle kinds of "Schorl," rocks that were frequently found in primitive and secondary mountains, were altered to lava and basalt in volcanic fires (Saussure, 1779, v. 1, p. 131).[17] Rocks of somewhat different compositions formed the different species of lava and basalt, but granite could not be the base of homogenous basalt because it contained infusible minerals. Saussure's thoughts about basalt evolved as his work continued, and as he read the reports of others.

The Italian Cosimo Alessandro Collini (1727–1806) was a well-traveled natural historian. In his 1781 publication, Collini was more concerned with the origin of mountains than with that of individual rocks. However, he remarked that some thought granite to be formed in the humid way, but the total lack of marine artifacts argued against this idea (Collini, 1781, p. 15). Collini noted that some naturalists believed in granite's origin by fire, while others changed "tour-à-tour," or took each possibility by turns (Collini, 1781, p. 16). He noted the idea of origin from solution followed by the effect of fire. Such uncertainty led to embarrassment and confusion in the study of the mineral kingdom. In a footnote (p. 16), Collini remarked that Buffon founded two of his systems on opposing principles. He suggested that it was difficult to know the relative ages of mountains, and thus the origin of those with different sorts of rocks:

Elles sont toutes sujettes à des doutes difficiles à résoudre; & toutes les fois qu'on a voulu établir des Régles générales à cet égard, on a été forcé d'admettre des exceptions qui les bornent trop, ou qui les détruisent. (Collini, 1781, p. 20)[18]

Collini argued about the extent of the fire of volcanoes, based on the sorts of rock formations he saw including columnar basalt, and found that neither very local fire nor the huge conflagration suggested by others could explain everything he saw. At best a few instances might be explained. He proposed a sequence for basalt origin that differed from those of others. Rather than it being a rock that was deposited from solution and then fused by local volcanic heat and subsequently cooled, he suggested that it began as a true fused lava. It was next decomposed by later vicissitudes but without altering its appearance, and then was transported and deposited by water (Collini, 1781, p. 36). That sequence helped him understand the parallel layers in which it is often found. It also could account for the approximate resemblance of columnar basalt to large crystals. Collini was well versed in the physical and chemical properties of minerals, as well as being a thoughtful and experienced observer in the field. Although he cited a great deal of both kinds of evidence in his discussions as shown throughout his 1781 publication, in response to those who felt they could explain the workings of volcanoes by chemistry, he remarked: "Mais la Chymie d'un Volcan peut & doit entièrement différer de la Chymie des hommes" (Collini, 1781, p. 49).[19] We can only agree with him.

This section demonstrates that theories of rock origin were frequently expressed before the seminal publications of Werner and Hutton. It is also obvious that most positions had already been explored, with both experiment and field evidence brought to bear on the problem. From this standpoint, the theories of Werner and Hutton appear almost anomalous. Hutton ascribed all to the effect of heat on unconsolidated "remains" of former worlds. Werner found that rocks said to be volcanic "show nothing which would hint at an origin through fire" (Werner, [1786] 1971, p. 86).

It is quite difficult to understand why the one-sided theories were advanced, and also why workers in the field were impelled to accept either of them. Of course, as noted earlier, not all geologists did argue for one or the other. There were other authors than those I have specifically mentioned who discussed these questions, although I believe most positions have been covered here. But the questions about, and attempts to explain, rock origin continued, whether driven by desire to "prove" one or the other theory, or simply to account for what had been observed. The 1780s were a decade when experimental design, techniques, and instrumentation were advancing rapidly, as was the understanding of chemical behavior. In a relatively few years, Lavoisier's understanding of chemical elements and John Dalton's of the

---

[16]Gohau, the editor of the 1971 edition, reminds us that this work predated Lavoisier's *Traite elémentaire de chimie* and still supported the idea of there being four elements. Thus, that idea about the acid was very wrong.

[17]"Roche de Corne" is hornstone, a variety of more brittle flint. An earlier usage sometimes referred to hornblende as "roche," or "Pierre de Corne." Schorl is black tourmaline, a common mineral in metamorphic rocks. Minerals as varied as hornblende and other amphiboles, augite and other clinopyroxenes, and olivine were sometimes classed as "schorls."

[18]"They are all subject to doubts that are difficult to resolve; and at the same time one wishes to establish general rules in this regard. One has been forced to admit exceptions that limit them too much, or that destroy them" (my translation).

[19]"But the chemistry of a volcano can and must be completely different than the chemistry of men" (my translation).

atomic nature of matter became part of the fabric of reasoning, which enabled new insights about reactivity, fusibility, solubility, and the formation of crystals to be developed.

Cronstedt had remarked that materials that showed the marks of fire could be seen where there was no evidence of a volcano. The editor of *An essay toward a system of mineralogy,* the 1788 English translation of Cronstedt's seminal work, included a footnote stating that Cronstedt had confounded "basalts" with "schoerls" (or "schorls," see note 17 above) (Cronstedt, 1788, v. 1, p. 912). We see that there was a good bit of rock and mineral misidentification for several reasons: lack of precise standards for identification; lack of agreement on characteristics of a particular mineral or rock; use of common names that differed in different regions; qualitative rather than quantitative measures of characters; and, not least, the absolute certainty with which many, if not most, workers identified rocks, which sometimes led their readers and followers to perpetuate errors. As stated previously, 1788 was also when one of the last more speculative theories of Earth, that of Whitehurst, was published. However, on inspection, its content is not noticeably more speculative than that in Hutton's first tract. Whitehurst cited Charles Marie de la Condamine's (1701–1744) and Hamilton's volcanic observations as one instance where there could be what he called "superficial alteration" of Earth's surface (Whitehurst, 1786 [1978], p. 87–88). In an interesting comment in 1788, Dolomieu noted that all bodies in nature could not remain constant, but carried within themselves their principles of destruction, as influences to both unify and destroy were always present (Dolomieu, 1788, p. 392). He noticed that the "humid way" operated on volcanic substances as soon as they were ejected. Recognition of the relatively rapid alteration of surface materials influenced thinking about original origin.

In 1790, Deluc took on the entirety of Hutton's first statement of his theory in a series of 40 points that were a distillation of his 1778 *Theory* as applied to Hutton. He went to great lengths to refute most of Hutton's assumptions about fusion. He only saw the marks of fusion on some, not all, of Earth's strata. In an interesting argument, he stated that substances that had been fused and then cooled to a solid state could be remelted with the same degree of heat. But then he noted that there were a number of things that did solidify with water, such as mortar and concretions, which could not then be melted. He argued with Hutton's conclusions about flint, namely, that nothing of such hardness could be seen to be produced by anything but fusion. Deluc then mentioned many field appearances of flint that were inconsonant with an origin from fusion, including its close association with chalk. Granite showed no mark of fusion that Deluc could discern. And if granite was between two strata that included "marine bodies," how could those have been retained in the heat of fusion? Deluc postulated six periods in the formation of Earth that were successive combinations of the actions of density and precipitation as well as the influence of "elastic fluids" or gases. He argued against Hutton's assumption that heat could raise the strata above the sea by stating that his observations of volcanoes didn't indicate the lifting

and breaking of large areas. In a telling comment about Hutton's objections to the efficacy of precipitation he remarked:

The only argument which you have opposed to that theory is, that *substances precipitated in a liquid, ought to be still soluble by that liquid:* but a moment's reflection would have made you discover, that, on the contrary, no substance can be *precipitated* in a *liquid,* while that *liquid* still retains the faculty of *dissolving* it; and in general, that the very modifications which produce the *precipitation* of substances in a *liquid,* are the cause of its incapacity of dissolving them again. (Deluc, 1790c, p. 220)[20]

A modern chemist can understand the foregoing statement without difficulty. While not all of Deluc's objections to Hutton's findings could withstand the test of time, none were completely implausible in the context of his current science, and some, like the quote above, are still acceptable today. In a second part of the paper, he addressed mainly geomorphological issues concerned with the surface appearance of Earth.

Despite Werner's objections to such a scheme, Beddoes had no difficulty with fusion being the origin of a wide variety of rocks or with different conditions that resulted in the formation of rocks as varied as basalt and granite. In 1791, Dolomieu differed from this view. After a discussion of how silica and calcareous substances could interact in solution, he stated:

Quel qu'ait pu être ce dissolvent, c'est avec M. de Saussure & M. de Luc que j'admets la précipitation comme première cause de la formation & la consolidation des plus anciens matériaux de nos montagnes. (Dolomieu, 1791b, p. 382)[21]

Even though Dolomieu, Saussure, Breislak, and others had great familiarity with active volcanoes and many field localities, the "war" between fire and water continued. In 1792, Dolomieu used arguments derived from solubility, affinity, and density to further discuss the order and method of strata deposition. He also engaged in a "discussion" with others of the time about how often the sea had covered the land, and what were its modes of advance and retreat. We will not discuss those here because they were dependent on observation only. Dolomieu rarely used the word "basalt" in his lengthy papers of 1794, instead referring to "prismatic" or "compact" lava. His extensive observations and experimentation, combined with references to the work of others, indicated to him that volcanic heat was not as intense as had been assumed, and that fusion characteristics of rocks and minerals depended more on the very long duration of more moderate heat (Dolomieu, 1794a, p. 117, 118). In those two papers of 1794, he maintained the primacy of deposition from water for the "original" rocks including granite, and their minor alteration by heat.

---

[20]David Oldroyd (2006, personal commun.) has pointed out that this is not a just criticism of Hutton. Hutton's process of crystallization from a fused substance can indeed be reversed by the addition of more heat. Deluc's process appears to be irreversible, in that a liquid from which matter has crystallized has changed so that it can no longer take the material up in solution, as we are familiar with in crystallizing salts out of a salt solution.

[21]"Whatever has been able to dissolve it, it is with M. de Saussure and M. de Luc that I admit precipitation as the prime cause of the formation and the consolidation of the most ancient materials of our mountains" (my translation).

In the first half of the 1790s, confusion reigned supreme. In La Métherie's 1792 translation of Bergman, the 27 point account of the origin of the currently seen Earth was repeated, along with the two difficulties he saw with Bergman's generally Neptunist reasoning. Saussure came to the conclusion that basalt might have two origins: by fire if associated with volcanic products; or, if there was no obvious relation to a volcano, it was of humid origin (Carozzi, 2000, p. 609). Despite Beddoes' allegation, Saussure and Dolomieu laid to rest the assertion that granite was the starting point for the formation of basalt, as they both, especially Dolomieu, through painstaking examination, recognized volcanic products that were of greater variety in their mineralogy than basalt. In other words, a lava derived from granite would be different from one derived from other rocks. Kirwan characterized Dolomieu as "the most sagacious, exact, and philosophical, of all of them" (Kirwan, 1794a, v. 1, p. 441). His "Examination of the supposed igneous origin of stony substances" (Kirwan, 1794b) has been credited with provoking Hutton into writing a full statement of his theory, which was published in 1795. In 1797, Kirwan attempted to explain the origin of our globe from Chaos, leading to present appearances. He promised to follow certain principles: investigations of past facts dependent on natural causes should adhere to laws of reasoning, causes should be supported by experience or approved testimony, and the power of a cause must be demonstrated by actual observation (Kirwan, 1797, p. 233).[22] Kirwan clearly didn't consider Hutton's *Theory* as the last word. In these publications in the latter part of the decade, Kirwan's references to experimental work that supported or could be interpreted as supporting his position in favor of aqueous origin increased. The authors he cited included most of those working in geology from many European countries, including Scandinavia. But he referred to "systems" opposed to his own:

There is another system which attributes not only to basalts, but to all stony substances, an igneous origin; it is that of Lazaro [*sic*] *Moro*, revived, and wonderfully improved, by Dr. Hutton, of Edinburgh, well known, by his excellent Essay on the Origin of Rain. This may be called the *Plutonic* system. (Kirwan, 1799a [1978], p. 455)

In print at least, Kirwan disagreed as a gentleman, although he proceeded to contest virtually all of Hutton's assertions.

La Métherie, despite his generally Neptunian stance, declared lava to be definitely a product of fire. He had more questions when it came to prismatic basalt, and wondered:

Sont-ils cristallisés par le feu?
Sont-ils cristallisés par l'eau? (La Métherie, 1795, v. 2, p. 56)[23]

He noted that many mineralogists said crystal form was due to cooling after fusion. But a number of them, including Bergman and "several celebrated German mineralogists" (La Métherie, 1795, v. 2, 57), assumed water had a role, and in consequence were called "Neptunists." Those who thought prismatic basalt originated as incandescent lava he called "Vulcanists." La Métherie requested that he be allowed to have an opinion between those of the other distinguished scholars. He didn't see how prismatic basalt could result from incandescent lava that had cooled slowly, as that would not produce regular forms. Rather, from crystals he had seen when some metals cooled, he thought it could result from a quick cooling, such as would occur in water (La Métherie, 1795, v. 2, p. 57–59). He later posed a series of 13 questions about the nature of volcanic fire, and noted that their solution would be a great day for the theory of Earth. He went on to investigate volcanic fire (La Métherie, 1795, v. 2, p. 388). As noted earlier, he summarized the theories of a great many other authors. Here was a very well-read, open-minded man, who found neither main theory compelling.

In 1796, at the end of the fourth volume of his *Voyages*, Saussure included "Agenda, or a Collection of Observations and Researches the Results of which may serve as the Foundation for a Theory of the Earth." The question must have been of wide interest as his lists of necessary items for such a study, which were titled as 23 projected chapters of a book, were reprinted in 1796 in the *Journal des Mines* and in 1799 in the *Philosophical Magazine*.[24] They ranged from "Astronomical Principles" through "Chemical and Physical Principles" to observations on various kinds of mountains, rivers, plains, rolled pebbles, earth strata, fissures, and valleys. Also included was a study of earthquakes as well as a list of instruments required to do proper geology. Of all of these projected chapters, the one of most interest in this book is the second, on chemical and physical principles. Among the many things to be investigated in that category were chemical affinities, precipitation, solution, and crystallization. Those were just the first of 18 suggestions on that list. He also included heat, gases, air pressure, mineralogy, electricity, magnetism, the origins and effects of acids and bases, the current temperature at the center of Earth, and interactions of all of those with earth materials. He observed that when traveling there were so many objects competing for his attention that it was easy to get distracted, and so he prepared an agenda of study topics. The lists for a theory of Earth were intended to be used in the same way, as guideposts for a most complex subject (Saussure, 1796, v. 4, p. 467–468). All the lists for the "agenda" were encyclopedic, which may explain why he never wrote his theory of Earth, finding the task too daunting.

After 1800

After Spallanzani's thorough investigation of fusion conditions for all sorts of rocks, minerals, and gases, volcanic and other-

---

[22]C.C. Gillispie has pointed out that Kirwan, who was a strict empiricist, felt he could trace present observations back to the origin of Earth in Mosaic terms (Gillispie, [1951] 1959, p. 50–53).

[23]"Are they crystallized by fire?
"Are they crystallized by water?" (my translation).

[24]I didn't attempt to see how often Saussure's list had been summarized or translated in other journals. These two were both complete, not summaries, and the one in *Philosophical Magazine* was a translation.

wise, published in 1795 and 1798, as well as his remarks about the variability of volcanic heat—all backed by his temperature observations and chemical analyses—he was hesitant to make overriding statements about fusion or solution. Via experiment, he had documented the fact that volcanic heat could vary greatly, and that much depended on the length of time that any given rock species was exposed to it. He noted that volcanic fires were often more powerful than any furnaces (Spallanzani, 1798, v. 4, p. 1–5). However, when discussing origin, he noted that it was important to know what the rock was that had been historically termed "basalt" and its place of origin. Some historical stones identified as basalt had no marks of fire on them (Spallanzani, 1798, v. 3, p. 193, 194). He couldn't agree with Werner and others that all columnar basalt was the result of deposition from solution because he had seen instances in which it was clearly connected to volcanic action: "It consequently appears that nature obtains the same effect by two different ways" (Spallanzani, 1798, v. 3, p. 196).

In 1801, Breislak pointed out that geologists who wanted to explain the formation of the globe by solution, precipitation, and crystallization had problems explaining the behavior of siliceous earth, which was a major part of many supposedly primitive substances. If they tried, he argued, "Il faut que, dans beaucoup de cas, ils renoncent aux idées fondamentales de la chimie, ou qu'ils imaginent dans la nature des agens qui n'y existent pas" (Breislak, 1801, v. 1, p. 180).[25] Since the globe had been fluid, and he recognized that crystallization was a widespread phenomenon, he felt that the "hypothèse de la fluidité ignée" was subject to the least difficulties (Breislak, 1801, v. 1, p. 156).

It is difficult to follow the web of arguments that ensued in the first quarter of the nineteenth century, mainly because there were so many accounts. Many historians have remarked that Playfair's defense of Hutton in less obscure language, along with Hall's fusion "proof," tipped the scales in Hutton's favor. But what was the content of the arguments of the time? Leopold von Buch mentioned the difficulty of determining the truth of Werner's theory of volcanic action, caused by burning coal seams, in the process of reporting what he had seen at many active eruption sites (Buch, 1802, v. 2, p. 166). He later asked the question "What is lava?" Playfair's defense of Hutton was published in 1802. But so was John Murray's *A comparative view of the Huttonian and Wernerian systems of geology*, which came to the opposite conclusion. Hall's experimentation on lava and calcium carbonates appeared in definitive publications in 1805 and 1812. However, the papers were read well before then, and there were earlier preliminary publications of his results. Thus it occurred that Kirwan's "Observations on the proofs of the Huttonian theory of the Earth, adduced by Sir James Hall," was read in 1800 and published in 1802. In it, he called Hall's fusion experiments "highly interesting" and suggested they would be "highly useful by suggesting new experi-

ments" (Kirwan, 1802a, p. 3). After an extensive discussion of Hall's comments on his generally unsuccessful granite fusion and crystallization, Kirwan critiqued the whin and lava experiments. Kirwan argued, with supporting references to others' experiments, that the changes Hall observed that after fusion, cooling, reheating, and slow cooling resulted from driving off the volatiles in the rocks, thus changing their composition. Kennedy's analyses had shown this to occur. Kirwan pronounced the support of Hutton's theory from Hall's work to be "none at all" (Kirwan, 1802a, p. 22). And further commented:

> I persist in thinking his [Hall's] experiments afford no confirmation of the high degrees of heat attributed to volcanoes, and still less to the many hypotheses gratuitously heaped on each other by Doctor Hutton, or to the volcanic origin of whins or traps, for the reasons already assigned. (Kirwan, 1802a, p. 27)

In 1803, Kirwan published a reply to Playfair entitled, "A reply to Mr. Playfair's reflections on Mr. Kirwan's refutation of the Huttonian theory of the Earth." In it, he objected to Playfair's use of personal invective to state Kirwan's differences with Hutton's theory. Much of it centered on Kirwan vis-à-vis Hutton on the question of "no vestige of a beginning, no prospect of an end." In defense of his own conjectures about origin, Kirwan asserted, "It was never pretended that Moses intended to write a treatise of geology" (Kirwan, 1803, p. 12). He also noted that there was clear evidence that rocks that were now hard had demonstrably been soft during their history. He concluded with a paragraph that is instructive for us now:

> Controversies managed as this has been by him [Playfair] and Dr. Hutton, whose favourite method of answering objections consists in depretiating [*sic*] or sneering at the understanding, and undermining the credit of their author, are a disgrace to philosophy, and sufficiently expose the weakness of the cause that obliges to have recourse to such experiments. (Kirwan, 1803, p. 13)

Even now, we still too often couch our objections in simple terms instead of exploring the web of reasoning that led to the differences.

Meanwhile, the arguments continued. Murray's *A comparative view of the Huttonian and Neptunian systems of geology* (1802) was more objective than Playfair's opus, and lacked the personal attacks. (See Fig. 11.4.) Despite the ingenuity and novelty of Hutton's system, Murray felt it to be "visionary and inconsistent with the phenomena of Geology" (Murray, [1802] 1978, p. iii). But he noted that Hutton avoided the hostility sometimes shown when controversial ideas were discussed, and was said to "rest the defense of the theory he supports on its intrinsic worth" (Murray, [1802] 1978, p. iv). Murray felt that Werner avoided hypotheses and used more induction. At the end of the volume, Murray praised the Huttonian system for its novelty and boldness, and noted that it attempted to account for the formation of successive worlds. But he found that too many assumptions were

---

[25]"It is necessary, in many cases, that they abandon fundamental ideas of chemistry, or that they imagine in nature agents that do not exist" (my translation).

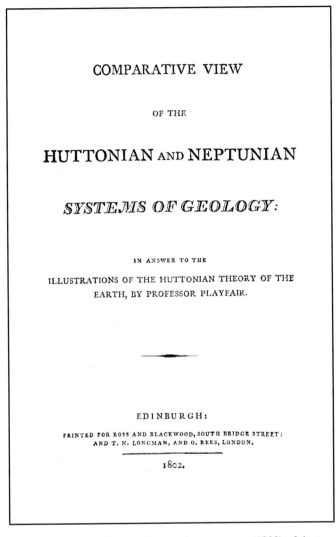

COMPARATIVE VIEW

OF THE

HUTTONIAN AND NEPTUNIAN

SYSTEMS OF GEOLOGY:

IN ANSWER TO THE

ILLUSTRATIONS OF THE HUTTONIAN THEORY OF THE
EARTH, BY PROFESSOR PLAYFAIR.

EDINBURGH:

PRINTED FOR ROSS AND BLACKWOOD, SOUTH BRIDGE STREET;
AND T. N. LONGMAN, AND O. REES, LONDON.

1802.

Figure 11.4. The title page of Murray's assessment (1802) of the two theories of rock origin.

involved, among them being the need for intense heat in Earth's interior. Neptunian theory contrasted with it by only being concerned with the present world. Even with what Murray viewed as deficiencies because of the current imperfect state of science, in general he contended that the Neptunian view was a

series of inductions, more or less perfect, referred to a common principle, and occasionally connected by a moderate and rational hypothesis. In a word, it may be considered as the commencement of a system which possesses the stability of truth and which time will extend and improve. (Murray, [1802] 1978, p. 255–256)

The tone of Murray's book was bound to impress thoughtful scientists.

In 1804, Gregory Watt reviewed Breislak's *Voyages physiques et lithologiques dans la Campanie* and mentioned the variety of volcanic localities in Italy. It is interesting to read his criticism of the French translation of Breislak's original Italian. Watt objected to Pommereul's (the translator's) conversion of Breislak's Italian measures into French meters "which arrogantly figure in the text, while the original expression is degraded to the notes" (Watt, 1804b, p. 30). He viewed larger units in the Italian system as "perfectly ridiculous" when put into French units down to centimeters. He also objected to Pommereul's rounding of figures so that distances were not at all accurate. Apart from that, he admired Breislak's fidelity in reporting volcanic phenomena; however, the problem of supplying heat for volcanism was not solved. Watt pointed out the difficulties with both assumptions of residual central heat and those of requiring sulfur, pyrites, or another substance for combustion. He spoke well of Hall's work, but clearly felt that there were many problems yet to be solved.

Another experimenter, Humphry Davy, was most concerned to draw conclusions only from facts and not speculation, religiously guided or not. He criticized Burnet, Whitehurst, Kirwan, and Deluc for using sacred writings to bolster their positions in theories of Earth (Siegfried and Dott, 1980, p. xxviii), although he gave them credit for using chemical and other experimental reasoning. Davy himself followed the path of observation and experiment. But according to Davy, neither Buffon's and others' speculations about comet collision with the Sun, nor Hall's controlled experimentation were sufficient for wide explanation for geological occurrences. Hall's experiments didn't account for too many instances such as granite, gems, or well-preserved organic remains, although Davy found plutonic explanations somewhat better (Siegfried and Dott, 1980, p. xxx–xxxi).

In 1805, Faujas noted the great difficulty of discussing theories. He identified and lauded the work of geologists from a number of countries, but focused on marine fossils rather than experimental work when speaking of theories of Earth. Later, Playfair evaluated Hall's experimental work objectively, being much impressed by it. He was informed about both the experiments on lava and whinstone and those on carbonates under pressure, and praised Hall for his ingenuity with instruments and his care in dealing with sources of error. Playfair particularly mentioned Hall's careful use of Wedgwood's pyrometer. But in a more measured response to Hutton's theory than in his earlier book, Playfair noted that what the experiments allowed was that carbonates *may* have been consolidated by fire, but that they *have actually been* so was a different proposition (Playfair, 1806, p. 30). With respect to the theory, he commented:

It is not for us, who pretend not to the character of geologists, and only venture to give an opinion here as on a matter of general science, to say whether the theory invented by Dr. Hutton, and so ably supported in the paper before us [Hall's], is in full possession of the advantage just stated. The time perhaps is not yet come when this question can be brought to a complete decision, and when philosophers can determine finally, whether the element of fire or of water is the consolidating power of the mineral kingdom. (Playfair, 1806, p. 31)

Playfair contended that theory was more likely to be acceptable if it was both supported by experiment and conformable with nature and then suggested a number of further experiments that would help clarify the questions being asked.

In 1807, William Richardson argued vehemently against the volcanic origin of basalt, and pointed out many of what he considered irreconcilable differences between basalt and lava. He also quoted Kirwan, Ferber, and John Strange (1732–1799) as saying they learned little about basalt from observing volcanoes. However, Cordier laid part of the blame for misinterpretation by the Neptunian school on poor sampling technique for the basalts that were observed by some of the northern Europeans. He felt they had sampled only the surface scoria, and not the interior of the flows, and, thus, that basalt characters were imperfectly developed (Cordier, 1808, p. 58 fn).

Robert Jameson clearly still supported the Neptunian view in 1808, as attested to by his book on Wernerian theory. His description of rocks continued to be accepted as did, for a while, their sequence of deposition. Of course, as field observations increased, the latter part was much amended. The more controversial part of his book was about rock origin. Jameson's book is credited with providing a "fuller expression [of Werner's rock origin] than in any of Werner's own publications" (Jameson, [1808] 1976, p. viii). Jameson's extensive travel, studies, and access through prominent people and positions led him to write this complete and influential book on the Wernerian theory of rock origin, about which he changed his mind only in the 1820s (Jameson, [1808] 1976, p. xviii). Like Werner and many others, he cited the evidence of Earth's original fluidity, as attested to by its spherical form. His opinion of Hall was that he had "published a series of curious experiments on the effects of heat on mineral substances; but the results do not apply to the present question [of rock origin]" (Jameson, [1808] 1976, p. 74). He believed that none of Hall's productions represented actual rocks of Earth's crust. He called Hutton "a man of unquestionable ingenuity, but very imperfectly skilled in mineralogy" (Jameson, [1808] 1976, p. 345). He objected to every part of Hutton's sequence for the deposition of primitive rocks, arguing that they had to be chemical and mechanical deposits. Jameson continued with his argument, rejecting the Huttonian view of the origin of each rock type in the sequence, not just those influenced by the role of heat. He was no more convinced by Playfair's statements than he was by Hutton's, particularly the Huttonian view of vein formation.

Deluc was also no admirer of Hutton's theory. After a discussion of the effects of heat on aqueous vapor and in carbonate calcining, Deluc stated baldly, "Dr. Hutton's hypothesis therefore receives no support from the experiments of Sir James Hall." Additionally, he wrote that fixed air rises quickly through water, rather than providing the pressure necessary for the elevation of strata. What he called "expansible fluids" couldn't be produced under the pressure of solid rock, which he interpreted Hutton's theory as meaning (Deluc, 1809, p. 366). He showed respect for Kirwan, who aided in the task of "pointing out the errors of the systems which are in opposition to the Book of *Genesis*" (Deluc, 1809, p. 367). However, while he thought Kirwan highly adept in chemistry and mineralogy, the essential parts of geology, those were not enough. Intensive study was required, in many countries, of the history of Earth as evidenced in the hills, mountains, valleys, plains, and coasts (Deluc, 1809, p. 368). From our vantage point, the disconnect felt between laboratory experiments and their application to the whole Earth is clearly expressed in Deluc's assessments.

In 1811, Pinkerton summarized the French mineralogist's A.J.F.M. Brochant de Villiers's (1772–1840) treatment of basalt origin (1801–1803). There were nine points in support of volcanic origin of basalt, two based on experiment, while the others included mineral or rock characters or stratigraphy. There were 13 points detailing the Neptunist position. Pinkerton cited Hall's experiments several times, but summarized their difficulties. Brochant de Villiers had hoped for fresh information before coming to a conclusion (Pinkerton, 1811, v. 1, p. 64). Pinkerton then set out Brongniart's third opinion, whose supporters he listed as Fortis, Dolomieu, Delrio, and Spallanzani. According to Brongniart, the third opinion was that basalt might have either kind of origin, and could be different in different places. Werner's work in mineralogy was universally respected, as initially were his stratigraphical observations, thus giving credence to some of his more sweeping pronouncements.

In 1812, Breislak reviewed the two theories, emphasizing the importance of close observation, temperatures, and experiment. Despite Hall's work, Breislak could accept neither Hutton's belief in the periodic renewal of continents nor the existence of intense heat at the bottom of the sea (Breislak, 1812, p. 115). He based his objections on actual measurements of temperatures both in the sea and at depth in Earth, and discussed the expansibility of fluids. However, he also could not accept the formation of granite by crystallization from solution, although he thought it might be possible with some rocks. He cited Dolomieu, Drée, and others for their work with granite fusion which he summarized. Breislak felt that experiments with granite fusion could not apply to large masses of materials (Breislak, 1812, p. 146).

By 1813, Bakewell pronounced:

I am inclined to think that the part of Dr. Hutton's theory which relates to the igneous origin of basaltic rocks, is as well established as the nature of the subject will admit of; other parts of the system are much less satisfactory. (Bakewell, 1813, p. 113)

He noted the difficulty of keeping the description of facts separate from the language of theory. Theories had their uses in connecting isolated facts and applying to nature (Bakewell, 1813, p. 114). He continued with a long exposition of details of the theories; and included most facets of them: marks of fusion, field position, solubility behavior, magnetic effects, etc.; and ended with the creation of Earth itself. But he reached no conclusions and looked to the future: "In the present state of geological science, facts are more wanted than speculations" (Bakewell, 1813, p. 330).

Despite the inconclusiveness of the discussion, numerous publications continued the arguments, many of which included a review of the history of the controversy. Except for a few workers wedded to one theory or the other, the great majority continued to list the points for and against each theory, using micro- to macro- evidence. Virtually all authors cited numerous experimental investigations, including analysis. The points made by the same, extensive list of participants were summarized again and again. In 1814, Daubuisson maintained that basalt couldn't originate from volcanic heat acting on any known rocks, and had never been and never could have been lava (Daubuisson, 1814, p. 118). In 1825, Brongniart summarized the new work by Cordier concerning his separation and microscopic examination of very small crystals in the matrix of volcanic productions of all kinds. Basically, all volcanics showed very similar mineral types and conditions, regardless of whether they were from lava, basalt, or scoria. Primary and secondary rocks did not show those characters (Cordier, 1815). After following many of the same topics as other books of its time, including most of the same observations, Kidd concluded his *A geological essay on the imperfect evidence in support of a theory of the Earth, deducible either from its general structure or from the changes produced on its surface by the operation of existing causes* (1815) with "positivist" language:

[T]he science of geology is at present so completely in its infancy as to render hopeless any attempt at successful generalization, and may therefore be induced to persevere with patience in the accumulation of useful facts. (Kidd, 1815, p. 269)

Water and fire origin were again discussed in the 1816 American edition of William Phillips' *An outline of mineralogy and geology*. He suggested several means of testing the theories in order to actually produce a quartz crystal from water solution, but repeated the same notes on the difficulties as had been provided 35 or more years earlier.[26] Later editions, both in England and in the United States, included more detailed chemistry. Also in the United States, in 1816, Cleaveland noted there were difficulties with the Neptunian position. However, he also called the Huttonian position "so ingeniously illustrated, but unsuccessfully supported by Professor Playfair" (Cleaveland, [1816] 1978, p. 593). He had more questions than answers on the subject of theory.

By 1818, Breislak had come to no new conclusions, and repeated his comments about mineral fusibility and the effects of sequential crystallization. Despite all the experimental and observational data about the effects of heat, he still felt aqueous origin was most likely for granite. Breislak noted the difficulties of scale: It was easier to explain things for small amounts of material than for the large masses of Earth. There was no solution for some time to come on the question of whether volcanic heat was local or the result of a whole-Earth phenomenon. Authors still made references to oxygen, and whether it was required for volcanic heat. It was suggested that because there were many products of volcanoes, perhaps there were many causes for their formation. Although consensus was growing on how basalt should be described, it was still necessary to determine if the word had a consistent meaning, and he commented: "La détermination de

ces roches forme aujourd'hui un des points les plus controversés dans la géologie" (Breislak, 1818, v. 3, p. 229).[27]

The well-traveled British botanist, physician, chemist, and geologist Charles B. Daubeny (1795–1867), although of a later generation than many people we've discussed, stated, after his extensive examination of the volcanoes of Auvergne:

Since, therefore, it seems to follow, that Fire and Water, although such opposite agents, have, in some instances, produced effects nearly, if not altogether identical, I do not see that the geologist who returns from Auvergne, persuaded that great part ... is of volcanic origin, ought to be accused of inconsistency, if he still hesitates as to the real origin of those rocks. (Daubeny, 1821, p. 314)

G.S. Mackenzie considered Hutton's theory, despite Playfair's help and the truth of some parts, to be "manifestly contradicted by facts" (Mackenzie, 1826, p. 84). However, he still referred positively to Hall's experiments. Excluding either water or fire led to difficulties. In his work on fluids present in mineral cavities, David Brewster felt it was necessary to try to see the relation of that work to the two competing theories, with the addition of the behavior of gases (Brewster, 1826a, p. 38).

As the nineteenth century progressed, new voices were added to the discussion, but there was no trace of consensus among the larger community of practicing geologists.[28] As the laboratory investigations continued, a kind of benchmark occurred with the publication of the first volume of the first edition of Lyell's *Principles of geology* in 1830, which of course drew on earlier work. While the individual field and laboratory investigations of a group of increasingly specialized practitioners continued, many of them having looked back at earlier work, Lyell summarized the history of geology to that point. He can thus serve as one of the last primary references for the eighteenth–nineteenth-century work, and the first of the secondary references about it. Some of his conclusions about the efficacy of Huttonian/Wernerian theory are stated in this chapter. In the next, the few arguments that he drew from experiment in his position of historian will be discussed.

Lyell's was a more complex statement of a theory of Earth than had come before. He had dragons to slay other than fusion or solution, and far more collected data, observation, and intelligence to accommodate. As is well known, he ended with a broadly cyclic, steady-state system, in which very gradual processes in the past were similar to those currently observable in operation, and had not differed from them in degree or kind. He discussed Hutton and Werner and the controversy between Neptunist and Plutonist positions. He repeated admiration for Werner's grasp of mineralogy and mining, and the devotion of his students, and noted as well that Werner had not traveled widely and denied the action of volcanoes in the production of basalt. About his general theory, Lyell remarked:

---

[26]This crystal growth had actually been seen in the course of silica analysis more than 30 years previously.

[27] "Today, the determination [of origin] of these rocks forms one of the greatest controversies in geology" (my translation).

[28]See Brongniart (1825), Daubuisson (1828), and Burat (1834–1835, v. 2 and 3), among many others.

His [Werner's] theory was opposed, in a two-fold sense, to the doctrine of uniformity in the course of nature; for not only did he introduce, without scruple, many imaginary causes supposed to have once effected great revolutions in the earth, and then to have become extinct, but new ones also were feigned to have come into play in modern times. (Lyell, [1830] 1990, v. 1, p. 58)

He further argued that despite the determination of the volcanic origin of basalt by many observers, "Werner by his dictum caused a retrograde movement," overturned the real theory, and "substituted for it one of the most unphilosophical ever advanced in all of science" (Lyell, [1830] 1990, v. 1, p. 59). Wernerians continued in their faith, however, and "the Vulcanists were not long in becoming contaminated with the same intemperate zeal" (Lyell, [1830] 1990, v. 1, p. 60). For Lyell, Hutton's theory was far more philosophic in that he only called upon causes that could currently be observed. Even so, he believed that "Its greatest defect consisted in the undue influence attributed to subterranean heat, which was supposed necessary for the consolidation of all submarine deposits" (Lyell, [1830] 1990, v. 1, p. 63). Lyell felt that Hutton had made the error of not supporting his claims of great antiquity for Earth with the use of organic remains, although one could argue that great age could be shown by the unconformities that Hutton observed. Reviewing the controversy in England, Lyell argued that, despite Playfair's eloquent defense,

[N]eptunianism and [religious] orthodoxy were now associated in the same creed; and the tide of prejudice ran so strong, that the majority were carried far away into the chaotic fluid, and other cosmological inventions of Werner. (Lyell, [1830] 1990, v. 1, p. 69)

Lyell suggested that after all the contention, a neutral school arose that paid no attention to the excesses of either theory, and instead dealt with observations that could advance geology in a better way. He went on to review the growing literature and, finally, in his three-volume *Principles of geology*, to arrive at his own theory.[29]

It is important to note that there was no decision on which of the theories was best for years to come, and there was never a definitive date when Hutton was seen to be generally correct with the exception of a relatively few misinterpretations. Lyell was certainly influential, but the picture was muddied wherever geology was spoken by the continued use of Wernerian terminology for lithostratigraphy that could imply Wernerian origin and time relation principles, whether intended or not. Investigations continued apace on all levels, with stratigraphy, in particular, coming more

into its own. For the laboratory and experimental work, a complex web was being woven. In no particular order, studies continued on pressure effects, the adequacy of sampling, fusion temperatures and tools and appearance of fused materials, solubility, gas identification and behavior, reactions, analysis/composition, affinities and disassociation, physical characters and identification, heat conduction, cooling rates, function and behavior of volatiles in solids, and specific gravity. The implications of compositional variation in single rock types were investigated. Also noted were stony deposits in water pipes, and glazing on the fused materials of several technologies. Instruments of all kinds were improved, and improved again, and there were new methods of approaching old problems with new instruments. There was significantly increased standardization of nomenclature, and far more stringent efforts to specifically identify rocks and minerals. Experimentation allowed a kind of control absent from field observation or theorizing because the process could be watched and repeated. Samples could be well characterized and variables controlled. The difficulty of extrapolating from the laboratory to nature was well recognized, but benchmarks such as fusion temperatures remained. It is safe to say that experimentation narrowed acceptable parameters and required theory statements to be fact-based, as well as being congruent with field evidence.

My contention has been that experiment and controlled or contrived observation was the third leg of the triad of field observation, theory, and experiment. It actually proved nothing, or very little, but pointed out numerous possibilities, and was a frequent component of the arguments of the geologists of the time. Despite its inconclusiveness on questions that were global in scale, it supplied reproducible facts about geological materials, and as such, I believe it was a major reason why neither Huttonian nor Wernerian theory was accepted. It also led to a plethora of theories about rock origin, in that each verifiable fact had to be accounted for. The position of experiment in geology continued and continues to be a subject of contention, although the methods pioneered in all sciences in the eighteenth century, such as attention to sampling technique, multiple trials, instrument calibration, care with calculations, etc., have become part of accepted experimental protocol. It both used extant instruments and contributed to the design of instruments and equipment that could function under the required conditions of pressure and temperature. Naturally, experiment could not address the increasing awareness of the enormity of time during which conditions applied. The same sorts of objections to the efficacy of experiment voiced then and, to some extent, now are currently echoed about the validity of computer models for large-scale phenomena. The fact that few historians of geology have discussed experimentation in detail has led to the impression that experimentation was not important to the science. I hope this book is a first step in remedying that impression.

---

[29]The study of Lyell, the influences on him, and his influence on others could be said to be nearly an "industry" in the history of geology. A few more recent references are Dolan (1998), Berggren (1998), and Young (1999).

The place of experiment

INTRODUCTION

One might argue whether a book on experiment in geology is necessary, given that there is general agreement that in the eighteenth-century, experimentation didn't change the course of early geology greatly. Experimentation might almost have been seen as an impediment to geological reasoning, since actions possible in laboratories were often not observed in nature. Conversely, the actions seen in nature, such as lava flowing down hillsides, seemed to be the product of forces so far in excess of humankind's puny efforts that attempting to draw conclusions from the latter was an exercise in futility. Rudwick has taken the view that collections and fieldwork had considerably more influence than experimentation until well into the nineteenth century, the role of the early laboratory being marginal (2005, p. 37–38). But there must be a beginning for everything, and the interweaving of experiment, thought, and other observations was an integral part of the growth toward truth in geological reasoning. Clearly, the practitioners of experiment and observation did not think they were wasting their time. As detailed in previous chapters, untold hours were devoted, sometimes at considerable risk and expense, to investigate the properties of the materials that make up Earth, as well as the way they behaved under conditions thought to exist in their regions.

At the turn of the nineteenth century, geology was not only popular, but in the minds of many, it was at least equal in importance to chemistry and physics, both practically and intellectually.[1] On the practical side, study of Earth made possible the discovery of useful substances: ores, coal, dyes, and reactive, preservative, medical, and fertilizing agents, to name a few. Experimental tests could approximately determine the quality of the substances. Intellectually, geology could arguably be said to be the key to the greatest adventure of all: the history of mankind and of Earth itself. But overarching theories of Earth increasingly gave way to objective summaries of earth material properties and detailed observation of earth processes. The then-new Geological Society of London rejected theorizing in favor of the patient accumulation of observable facts. Many of those facts were from fieldwork, but a recent paper analyzing oral papers given at the Geological Society has shown that practical and technological work contributed more than previously thought (Veneer, 2008). Members had economic and mining interests, and mineral analysis and instrument discussion were relatively common. The official objectivity didn't prevent speculation or adventurous thinking; witness the intense efforts to understand the placement and changes within fossils, or the puzzles of stratigraphy. Then and now, the issue of geologic time occupied some of the best minds. But at this time of ferment, experiment acted as a brake on the premature choice of theory. Even if in most cases it didn't provide specific answers to many fundamental geological questions, it undoubtedly raised possibilities and opened new avenues. Just one example of that would be Grignon's demonstration that crystals could form from a means other than liquid solution.

EXPERIMENT/OBSERVATION

In this book, the word "experiment" has been employed to refer to what might be called "regulated" observation[2] and instrumentally aided observation, as well as the later meaning of a sequence of actions designed specifically to test a hypothesis with controlled variables except for the one under question. The recalcitrance of geology to the variables of time and total mass has been noted by most practitioners. In his address to the annual general meeting of the Geologists' Association of London on 1 February 1889, the outgoing president, Frederick William Rudler (1840–1915), selected experimental geology as the topic for his presidential address. He chose it because he "sought to avoid all topics which have been brought forward by previous occupants of this chair," which implied that it had not previously been considered a topic of prime importance (Rudler, 1889, p. 69). Rudler liked the definition that Sir John Herschel (1792–1871) gave for "experiment," which he defined as "active observation" (p. 70). Rudler thought most of geology consisted of passive observation, but experimentation occurred when the geologist "introduces artificial conditions to aid in the observation of nature" (p. 70). Even drilling into a formation could fit within this definition, as witnessed in the stratigraphic work of William Smith and Greenough. But unlike laboratory experiment, the field geologist couldn't adjust the conditions under which his observations took place. He could record the temperature, barometric pressure, and other accessible variables.[3] But duplicating the time period was impossible. In an interesting "take," Rudler discussed

---

[1] We recall that the three (or more) sciences were not separated from each other at that time.

[2] The word "observation" covers a lot of territory. I have tried to emphasize some kind of imposition of order over and above field observation that simply recorded what was seen. Apart from the use of instruments, it might have included notes taken at the same time in a sequence of days or collection of gases from a fumarole on a regular basis.

[3] Hall estimated the pressure of seawater at depth, where consolidation in the Huttonian sense might occur (Hall, 1812, p. 184).

the deposition of minerals on walls built by the Romans around hot springs. The resulting mineralization had occurred in about 20 centuries. Rudler felt that this situation justified application of the knowledge of the historical process to explanation of a geological occurrence (Rudler, 1889, p. 73). He continued with a discussion of what conclusions could be drawn from the synthesis of minerals in the laboratory, having provided a historical review of those attempts. He pointed out that just because some minerals had been synthesized under particular conditions, it didn't rule out their being produced under different conditions. In the end, he argued that experiment needed to be guided by hypothesis:

The imaginative faculty which can create a rational hypothesis—that is, a hypothesis conformable to ascertained facts—thus becomes a quality of the first importance to the geological experimentalist. (Rudler, 1889, p. 103)

So when:

[T]he waters of experimental verification have percolated through the sands of speculation the ground beneath our feet acquires firmness and fixity. If some of our favourite theories are dissolved away others are consolidated, and we have the satisfaction of seeing what were previously mere probabilities crystallize into certainties. (Rudler, 1889, p. 91–92)

In his discussion, Rudler displayed a thorough understanding of the place of experiment in geology. His lucid defense of geological experiment should be read now by both geologists and historians of geology.

IDEAS IN FLUX

Nineteenth-century geologists, such as Lyell and Bakewell, who wrote on the history of geology gave experiment a minimum of space, limited largely to mention of the work of Hall and Cordier, with some reference to Saussure.[4] Later Kobell ([1864] 1965), as a practicing mineralogist, thoroughly reviewed methods and instrumentation applied to mineral characterization as they appeared in the discovery timeline. Prestwich (1895) discussed controversial questions in geology and noted that there were usually two sides to the questions of whether there was uniformity of both kind and degree, or just of kind, in geological processes. He cited experimental evidence that sometimes went back to the eighteenth century. These books are better references for consideration of the role of early experimentation in geoscience than are many later works.

Few of the general standard histories of geology after that time through the twentieth century discussed experiment at any length. Writing in the "hero" mode, authors have given three different researchers credit for being the "first experimental petrologist" or the "first experimental geologist": Spallanzani, Saussure,

and Hall. This designation is less important than the specific attention given to their work. But, sometimes, an author is betrayed by an incomplete reading of sources or, perhaps, by too hasty a conclusion. In my survey of histories of geology and related papers, I have encountered three statements that made claims about Hall's work that Hall himself had not made. Porter wrote: "Hall in particular persuasively demonstrated in a distinguished series of experiments the interconvertibility of glassy and basaltine materials, and the fusibility of granite" (Porter, 1977, p. 175). But Hall never claimed to have fused granite, and in fact delayed further experimentation because of his lack of success with it. Carozzi's translation of Gohau stated, "Thanks to James Hall's laboratory demonstration, the igneous origin of granite was accepted rather quickly, and geology became Huttonian (Gohau, 1990, p. 121). A distinguished geologist made the same error, saying that Hall's fusion of granite was successful (Pitcher, 1993, p. 2). In contrast, as we have seen, acceptance of granite's igneous origin was anything but quick, and for a long while it depended far more on field observation than on laboratory results. Porter also discussed the general lack of attention paid to geology in histories of science. He pointed out that often new and specific data in geology tended more to confuse matters than to clarify them: "[F]resh data were an irritant, provoking new problems, and new domains of conceptualization, rather than simply confirming or refuting old theories" (Porter, 1980, p. 311). Porter pointed out other instances where more data on large-scale phenomena such as glaciation led to uncertainty rather than the opposite. For rock origin, showing that a rock such as basalt might have an igneous origin did not help to explain limestone, and even a plausible, perhaps low-temperature, origin for limestone might not explain all the features of marble.

FITTING IT IN

Space specifically designed for experiment became a feature of science in the seventeenth century with the work of Boyle. From that time, experiment has been called "a systematic means of generating natural knowledge" (Shapin and Schaffer, 1985). But as has been noted, in geology, there is not a clear demarcation between a glass-, ceramics-, or metalworker trying out a new combination of earth substances and a scientist forming hypotheses and testing theory. A continuum of endeavor lies between the two. Ideally we would see what Oldroyd (1986) has called "the arch of knowledge," hypothesis formulation and testing with inductive and deductive "legs." Essentially, the upward leg of the "arch" was formed of facts, phenomena, or data, which rose by induction from the facts to the top, which consisted of scientific principles, or even theory. From the principles thus deduced, one could deduce further facts, that process being the other, descending, leg of the arch, which facts could then be checked out experimentally (Oldroyd, 1986, p. 363). This is a simple statement of how science should work with respect to induction and deduction, a matter that has been much discussed by philosophers of science. It would be well past the beginning of the nineteenth

---

[4]It has been suggested that while experiment was common to all the sciences, the use of fieldwork enabled geology to be more distinctive (K. Taylor, 2008, personal commun.).

century before the "arch" would be anywhere near being built in geological experimentation, and then only in a few examples, the barriers being those of space and time.

In his writing about science, William Whewell (1794–1866) distinguished between two sorts of evidence: There was the kind of evidence that established a necessary truth, for Whewell, intuition of its self-evidence, which was different from the kind of evidence—observation or experiment—that established an empirical proposition (Butts, 1968, p. 10). The geology we have been discussing surely depends on both sorts of evidence. But in reading histories of science, it appears that the two types are not always delineated or distinguished from one another. We have been discussing the second kind in the main. Whewell did distinguish between descriptive and physical geology, which he refined as phenomenal and theoretical, or etiological, geology (Whewell, 1858, v. 2, p. 263). Description of causes in geology he found to result partly from calculation and reasoning, but observation of phenomena was also necessary (Whewell, 1858, v. 2, p. 279). He noted how Hall's experiments underpinned theories that could not be directly tested. With respect to the contest between what he called "Neptunian" and "Plutonian" or "Vulcanian" geology, he remarked:

[A]ll that is to remain as permanent science in each of these systems must be proved by the examination of many cases and limited by many conditions and circumstances. Theories so wide and simple, were consistent only with a comparatively scanty collection of facts, and belong to the early stage of geological knowledge. In the progress of the science, the 'theory' of each part of the earth must come out of the examination of that part, combined with all that is well established concerning all the rest; and a general theory must result from the comparison of all such partial theoretical views. (Whewell, 1857, v. 3, p. 503–504)

He believed that patient accumulation of observations was still needed to determine geological causes.

A pattern emerges. The experimental and observational/instrumental work discussed in most histories of geology concerned with the eighteenth and early nineteenth centuries falls into only a very few categories, and in general it was not used as a major part of argument. With some exceptions, as noted, rather than a comprehensive report on experiment, or "tweaked" observation, as an integral part of the history of geology, in most cases only those cases that impinge directly on the portion of geology under discussion are cited. Under a broad umbrella of what might be considered were: Steno and his tectonic diagrams and studies of crystal forms; Nicolas Lemery's (1645–1715) "volcano"; Buffon's cooling experiments; Sir James Hall's work on basalt formation, carbon dioxide retention with heating under pressure, and folding due to directed pressure; modes of rock consolidation; fusion conditions; temperatures of surface water and Earth, and the same at depth; and volatile contents of water and lava. Few histories mention all, or even a majority, of those. The scope of mineralogical investigations with respect to theory was questioned. For example, Henry Clifton Sorby's (1826–1908) work was not always seen as advancing geology, and was referred to by some as "looking at mountains with a microscope" (Geikie, 1895, p. 343). However, despite the limits of any one investigation, the insights gained did enter into wider theorizing.

The difficulty of using the usual experimental method of science in a historical science has been remarked on by many. Although he was referring to Darwin's methodology, Gould's comment about natural historical sciences fits geology well: "Because you can't do experiments, you must be able to use the results of history to explain nature" (Gould, 1989, p. 1000). Kirwan addressed this problem in 1799:

In the investigation of past facts dependent on natural causes, certain laws of reasoning should be adhered to. The first is that no effect shall be attributed to a cause whose *known* powers are inadequate to its production. The second is, that no cause should be adduced whose existence is not proved either by actual experience or approved testimony. ... The third is that no powers should be ascribed to an alleged cause but those that it is known by actual observation to possess in appropriated circumstances. (Kirwan, 1799a, p. 1–2)

Of course, neither Kirwan nor Werner complied with this ideal situation, Kirwan being "biblical" in some instances. Even Hall was convinced that his observation of basalt fusion and cooling could be extended by analogy to lava flows that he had seen in the field (Hall, 1805b, p. 66). He relied on Kennedy's chemical analyses to support his fusion/cooling experiment that showed the identity of whinstone and lava. Before his work with carbonates, he noted nodules of calcareous spar in porous lava, and proposed that Hutton's supposition of pressure could explain the anomaly despite the heat applied. Hall wrote:

He [Hutton] rested his belief of this influence on analogy; and on the satisfactory solution of all the phenomena, furnished by this supposition. It occurred to me, however, that this principle was susceptible of being established in a direct manner by experiment, and I urged him to make the attempt; but he always rejected this proposal, on account of the immensity of the natural agents. (Hall, 1812, p. 75–76)

The carbonate experiments did support the suggestion that pressure confined the carbon dioxide, despite the difference in scale. Hall also investigated whether the presumed fuel for basalt fusion, coal or organic substances, could be brought to a sufficiently high temperature. That was a game attempt, but he confessed he could not determine whether organic material or a (hypothetical) great central fire was the actor. Hall deserves credit for having the temerity to attempt to harness the forces of nature in a laboratory, even though few attempted to repeat his work. Seasoned observers of the time found that correlating the field evidence with Hall's work was confusing.

JUDGEMENT NOW

None of this experimentation or argument addressed many of the problems now discussed by philosophers of science with respect to historical and current experimentation. The supposition that we can re-create the work of the eighteenth or nineteenth centuries by using instrumentation of that period or replicas of it has not proved to be easy.[5] When attempted with, for example, Michael Faraday's (1791–1867) work, the importance of both the instrumentation and its manipulation became apparent (Gooding, 1990).

Historians and philosophers have not always been in agreement about what the relations of earlier efforts were to an accurate depiction of nature, this topic in itself having generated its own literature.[6] Gooding et al. (1989) have, for example, remarked:

> Experiment has many uses apart from supporting or refuting knowledge claims: active observation, invention, the construction of models, imitation of natural phenomena, or the design of instruments to extend the senses. Not the least important use is the provision of evidence for rival philosophies of science, even if this evidence usually relies on a cursory account of experimental work. (Gooding et al., 1989, p. xv)

And things have become much more complex in the twenty-first century. As instruments have become incomparably more intricate, experiments are more and more removed from nature. The debates of the eighteenth century and Hutton's doubts about whether a laboratory could provide truth about nature pale beside particle accelerators, protein sequencers, electron microscopes, and tomography, to mention just a few. How many "layers" are there between nature and a cluster of points on a graph purporting to show basalt affinity with one or another tectonic regime? The way in which data are presented has evolved into entirely new graphic representations over nearly 200 years, as well shown by Richard Howarth (2002).

We can easily follow Sir James Hall's calculations of carbonate density before and after rocks' subjection to high pressures as recorded in his laboratory notebooks, from which he drew some of his conclusions. The increasing use of mathematics in the sciences continued from its inception in the early seventeenth century. But what do many of us know about unintentional biases introduced by computer data-crunching now? Rather than making experiments more transparent to reasoning, our sophisticated instruments and computer models may veil them in complexity. The range of knowledge required to be a "scientist" has expanded enormously, and the implications of computer models and data processing are not always understood even by those who use them. There are, of course, standard methods for determining the "fit" of data. But these multiple problems quickly become more of a problem of philosophy than of history as evidenced in the work of Hacking (1983), Brown (1987), and others. They have not been addressed in this book, which discusses the tools, procedures, thoughts, and opinions about rock origin of those who worked in the eighteenth and early nineteenth centuries. As mentioned earlier, also not addressed is the entire and valid question

of the social import of the web of workers, thinkers, and philosophers that comprised and comprise the scientific enterprise. The aim here has been to highlight a portion of the reasoning process that has hardly been addressed in most histories of the beginning of geology, yet which, to my thinking, constitutes at least a part of the sturdy feet of Oldroyd's arch.

CONCLUSION

So, what function did experiment have, and what larger themes did the experiments address within the nascent science of geology? It seemed to begin in the practical realm. What mineral substance could flavor or preserve food? Which would be an emetic? What substances were safe for food storage? Which rock could yield a metal? What could surface color or structure show about what was underground? Which rocks and minerals would provide pigments? What rock or combination of minerals could be used to enclose a fire? Which could help someone recover from a fever? Could expensive jewels be duplicated? What kind of soil or soil addition helped food plant growth? As we've seen in the sixteenth- and seventeenth-century mining treatises, the practical identification of many useful classes of substances was well advanced, as were ways to combine and/or treat them to make desired articles. Combination was not well understood, although advantageous components might be, as shown in the very early understanding of how to make different kinds and colors of glass. In all of the practical work, the method of "try it and see what happens" (which we still use today, especially in solving computer problems!) was much in use.

It should be noted that "interference" with nature was not nearly so extensive 200 years ago as that referred to in discussions of experiment today. Basically, phenomena were observed, sometimes with instrumentation that allowed closer or more quantitative observation, but there was virtually no creation of phenomena as, for example, a cloud chamber allows. And as opposed to the view of many works on experiment in the final decades of the twentieth century, observations were not all theory laden (Lenoir, 1997, p. 22).[7] Neptunism and Vulcanism/Plutonism were such overriding theories, with so little direct evidence, that only the possible was investigated. Those who were surer of their theories, for example, Hutton and Werner, were not laboratory experimenters with respect to testing geological theory. And as frequently noted, groups such as the Geological Society of London specifically rejected theory in favor of field observations.[8]

Not surprisingly, experimental work in geology at the end of the eighteenth century parallels Holmes's (1985) description of that of Lavoisier in respiration, although the theory involved

---

[5]In an attempt to show a simple Daltonian weight ratio for a reaction, I once cleared a laboratory of innocent inorganic chemistry students because the twentieth-century vial of "pure" iron filings was contaminated with sulfur, leading to unexpected production of hydrogen sulfide.

[6]We won't consider that literature here as it lies more in the area of philosophy than history of science.

[7]Lenoir mentions I. Hacking (1936–) as an author who supports the position that experiment has a life of its own apart from theory. One could argue that all observations were theory laden, because otherwise, what would be the point of doing them?

[8]As Laudan (1987) has pointed out, in Europe there was a stronger tradition of meticulous mineralogical and chemical work, and in England there were Neptunian exceptions such as Robert Jameson.

in geology was of far larger scale and even more elusive. Unrelated experiments might give clues to promising modes of inquiry. Rather than fixed on a goal, an investigation was often open-ended, with suggestions coming from the work itself rather than prior theory (Lenoir, 1997, p. 10). Apparatus was not necessarily designed for a specific purpose, but drawn from other areas. Hall adapted Rumford's pressure apparatus (a gun barrel) from Rumford's experiments on the power of gunpowder among other parts of a web of investigation of pressure effects. But the late eighteenth-century and early nineteenth-century experimenters identified many questions of continuing interest in geology, and the work of solving the problems of mineral and rock formation, identification, and synthesis continued, on an ever more sophisticated level.

However, as the work continued, as Fritscher has pointed out, even in the closing decades of the nineteenth century, metamorphism was studied by neither experiment nor chemistry, but instead, the petrographic microscope was employed to seek for petrographic and stratigraphic relations (Fritscher, 2002, p. 146). In the early decades of the twentieth century, the application of chemical equilibrium theory to mineral assemblages still suffered from a lack of proper instruments and methods with which to study it (Fritscher, 2002, p. 159).[9] There is a dichotomy between what P. Galison (1997) has called the "homomorphic" and the "logic" styles of investigation. Although his work referred to twentieth-century physics, that thinking can enlighten our view of much earlier geological experimentation. The first instance referred to a full reproduction of natural processes, which was definitely what eighteenth- and early nineteenth-century experimenters were attempting, although they knew they fell short. At that time, there was far less of the second, which is basically the statistical use of a great deal of data (Fritscher, 2002, p. 160).

And so, to which of the roots of geology did experimentation contribute? Rudwick's *Bursting the limits of time* (2005) has traced the multiple strands of geology as they arose from the previous era of geotheories. Perhaps the first general statement of the role of the "geosciences," as we now term them, came with the publication of Saussure's *Agenda* in 1796 (v. 4). The multiple facets included the double necessity of both indoor, meaning museum and laboratory, and outdoor or fieldwork, in this highly empirical science (Rudwick, 2005, p. 346). At that time, both natural history with its description and classification, as well as natural philosophy, which included mathematical relations and causal theories of natural phenomena, were employed. A major strand of geology, mineralogy (Laudan, 1987), with its extensive laboratory investigation, was considered part of natural history. What distinguished it from pure chemistry was the naturally occurring origin of its subject rather than the compounds or operations themselves. The earlier primary thrust of mineralogy had been collecting, describing, and classifying. In the early

years, the word "fossil" was used both for remnants or records of organisms, as well as in the sense of anything that was found in Earth. But the process of identification included the extensive methods and equipment as I've discussed in earlier chapters, whether applied in a laboratory or in the field with traveling kits. Complexity was added when combination of mineral assemblages into rocks was considered.

Physical geography, or the description of large features of Earth, was another strand. Description of a mountain range, volcano, or river valley, as seen in fieldwork, would not seem related to our modern concept of experiment, and was indeed characterized as natural history. But as stated in earlier chapters, instruments were devised to quantify phenomena observable in the field, such as temperatures in various situations, barometric pressure, gravitational attraction, and other parameters. Experimenters continued to use instruments to measure and quantify the angle of beds, and the humble hammer became the emblem of the geologist. Another strand, the miners' discipline of geognosy, focused on the third dimension. It also required instrument use and tests, although theory was eclipsed by the need for physical description of underground structure to aid in finding commercially useful resources. The practical tests employed led later to chemical clarification useful in conjectures about mineral and rock origin. Additionally, fusion conditions and observations of ores, slags, and metal crystals fed directly into understanding of geological origin conditions. All three of these natural history strands became increasingly quantifiable, and raised questions that might be tested by experiment.

The last of the four strands identified by Rudwick was the one that belonged more to natural philosophy than natural history. He said:

> Rather than describing and classifying, it used the natural-history sciences as raw material for detecting the regularities or "natural laws" underlying the observable occurrence of terrestrial features and processes, with the ultimate goal of determining their physical *causes*. (Rudwick, 2005, p. 99)

This activity is closer to the information gathering and testing that we associate with experiment. *The world in a crucible: Laboratory practice and geological theory at the beginning of geology* has emphasized the search for the origins of minerals and rocks. The attempt to find causal agents was a major part of that, and appears repeatedly on all scales, including that applied to the origin of rocks, the violence of earthquakes and volcanoes, and the abiding question of the source of Earth's heat, the location of fossils, and the deposition of ores.[10] Forces or activity might be postulated, as well as the question of the time during which forces might be applied. In all of this study, I have not seen so much of the separation of those different strands of geology as Rudwick discusses, but rather I see the unity of their contributions to the science that geology became. In all, the methods of science and

---

[9]Fritscher mentioned that this is discussed in detail in Yoder (1980) and Geschwind (1995).

[10]I have not considered questions of surface forms and erratic blocks seemingly unconnected with their surroundings.

the experimental spirit permeated questions about Earth. I would argue that it was much more important than Rudwick has indicated, brought into play in one form or another in all of his four areas, and indeed could be said to be the unifying principle that makes geology into a science. The nascent science of geology never lacked for questions, which situation is still familiar to us in the twenty-first century.

EPILOGUE

Earth is a huge and unwieldy object. It cannot be manipulated as a whole by humans, nor can its past be directly investigated. There is the perception in some quarters that geology is a derivative science, dependent on the major sciences of chemistry, physics, and biology for its objective truths. But to its practitioners, geology has a center and mission, an intrinsic identity, which is aided by, though far from limited by, the tools provided by those other sciences. It is perhaps the best example of an interdisciplinary endeavor, yet it possesses a core value of its own. Unlike the other sciences, it is concerned with the great necessities of both life (and spirit). Soil, water, the permanence of formations and sediment, resource location, and geological hazards continue to be of essential interest. What other science, with the exception of some parts of astronomy, considers the history of the Earth we inhabit; what it might have been like; and what it may turn into in the near or distant future?

There are vast reaches of time, materials, and processes so large as to be imperfectly understood, all most emphatically not under human control. One must be comfortable with the probable and the possible, not the sure proof. It is not only the fieldwork which makes geologists the adventurous souls that they are, but the necessity of grappling with problems that often cannot be neatly delineated in space or time. One thing that attests to the importance of early experimental methods in geology is its legacy. There has never been any break in the progress of experimentation. Each of the topics discussed in this book was followed through the next centuries with improved methods, advanced instrumentation, and more sophisticated mathematical analysis and graphic representation. One may even see the Mars rovers as distant descendants of the processes begun so long before. Geological experiment in the eighteenth century helped define what experiment could be and do in this science of ours.

Thus, the science of geology still has need of its multiple branches and ways of working. Because of it inconclusiveness, historians, particularly in the twentieth century and beyond, have not given geological experimentation its due. It was a far richer endeavor, and far more part of the overall fabric of geology than historians of that science have previously recognized.

I believe in scientific inquiry for its own sake. I think the history of science gives ample examples that pure investigation has enormous benefit. ... I can't tell you what this might be good for, but learning about nature is important. And lovely things turn up.
—James A. Van Allen, quoted in his obituary
in *The Washington Post*, 10 August 2006

References Cited

REFERENCES BEFORE 1840

Agricola, G., [1546] 1955, De natura fossilium (translated by Bandy, M.C., and Bandy, J.A.): Geological Society of America Special Paper 63.

Agricolae, G., 1556, De re metallica: Basileae.

Agricola, G., [1556] 1950, De re metallica (translated by Hoover, H.C., and Hoover, L.C.): New York, Dover Publications.

Anonymous, 1791, Observations on the affinity between basaltes and granite: Monthly Review, v. 6, 2nd ser., p. 139–142.

Arderon, W., 1748, Part of a letter from Mr. William Arderon F.R.S. to Mr. Henry Baker F.R.S. concerning the formation of pebbles: Philosophical Transactions of the Royal Society of London, 44, p. 467–71.

Argenville, A.J.D. d', 1742, L'histoire naturelle éclaircie dan deux de ses parties principales, la lithologie et la conchyliologie: Paris, Chez de Bure l'Aîné.

Argenville, A.J.D. d', 1755, L'histoire naturelle éclaircie dans une de ses parties principales, l'oryctologie: Paris, Chez de Bure l'Aîné.

Bakewell, R., 1813, An introduction to geology: London, J. Harding.

Bakewell, R., 1819, An introduction to mineralogy: London, Longman, Hurst, Rees, Orme, and Brown.

Bakewell, R., [1833] 1978, An introduction to geology: New Haven, H. Howe & Co.

Bayen, P., 1779, Examen chymique de diférentes pierres, Quatrième partie: Observations sur la Physique, v. 14, part II, p. 446–55.

Bayen, P., 1797, Opuscules chimiques de Pierre Bayen: Paris, Chez A.J. Dugour et Durand.

Beddoes, T., 1791, On the affinity between basalts and granite: Philosophical Transactions of the Royal Society of London, v. 81, p. 48–70.

Bergman, T.O., 1784a, Manuel du minéralogiste, ou sciagraphie du règne minéral, distribué d'après l'analyse chimique (translated by Mongez): Paris, Cuchet.

Bergman, T.O., 1784b–1791, Physical and chemical essays: London, J. Murray, 3 vols.

Bergman, T.O., 1792, Manuel du minéralogiste; ou sciagraphie du règne mineral: Paris, Cuchet, 2 vols. New edition by La Métherie, J.-C. de, ed.

Biringuccio, V., 1540, De la pirotechnia libri x: Venice, Per Venturino Rossinello ad instantia di Curt Nauo.

Biringuccio, V., [1540] 1990, The pirotechnia of Vanoccio Biringguccio (translated by Smith, C.S., and Gnudi, M.T.): New York, Dover.

Black, J., 1794, An analysis of the water of some hot springs in Iceland: Transactions of the Royal Society of Edinburgh, v. 3, p. 95–126. (read in 1791)

Boccone, P., 1674, Recherches et observations naturelles: Amsterdam, Chez Jean Jansson A. Waesberge.

Boyle, R., [1672] 1972, An essay about the origine and virtues of gems: New York, Hafner Publishing Co.

Breislak, S., 1792, Essaies mineralogiques sur la Solfatara de Pouzzole (translated by F. De Pommereul): Naples, R. Giaccio.

Breislak, S., 1801, Voyages physiques et lythologiques dans la Campanie: Paris, Dentu, 2 vols.

Breislak, S., 1812, Introduction à la géologie (translated by Bernard, J.J.B.): Paris, Chez J. Klostermann.

Breislak, S., 1818, Institutions géologiques: Milan, à l'imprimerie imperiale et royale, 3 vols.

Brewster, D., 1826a, On the existence of two new fluids in the cavities of minerals: Transactions of the Royal Society of Edinburgh, v. 10, part 1, p. 1–41.

Brewster, D., 1826b, On the refractive power of the two new fluids in minerals: Transactions of the Royal Society of Edinburgh, v. 10, part II, p. 407–428.

Brochant de Villiers, A.J.F.M., 1801–1803, Traité élémentaire de mineralogie: Paris, Chez Villier.

Brongniart, A., 1825, Introduction à la minéralogie ou exposé des principes de cette science: Paris, F.G. Levrault.

Buch, L. von, 1802, Geognostische Beobachtungen auf Reisen durch Deutschland und Italien, v. 2: Berlin, Haude und Spener.

Buch, L. von, 1809, Geognostische Beobachtungen auf Reisen durch Deutschland und Italien: Berlin, Haude und Spener, 2 vols.

Buffon, G.-L. Leclerc, Comte de, 1749, Histoire naturelle: Paris, Imprimerie royale.

Buffon, G.-L. Leclerc, Comte de, 1774, Histoire naturelle, générale et particulière: Supplement, v. 1: Paris, Imprimerie royale.

Buffon, G.-L. Leclerc, Comte de, [1778] 1971, Des époques de la nature: Paris, Éditions rationalistes. Introduction by G. Gohau.

Burat, A., 1834–1835, Traité de géognosie: Paris, F.G. Levrault, 3 vols.

Burnet, T., 1684, The sacred theory of the Earth: London, R. Norton for Walter Kettilby.

Burnet, T., [1684] 1965, The sacred theory of the Earth: Carbondale, Illinois, Southern Illinois University Press.

Cadet, C.-L., 1803, Dictionnaire de chimie: Paris, Imprimerie de Chaignieau aîné, 2 vols.

Cleaveland, P., [1816] 1978, An elementary treatise on mineralogy and geology: New York, Arno Press.

Collini, C.A., 1781, Considerations sur les montagnes volcaniques: Mannheim, Germany, Schwan.

Cordier, L., 1807, Recherches sur différens produits des volcans: Journal des Mines, v. 21, no. 124, p. 249–260.

Cordier, P.-L.-A., 1808, Suite des recherches sur différens produits des volcans: Journal des Mines, v. 23, p. 55–74.

Cordier, P.-L.-A., 1815, Sur les substances minérales, dites en masse, qui servent de base aux roches volcaniques: Journal des Mines, v. 38, p. 383–394.

Cordier, P.-L.-A., 1816. See Cordier, 1868.

Cordier, P.-L.-A., 1828, Essay on the temperature of the interior of the Earth: Amherst, John and Charles Adams.

Cordier, P.-L.-A., 1868, Descriptions des roches composant l'écorce terrestre et des terrains cristallin, rédige d'apres la classification, les manuscrits inédit et les leçons publiques (C.C. d'Orbigny edition): Paris, Chez Savy and Chez Donod. (This book includes Cordier, P.-L.-A., 1816, Mémoire sur les substances minérales dites en masse, etc.: Journal de Physique, v. 83, p. 135–163, 285–307, 352–386.)

Cramer, J.A., [1739] 1741, Elements of the art of assaying metals: (Elementa Artis Docimasticae) (translated from Latin by C. Mortimer): London, Tho Woodward.

Cronstedt, A.F., 1770, An essay towards a system of mineralogy: London, E. & C. Dilly.

Cronstedt, A.F., 1788, An essay towards a system of mineralogy (edited by Magellan, J.H.): London, Dilly, 2 vols.

Cuvier, G., [1817] 1978, Essay on the theory of the Earth (translated by Jameson, R.): New York, Arno.

Darcet, J., 1766 (v. 1), 1771 (v. 2), First and second Memoires sur l'action d'un feu egal, violent, et continué, pendant plusieurs jours sur un grand nombre de terres, de pierres & chaux métalliques: Paris, P.G. Cavelier.

Darcet, J., Fourcroy, A.F. de, and Berthollet, C.L., 1791, Rapport sur un ouvrage de M. Loysel, qui a pour titre: Essai sur les Principes de l'Art de la Verrerie: Annales de chimie, v. 9, p. 113–237, 235–260.

Daubeny, C., 1821, On the ancient volcanoes of the Auvergne: Edinburgh Philosophical Journal, v. 4, no. 8, p. 300–315.

Daubuisson, J.F., 1814, An account of the basalts of Saxony (translated by Neill, P.): Edinburgh, A. Constable and Co.

Daubuisson, J.F., 1828, Traité de géognosie: Paris, Chez F.G. Levault, 2 vols.

Davy, H., [1805] 1980, Humphry Davy on geology (edited by Seigfriend, R., and Dott, R.H.): Madison, University of Wisconsin Press.

Delamétherie, J.-C. See La Métherie, J.-C de.

Deluc, J.-A., 1790a, Neuvième lettre de M. Deluc à M. Delamétherie, sur les substances terrestres: Journal de Physique, v. 37, p. 290–307.

Deluc, J.-A., 1790b, Onzième lettre de M. Deluc à M. Delamétherie: Sur la formation des couches calcaires et leurs premiéres catastrophes, & sur les eruptions volcaniques: Journal de Physique, v. 37, p. 441–459.

Deluc, J.-A., 1790c, First letter to Dr. James Hutton, F.R.S.: Monthly Review, v. 2, p. 206–227.

Deluc, J.-A., 1809, An elementary treatise of geology: London, R.C. and J. Rivington.

Dolomieu, D. de, 1788, Mémoire sur les Iles Ponces et catalogue raisonné des produits de l'Etna: Paris, Chez Cuchet.

Dolomieu, D. de, 1791a, Communiqués à messieurs les naturalistes, qui font le voyage de la Mer du Sud: Observations sur la Physique, v. 39, p. 310–317.

Dolomieu, D. de, 1791b, Mémoire sur les pierres composées et sur les roches: Observations sur la Physique, v. 39, p. 374–407.

Dolomieu, D. de, 1794a, Distribution méthodique de toutes les matiéres dont l'accumulation forme les montagnes volcaniques: Journal de Physique, v. 44, p. 102–125. (An 2, Pluviose = January 1794.)

Dolomieu, D. de, 1794b, Mémoire sur les roches composées en géneral ...: Journal de Physique, v. 44, p. 175–200. (An 2, Ventose = February 1794.)

Drée, E. de, 1808, Sur un nouveau genre de liquéfaction ignée qui explique la formation des laves lithoïdes: Journal des Mines, v. 139, p. 33–70.

Ercker, L., [1580] 1951, Treatise on ores and assaying: Chicago, University of Chicago Press.

Faujas, B., de Saint-Fond, 1805, Essai de géologie, ou mémoires pour servir à l'histoire naturelle du globe: Paris, Chez Levrault, Schoell et Compagnie.

Fortis, A., 1802, Mémoires pour servir à l'histoire naturelle: Paris, J.J. Fuchs, 2 vols.

Fourcroy, A.F. de, 1789, Elémens d'histoire naturelle et de chimie (third edition): Paris, Cuchet, 5 vols.

Fourmy, 1803, Sur le thermométres en terre cuites, appelés en France pyrométres: Journal des Mines, v. 14, p. 423–437.

Fourmy, 1810, Sur le pyrométres, ou thermométres en terres cuites: Journal des Mines, v. 28, p. 427–442.

Garnier, P., 1693, Histoire de la baguette de Jacques Aimar: Paris, Chez Jean-Baptiste Langlois.

Gazeran, I., 1800, Sur la fabrication des boules pyrométriques de Wedgwood: Annales de Chimie, v. 36, p. 100–104.

Gehler, J.K., 1757, De characteribus fossilium externis: Leipzig.

Gellert, C.E., 1776, Metallurgic chymistry: Being a system of mineralogy in general: London, Becket, Gillet-Laumont, F.P.N.

Gesner, C., 1565, De rerum fossilium: Tiguri.

Gesner, K., 1576, Euonymous: The newe jewell of health (translated by Baker, G.): London, Henrie Denham.

Gray, S., 1733, A letter to Cromwell Mortimer, M.D. Secr. R.S. containing several experiments concerning electricity: Philosophical Transactions of the Royal Society of London, v. 37, p. 18–44. (Paper written in 1731.)

Gray, S., 1735, Experiments and observations upon the light that is produced by communicating electrical attraction to animal or inanimate bodies: A letter from Mr. Stephen Gray, F.R.S. to Cromwell Mortimer, Philosophical Transactions of the Royal Society of London, v. 39, p. 16–24.

Gray, S.F., 1828, The operative chemist: Being a practical display of the arts and manufactures which depend on chemical principles: London, Hurst, Chance and Co.

Greenough, G.B., [1819] 1978, A critical examination of the first principles of geology: New York, Arno.

Griffin, J.J., 1827, A practical treatise on the use of the blowpipe in chemical and mineral analysis: Glasgow, R. Griffin & Co.

Grignon, P.C., 1775, Mémoires de physique sur l'Art de fabriquer le fer, etc.: Paris, Chez Delalain.

Guyton de Morveau, L.B. For correspondence with Kirwan, *see* Grison et al., eds., 1994.

Guyton de Morveau, L.B., 1810a, Sur la pyrométrie: Annales de Chimie, v. 73, p. 254–263.

Guyton de Morveau, L.B., 1810b, De l'extrait de l'essai de pyrométrie, du pyrométrie à piéces d'argile: Annales de Chimie, v. 74, p. 18–46.

Guyton de Morveau, L.B., 1810c, De la seconde partie de l'extrait de l'essai de pyrotechnie, etc.: Annales de Chimie, v. 74, p. 129–152.

Guyton de Morveau, L.B., 1811, De l'effet d'une chaleur égale, longtems continuée sur les pièces pyrométriques d'argile: Annales de Chimie, v. 78, p. 73–85.

Guyton de Morveau, L.B., 1814a, Suite de l'essai de pyrométrie, correction de la table de Wedgwood: Annales de Chimie, v. 90, p. 113–137.

Guyton de Morveau, L.B., 1814b, Suite de l'essai de pyrométrie, correction, etc.: Annales de Chimie, v. 90, p. 225–238.

Haidinger, K.W. von, 1845, Handbuch der bestimmenden Mineralogie: Wien, Bei Braumüller & Seidel.

Haidinger, W., 1825, Treatise on mineralogy or the natural history of the mineral kingdom by Frederick Mohs, translated from the German with considerable addition: Edinburgh, A. Constable & Co, 3 vols.

Hall, B., 1834, Notice of a machine for regulating high temperatures: Proceedings of the Geological Society of London, v. 1, p. 478–479.

Hall, J., 1794, Observations on the formation of granite: Transactions of the Royal Society of Edinburgh, v. 3, p. 8–12. (read in 1790)

Hall, J., 1805a, Ms. 5020, Laboratory notebook, National Library of Scotland, Edinburgh.

Hall, J., 1805b, Experiments on whinstone and lava: Transactions of the Royal Society of Edinburgh, v. 5, p. 43–76. (read in 1798)

Hall, J., 1812, Account of a series of experiments, shewing the effects of compression in modifying the action of heat, v. 6: Transactions of the Royal Society of Edinburgh, p. 71–185. (read in 1805)

Hall, J., 1815a, On the vertical position and convolutions of certain strata, and their relation with granite: Transactions of the Royal Society of Edinburgh, v. 7, p. 79–108. (read in 1812)

Hall, J., 1815b, On the revolutions of the Earth's surface, Part I: Transactions of the Royal Society of Edinburgh, v. 7, p. 139–167. (read in 1812)

Hall, J., 1815c, On the revolutions of the Earth's surface, Part II: Transactions of the Royal Society of Edinburgh, v. 7, p. 169–211. (read in 1812)

Hall, J., 1826, On the consolidation of the strata of the Earth: Transactions of the Royal Society of Edinburgh, v. 10, p. 314–329.

Hamilton, W., 1772, Observations on Mount Vesuvius, Mount Etna, and other volcanos: London, T. Cadell.

Haüy, R.J., 1801, Traité de minéralogie: Paris, Chez Louis, 5 vols.

Haüy, R.J., 1817, Traité de caractèrs physiques des pierres précieuses: Paris, Courcier.

Henckel, J.F., 1757, Pyritologia, or, a history of the pyrites: London, A. Millar.

Henkel, J.F., 1744, Kleine mineralogische und chymische Schrifften: Dresden, Zimmermann.

Henry, W., 1814, The elements of experimental chemistry (third American edition from the sixth English edition): Boston, Thomas & Andrews, 2 vols.

Hill, J., 1748, A general natural history: London, Thomas Osborne, 3 vols.

Hutton, J., [1785] 1973, System of the Earth (reprint): New York, Hafner Press.

Hutton, J., [1788] 1973, Theory of the Earth: Transactions of the Royal Society of Edinburgh, v. 1, p. 209–304. (read in 1785)

Hutton, J., 1794, Observations on granite: Transactions of the Royal Society of Edinburgh, v. 3, p. 77–85. (read in 1790)

Hutton, J., 1795, Theory of the Earth with proofs and illustrations: Edinburgh, printed for Cadell and Davies, 2 vols.

Hutton, J., [1795] 1972, Theory of the Earth with proofs and illustrations: New York, Verlag von J. Cramer, 2 vols.

Jameson, R., [1808] 1976, The Wernerian theory of the Neptunian origin of rocks: New York, Hafner Press.

Jameson, R., 1816, A system of mineralogy: Edinburgh, Constable, Longman, 3 vols.

Jameson, R., 1817, A treatise on the external, chemical, and physical characters of minerals: Edinburgh, A. Constable and Company.

Keill, J., 1698, An examination of Dr. Burnet's Theory of the Earth, together with some remarks on Mr. Whiston's new Theory of the Earth: Oxford, printed at the theater.

Keir, J., 1776, On the crystallizations observed in glass: Philosophical Transactions of the Royal Society of London, v. 66, p. 530–542.

Kennedy, R., 1805, A chemical analysis of three species of whinstone and two of lava: Transactions of the Royal Society of Edinburgh, v. 5, p. 76–98. (read in 1798)

Kidd, J., 1809, Outline of mineralogy: Oxford, N. Bliss.

Kidd, J., 1815, A geological essay on the imperfect evidence in support of a theory of the Earth, deducible either from its general structure or from the changes produced on its surface by the operation of existing causes: Oxford, University Press.

Kircher, A., 1669, The vulcanos, or burning and fire-vomiting mountains: London, J. Darby for John Allen.

Kirwan, R., 1784, Elements of mineralogy (first edition): London, printed for P. Elmsly.

Kirwan, R., 1794a, Elements of mineralogy (second edition): London, 2 vols.

Kirwan, R., 1794b, Examination of the supposed igneous origin of stony substances: Transactions of the Royal Irish Academy, v. 5, p. 51–81. (read in 1793)

Kirwan, R., 1797, On the primitive state of the globe and its subsequent catastrophe: Transactions of the Royal Irish Academy, v. 6, p. 233–308. (read in 1796)

Kirwan, R., [1799a] 1978, Geological essays: New York, Arno.

Kirwan, R., 1799b, An essay on the analysis of mineral waters: London, D. Bremner.

Kirwan, R., 1802a, Observations of the proofs of the Huttonian theory of the Earth adduced by Sir James Hall: Transactions of the Royal Irish Academy, v. 8, p. 3–27. (read in 1800)

Kirwan, R., 1802b, An illustration and confirmation of some facts mentioned in an essay on the primitive state of the globe: Transactions of the Royal Irish Academy, v. 8, p. 29–34. (read in 1800)

Kirwan, R., 1803, A reply to Mr. Playfair's reflections on Mr. Kirwan's refutation of the Huttonian theory of the Earth: Philosophical Magazine, v. 14, p. 3–17.

Klaproth, M.H., 1801, Analytical essays towards promoting the chemical knowledge of mineral substances: London, T. Cadell & W. Davies, 2 vols.

Klaproth, M.H., 1802, Analyse de basalte: Journal des Mines, v. 13, p. 123–134.

La Métherie, J.-C. de, 1795, Théorie de la terre: Paris, Chez Maradan, 3 vols.

Lavoisier, A., 1789, Traité elémentaire de chimie: Paris, 2 vols.

Lavoisier, A., [1790] 1965, Elements of chemistry (translated by Kerr, R.): Edinburgh, Creech.

Loysel, P. See Darcet et al., 1791, for comments on Loysel.

Loysel, P., [1800] 2003, Essai sur l'art de la verrerie, Parts 1 & 2: An 8, 1798–99?: l'Institute des Sciences et des Arts. Posted at http://hdelboy.club.fr/loysel_1.html (accessed 13 February 2009).

Lyell, C., [1830] 1990 (v. 1); [1832] 1991 (v. 2); [1833] 1991 (v. 3), Principles of geology: Chicago, University of Chicago Press.

Mackenzie, G.S., 1815, An account of some geological facts observed in the Faroe Islands: Transactions of the Royal Society of Edinburgh, v. 7, p. 213–227.

Mackenzie, G.S., 1826, On the formation of chalcedony: Transactions of the Royal Society of Edinburgh, v. 10, p. 82–104.

Macquer, P.J., 1749, Elemens de chymie theorique: Paris, J.-T. Herissant.

Macquer, P.J., 1766, Dictionnaire de chymie (first edition): Paris, Lacombe, 2 vols.

Magnus, A., [~1262] 1967, Book of minerals [De mineralibus] (translated by Wyckoff, D.): Oxford, Clarendon Press.

Meinecke, J.L.G., 1808, Lehrbuch der Mineralogie mit Beziehung auf Technologie und Geographie: Halle, Hemmerde und Schwetschke.

Miché, A., 1803, Fait à la conférence des mines au nom d'une commission, sur le pyrométre de Wedgwood: Journal des Mines, v. 14, p. 42–49.

Mohs, F., 1812, Versuch einer Elementar-Methode zur naturhistorischen Bestimmung and Erkennung der Fossilien: Vienna.

Mohs, F., 1820, Of the classes, orders, genera, and species; Or, the characteristics of the natural history system of mineralogy: Edinburgh, printed for W. and C. Tait.

Mohs, F., 1825, Treatise on mineralogy, or the natural history of the mineral kingdom (translated by Haidinger, K.W. von): Edinburgh, printed for Archbald Constable and Co., 2 vols.

Moro, Antonio-Lazzaro, 1740, Dei Crostacei e degli Altri Corpi Marini che si Trovano sui Monti: Venice, Italy, S. Monti.

Mortimer, C., 1747, A discourse concerning the usefulness of thermometers in chemical experiments: Philosophical Transactions of the Royal Society of London, v. 44, p. 672–695.

Murray, J., 1801, On Mr. Wedgwood's pyrometer: Philosophical Magazine, v. 9, p. 153–158.

Murray, J., [1802] 1978, A comparative view of the Huttonian and Neptunian systems of geology: New York, Arno.

Murray, J., 1806–1807, A system of chemistry: London, Edinburgh, 4 vols.

Murray, J., 1815, On the diffusion of heat at the surface of the Earth: Transactions of the Royal Society of Edinburgh, v. 7, p. 411–434.

Murray, J., 1818, A general formula for the analysis of mineral waters: Transactions of the Royal Society of Edinburgh, v. 8, p. 259–279.

Mushet, D., 1799, Description of an assay-furnace, with an apparatus for measuring the degree of heat required: Philosophical Magazine, v. 4, p. 255–259.

Orbigney, C.C. d'. See Cordier, 1868.

Otter, W., 1825, The life and remains of Edward Daniel Clarke: London, George Cowie and Co., 2 vols.

Papin, D., 1681, A new digester or engine for softening bones: London, printed by J.M. for Henry Bonwicke.

Papin, D., [1681] 1966, A new digester or engine for softening bones (facsimile reprint): London, Dawsons of Pall Mall.

Papin, D., 1687, A continuation of the new digester of bones: London, Joseph Streater.

Phillips, W., 1815, An outline of mineralogy and geology: London, William Phillips.

Phillips, W., [1816] 1978, An outline of mineralogy and geology: New York, Arno.

Pinkerton, J., 1811, Petralogy, a treatise on rocks: London, printed for White, Cochrane, & co, 2 vols.

Playfair, J., [1802] 1964, Illustrations of the Huttonian theory of the Earth: New York, Dover.

Playfair, J., 1805, Biographical account of the late Dr. James Hutton, F.R.S. Edinburgh: Transactions of the Royal Society of Edinburgh, v. 5, p. 39–99.

Playfair, J., 1806, An account of a series of experiments, shewing the effects of compression in modifying the action of heat: Edinburgh Review, v. 9, 1931.

Pott, J.H., [1746] 1753, Lithogéognosie ou examen chymique des pierres et des terres en général: Paris, Herissant.

Pott, J.H., 1757, Chymische Untersuchungen: Berlin, Voss.

Presl, K.B., 1834. Anleitung zum Selbstudium der Oryktognosie: Prague, Gottlieb Haase, Söhne.

Priestley, J., [1767] 1948, Electricity, in Boynton, H., ed., The beginnings of modern science: Roslyn, New York, Walter J. Black, p. 281–283.

Priestley, J., [1790] 1970, Experiments and observations on different kinds of air: New York, Kraus Reprint, 3 vols.

Réaumur, R.-A.F. de, 1722, L'art de convertir le fer forgé en acier, et l'art d'adoucir le fer fondu, ou de faire des ouvrages de fer fondu aussi finis que de fer forgé: Paris.

Richardson, W., 1807, Arguments against the volcanic origin of basalt: Nicolson's Journal, new series, v. 16, p. 277–290.

Romé de l'Isle, J.-B.L., 1772, Essai de cristallographie: Paris, Chez P. Fr. Didot le jeune.

Romé de l'Isle, J.-B.L., 1781, L'action du feu central démontrée nulle à la surface du globe (second edition): Stockholm/Paris, Chez P. Fr. Didot le jeune.

Romé de L'Isle, J.-B.L, 1783, Cristallographie, ou description des formes propres à tous les corps du règne minéral, dans l'état de combinaison saline, pierreuse ou métallique: Paris, Imprimerie de Monsieur.

Rumford, B.T., 1797, Experiments to determine the force of fired gunpowder: Philosophical Transactions of the Royal Society of London, v. 87, p. 222–292.

Rumford, B.T., [1797] 1870, The complete works of Count Rumford: Boston, American Academy of Arts and Sciences, 4 vols.

Saussure, H.-B. de, 1779 (v. I), 1786 (v. II), 1796 (v. III and v. IV), Voyages dans les Alpes (4 vols., I, Neuchâtel: Samuel Fauche, 1779; II, Genève: Barde et Manget, 1786; III and IV, Neuchâtel: Fauche-Borel, 1796).

Saussure, H.-B. de, 1785, Sur l'usage du chalumeau: Journal de Physique, v. 26, p. 409–413.

Saussure, H.-B. de, 1791, D'un cyanomètre, ou d'un appareil destiné à mesurer l'intensité de la couleur bleue de ciel: Journal de Physique, v. 38, p. 199–208.

Saussure, H.-B. de, 1794, Nouvelle recherches sur l'usage du chalumeau: Journal de Physique, v. 45, p. 3–44.

Saussure, H.-B. de, 1803, Voyages dans les Alpes (second edition): Neuchatel, Chez Louis Fauche-Borel, 4 vols.

Scherer, A.N., 1799, Sur la pyromètre de Wedgwood: Annales de Chimie, v. 31, p. 171–176.

Schmeisser, J.G., 1795, A system of mineralogy formed chiefly on the plan of Cronstedt: London, Dilly, 2 vols.

Sherley, T., [1671] 1978, A philosophical essay declaring the probable causes whence stones are produced in the greater world: New York, Arno.

Spallanzani, L., 1795, Voyages dans les deux Sicilies: Berne, Chez Emaluel Haller, 2 vols.

Spallanzani, L., 1798, Travels in the two Sicilies and some parts of the Appenines: London, G.G. and J. Robinson, 4 vols.

Thomson, T., 1803, A new system of chemistry: Philadelphia, printed for T. Dobson.

Thomson, T., [1830–1831] 1975, The history of chemistry: New York, Arno, 2 vols.

Thomson, T., 1836, Outlines of mineralogy, geology, and mineral analysis: London, Baldwin & Cradock, 2 vols.

Tschirnhaus, E.W., 1702, Effets des verres brulans de trois ou quatre pieds de diamètre: Histoire de l'Académie Royale des Sciences for 1699, p. 90–94.

Valmont de Bomare, J.C., 1762, Minéralogie, ou nouvelle exposition du règne minéral avec un dictionnaire nomenclateur et des tables synoptiques: Paris, Vincent, 2 vols.

Vauquelin, N.L., 1799, Reflections on the quality of earthen-ware, and the results of the analysis of some earths and common kinds of earthenware: Philosophical Magazine, v. 5, p. 288–290.

Wallerius, J.G., [1747] 1753, Minéralogie; ou description general des substances du règne minéral: Paris, Drand, Pissot, 2 vols.

Warren, E., 1690, Geologia: Or a discourse concerning the Earth: London, R. Chiswell.

Warren, E., [1690] 1978, Geologia: Or, a discourse concerning the Earth before the Deluge: New York, Arno.

Watt, G., 1804a, Observations on basalt: Philosophical Transactions of the Royal Society of London, part II, v. 94, p. 279–314. (read in 1804)

Watt, G., 1804b, Watt's review of Breislak's *Voyage physique et lithologique*, etc.: Edinburgh Review, v. 4, p. 26–42.

Wedgwood, J., [1782] 1809a, An attempt to make a thermometer for measuring the higher degrees of heat from a red heat up to the strongest that vessels made of clay can support: Abridged Philosophical Transactions of the Royal Society of London, v. 15, p. 278–290.

Wedgwood, J., [1784] 1809b, An attempt to compare and connect the thermometer for strong fire with the common mercurial ones: Abridged Philosophical Transactions of the Royal Society of London, v. 15, p. 571–586.

Wedgwood, J., [1786] 1809c, Additional observations on making a thermometer for measuring the higher degrees of heat: Abridged Philosophical Transactions of the Royal Society of Edinburgh, v. 16, p. 136–145.

Werner, A.G., [1774] 1962, On the external characters of minerals [Von den äusserlichen Kennzeichen der Fossilien] (translated by Carozzi, A.V.): Urbana, University of Illinois Press.

Werner, A.G., [1786] 1971, Short classification and description of the various rocks (translated by Ospovat, A.M.): New York, Hafner.

Whiston, W., [1696] 1978, A new theory of the Earth: London, printed by R. Roberts.

Whitehurst, J., 1778, An inquiry into the original state and formation of the Earth: London, printed by J. Cooper.

Whitehurst, J., [1786] 1978, An inquiry into the original state and formation of the Earth: New York, Arno.

Withering, W., [1782] 1809, An analysis of two mineral substances: Abridged Philosophical Transactions of the Royal Society of London, v. 15, p. 290–294.

Woodward, J., 1695, An essay toward a natural history of the Earth: London, Wilkin.

Woodward, J., [1695] 1978, An essay toward a natural history of the Earth and terrestrial bodies: New York, Arno.

REFERENCES AFTER 1840

Adams, F.D., [1938] 1954, The birth and development of the geological sciences: New York, Dover.

Agnew, D.C., 1998, Instruments, gravity, *in* Good, G.A., ed., Sciences of the Earth: An encyclopedia of events, people, and phenomena, v. 2: New York, Garland, p. 453–455.

Albrecht, H., and Ladwig, R., eds., 2002, Abraham Gottlob Werner and the foundation of the geological sciences, Freiberger Forschungshefte D207: Freiberg, Technische Universität Bergakademie Freiberg.

Ambler, L.T., 1969, Early science at Harvard: Innovators and their instruments 1765–1865: Cambridge, Massachusetts, Fogg Art Museum, Harvard University.

Anderson, R.G.W., 1978, The Playfair collection and the teaching of chemistry at the University of Edinburgh, 1713–1858: Edinburgh, The Royal Scottish Museum.

Anderson, R.G.W., 1982, Joseph Black, and outline biography, *in* Simpson, A.D.C., ed., Joseph Black, 1728–1799: A commemorative symposium: Edinburgh, The Royal Scottish Museum, p. 7–11.

Anderson, R.G.W., 1985, Instruments and apparatus, *in* Russell, C.A., ed., 1985, Recent developments in the history of chemistry: London, The Royal Society of Chemistry, p. 217–237.

Anderson, R.G.W., 1998, Furnaces, *in* Bud, R., and Warner, D.J., eds., 1998, Instruments of Science: An historical encyclopedia: New York, Garland Publishing, p. 251–253.

Anderson, R.G.W., 2000, The archeology of chemistry, *in* Holmes, F.L., and Levere, T.H., eds., Instruments and experimentation in the history of chemistry: Cambridge, Massachusetts Institute of Technology Press, p. 5–34.

Anonymous, 1900, A treatise on chemistry and chemical analysis, v. 2: Scranton, The Colliery Engineer Co.

Atterbury, P., ed., 1982, The history of porcelain: London, Orbis Publishing.

Barrett, W.R., and Besterman, T., [1926] 1968, The divining rod, an experimental and psychological investigation: London, Methuen.

Bascom, F., 1927, Fifty years of progress in petrography and petrology: 1876–1926, *in* Mathews, E.B., ed., Fifty years' progress in geology, 1876–1926: Baltimore, Johns Hopkins Press.

Beretta, M., 1993, The enlightenment of matter: Canton, Massachusetts, Science History Publications.

Beretta, M., 1998a, Mineralogy: Disciplinary history, *in* Good, G.A., ed., Sciences of the Earth: An encyclopedia of events, people, and phenomena, v. 2: New York, Garland Publishing Co., p. 578–582.

Beretta, M., 1998b, Minerals and crystals, fifteenth to eighteenth centuries, *in* Good, G.A., ed., Sciences of the Earth: New York, Garland Publishing, p. 582–585.

Berggren, W.A., 1998, The Cenozoic era: Lyellian (chrono)stratigraphy and nomenclature reform at the millennium, *in* Blundell, D.J., and Scott, A.C., eds., Lyell: The past is the key to the present: Geological Society of London Special Publication 143, p. 111–132.

Bertucci, P., 2004, Cavallo, Tiberius, *in* Dictionary of national biography, v. 10: Oxford, Oxford University Press, p. 590–591.

Birembaut, A., 1970, Arnould Carangeot, *in* Gillispie, C.C., ed., Dictionary of scientific biography, v. 2: New York, Scribner, p. 61–62.

Boklund, U., 1971, Axel Fredrik Cronstedt, *in* Gillispie, C.C., ed., Dictionary of scientific biography, v. 3: New York, Scribner, p. 473–474.

Boklund, U., 1975, Scheele, Carl Wilhelm, *in* Gillispie, C.C., ed., Dictionary of scientific biography, v. 12: New York, Scribner, p. 143–150.

Bowen, N.L., [1928] 1956, The evolution of the igneous rocks: New York, Dover Publishing.

Boyer, C.B., [1968] 1985, A history of mathematics: Princeton, New Jersey, Princeton University Press.

Boynton, H., ed., 1948, The beginnings of modern science: Roslyn, New York, W.F. Black.

Brock, W.H., 1993, The Norton history of chemistry: New York, W.W. Norton & Co.

Brown, H.I., 1987, Observation and objectivity: Oxford, Oxford University Press.

Brush, S.G., 1965, Kinetic theory, v. 1: Pergamon Press, Oxford.

Brush, S.G., 1966, Kinetic theory, v. 2: Pergamon Press, Oxford.

Brush, S.G., 1976, The kind of motion we call heat: A history of the kinetic theory of gases in the 19th century: Amsterdam, North-Holland Publishing Co.

Brush, S.G., 1979, Nineteenth-century debates about the inside of the Earth: Solid, liquid, or gas?: Annals of Science, v. 36, p. 225–254.

Brush, S.G., 1982, Chemical history of the Earth's core: Eos (Transactions, American Geophysical Union), v. 47, p. 1185–1188.

Brush, S.G., 1983, Statistical physics and the atomic theory of matter: Princeton, New Jersey, Princeton University Press.

Brush, S.B., 1984, Inside the Earth: Natural History, v. 2, p. 26–34.

Brush, S.G., 1996, v. 1, Transmuted past: The age of the Earth and the evolution of the elements from Lyell to Patterson; v. 2, A history of modern planetary physics: Cambridge, Cambridge University Press.

Buchanan, P.D., 1998, Balances, hydrostatic, *in* Bud, R., and Warner, D.J., eds., Instruments of science: An historical encyclopedia: New York, Garland Publishing Co., p. 49–50.

Bud, R., and Warner, D.J., eds., 1998, Instruments of science: An historical encyclopedia: New York, Garland Publishing Co.

Buntebarth, G., 1998a, Geophysics: Disciplinary history, *in* Good, G.A., ed., Sciences of the Earth: An encyclopedia of events, people, and phenomena, v. 1: New York, Garland Publishing Co., p. 377–380.

Buntebarth, G., 1998b, Heat, internal, eighteenth and nineteenth centuries, *in* Good, G.A., ed., Sciences of the Earth: An encyclopedia of events, people, and phenomena, v. 2: New York, Garland Publishing Co., 409–412.

Burchard, U., 1998, History of the development of the crystallographic goniometer: The Mineralogical Record, v. 29, p. 517–583.

Burchfield, J., [1975] 1990, Lord Kelvin and the age of the Earth: Chicago, University of Chicago Press.

Burke, J.G., 1966, Origins of the science of crystals, Berkeley: University of California Press.

Burke, J.G., 1974, Friedrich Mohs, *in* Gillispie, C.C., ed., Dictionary of scientific biography, v. 9: New York, Scribner, p. 447–449.

Burnett, J., 1998, Thermometer, *in* Bud, R., and Warner, D.J., eds., Instruments of science: An historical encyclopedia: New York, Garland Publishing Co., p. 615–618.

Butts, R.E., 1968, William Whewell's theory of scientific method: Pittsburgh, University of Pittsburgh Press.

Campbell, W.A., 1985a, Analytical chemistry, *in* Russell, C.A, ed., 1985, Recent developments in the history of chemistry: London, Royal Society of Chemistry, p. 177–190.

Campbell, W.A., 1985b, Industrial chemistry, *in* Russell, C.A., ed., Recent developments in the history of chemistry: London, Royal Society of Chemistry, p. 238–252.

Caneva, K.L., 2001, The form and function of scientific discoveries: Washington, D.C., Smithsonian Institution Libraries.

Carozzi, A.V., 1975, Horace Bénédict de Saussure, *in* Gillispie, C.C., ed., Dictionary of scientific biography, v. 12: New York, Scribner, p. 119–123.

Carozzi, A.V., 1989, Forty years of thinking in front of the Alpes: Saussure's (1796) unpublished theory of the Earth: Earth Sciences History, v. 8, no. 2, p. 123–140.

Carozzi, A.V., 2000, Manuscripts and publications of Horace-Bénédict de Saussure on the origin of basalt (1772–1797): Geneva, Éditions Zoé.

Chaldecott, J.A., 1968, Scientific activities in Paris in 1791: Annals of Science, v. 24, p. 21–52.

Chaldecott, J.A., 1972a, Hans Loeser's metallic thermometers of 1746 and 1747: Annals of Science, v. 28, p. 87–100.

Chaldecott, J.A., 1972b, The platinum pyrometers of Louis Bernard Guyton de Morveau: Annals of Science, v. 28, p. 347–368.

Chaldecott, J.A., 1975, Josiah Wedgwood (1730–95) scientist: British Journal for the History of Science, v. 8, part I, p. 1–16.

Challinor, J., 1971, Faujas de Saint-Fond, Barthélemy, *in* Gillispie, C.C., ed., Dictionary of scientific biography, v. 4: New York, Scribner, p. 548–549.

Chester, D.K., Duncan, A.M., Guest, J.E., and Kilburn, C.R.J., 1985, Mount Etna: The anatomy of a volcano: Stanford, California, Stanford University Press.

Chinese Academy of Sciences, 1983, Ancient China's technology and science: Beijing, Foreign Press.

Coley, N.G., 1985, Chemistry to 1800, *in* Russell, C.A., ed., Recent developments in the history of chemistry: London, Royal Society of Chemistry, p. 49–76.

Copeland, R., 1995, Wedgwood ware: Buckinghamshire, Shire Publications.

Corapcioglu, M.Y., 1998, Water wells, *in* Good, G.A., ed., Sciences of the Earth: An encyclopedia of events, people, and phenomena, v. 2: New York, Garland, p. 835–836.

Crawforth-Hitchens, D.F., 1998, Balance, general, *in* Bud, R., and Warner, D.J., eds., Instruments of science: An historical encyclopedia: New York, Garland Publishing Co., p. 47–49.

Daumas, M., [1953] 1972, Scientific instruments of the seventeen and eighteenth centuries (translated by Holbrook, M.): New York, Praeger.

Davison, C., [1927] 1978, The founders of seismology: New York, Arno.

Dawson, J.B., 1992, First thin sections of experimentally melted igneous rocks: Sorby's observations on magma crystallization: Journal of Geology, v. 100, p. 251–257.

Dean, D.R., 1997, James Hutton in the field and in the study: Delmar, New York, Scholars' Facsimiles and Reprints.

Debus, A.G., 1967, Fire analysis and the elements in the sixteenth and the seventeenth centuries: Annals of Science, v. 23, 127–147.

den Tex, E., 1996a, Clinchers of the basalt controversy: Empirical and experimental evidence: Earth Sciences History, v. 15, p. 37–48.

Dolan, B., 1998, Representing novelty: Charles Babbage, Charles Lyell, and experiments in early Victorian geology: History of Science, v. 36, p. 299–327.

Dolman, C.E., 1975, Lazzaro Spallanzani, *in* Gillispie, C.C., ed., Dictionary of scientific biography, v. 12: New York, Scribner, p. 553–567.

Donovan, A., 1978, James Hutton, Joseph Black and the chemical theory of heat: Ambix, v. 25, p. 176–190.

Donovan, A., 1996, Antoine Lavoisier: Science, administration, and revolution: Cambridge, Cambridge University Press.

Dörries, M., 1998, Torsion balance, *in* Bud, J., and Warner, D.J., eds., Instruments of science: New York, Garland Publishing Co., p. 626–628.

Drake, E.T., 1996, Restless genius: Robert Hooke and his earthly thoughts: Oxford, Oxford University Press.

Eklund, J., 1975, The incompleat chymist: Being an essay on the eighteenth-century chemist in his laboratory, with a dictionary of obsolete chemical terms of the period: Smithsonian Studies in History and Technology 33: Washington, D.C., Smithsonian Institution Press.

Ellenberger, F., 1984, Louis Cordier (1777–1861), initiateur de l'étude microscopique des laves: Earth Sciences History, v. 3, p. 44–53.

Ellis, A.J., 1917, The divining rod: A history of water witching: U.S. Geological Survey Water Supply Paper 416 (1957 reprint): Washington, D.C., Government Printing Office.

Eyles, V.A., 1961, Sir James Hall, Bt. (1761–1832): Endeavor, v. 20, p. 210–216.

Eyles, V.A., 1963, The evolution of a chemist: Annals of Science, v. 19, p. 153–183.

Eyles, V.A., 1972, James Hall, Jr., *in* Gillispie, C.C., ed., Dictionary of scientific biography, v. 6: New York, Scribner, p. 53–56.

Ferchl, F., and Süssenguth, A., 1936, Kurzgeschichte der Chemie: Mittenwald, Arthur Nemayer Verlag.

Ferchl, F., and Süssenguth, A., 1939, A pictorial history of chemistry: London, Wm. Heinemann.

Ferrari, G., 1998a, Instruments, seismic, *in* Good, G.A., ed., Sciences of the Earth: An encyclopedia of events, people, and phenomena, v. 1: New York, Garland Publishing Co., p. 462–470.

Ferrari, G., 1998b, Seismograph, *in* Bud, R., and Warner, D.J., eds., Instruments of science: New York, Garland Publishing Co., p. 528–530.

Findlen, P., [1994] 1996, Possessing nature: Museums, collecting, and scientific culture in early modern Italy: Berkeley, University of California Press.

Frängsmyr, T., Heilbron, J.L., and Rider, R.E., eds., 1990, The quantifying spirit in the eighteenth century: Berkeley, University of California Press.

Freshfield, D.W., and Montagnier, H.F., 1920, The life of Horace Bénédict de Saussure: London, Edward Arnold.

Fritscher, B., 1988, Die "James Hall-Sammlung" in Keyworth: Berichte zur Wissenschaftsgeschichte, v. 11, p. 27–34.

Fritscher, B., 1991, Vulkanismusstreit und Geochemie: Die Bedeutung der Chemie und des Experiments in der Vulkanismus-Neptunismus-Kontroverse: Stuttgart, Franz Steiner Verlag.

Fritscher, B., 2002, Metamorphism and thermodynamics: The formative years, *in* Oldroyd, D.R., ed., The Earth inside and out: London, Geological Society of London, p. 143–156.

Fritscher, B., and Henderson, F., eds., 1998, Toward a history of mineralogy, petrology, and geochemistry: Proceedings of the International Symposium on the History of Mineralogy, Petrology, and Geochemistry, Munich, 1996: Heft 23: Munchen, Institut fur Geschichte der Naturwissenschaften.

Fruton, J.S., 2002, Methods and styles in the development of chemistry: Philadelphia, American Philosophical Society.

Galison, P., 1997, Image and logic: A material culture of microphysics: Chicago, Chicago University Press.

Geikie, A., 1895, Memoire of Sir Andrew Crombie Ramsay: London, Macmillan.

Geikie, A., [1905] 1962, The founders of geology: New York, Dover Publications.

Geschwind, C.-H., 1995, Becoming interested in experiments: American igneous petrologists and the Geophysical Laboratory, 1905–1965: Earth Sciences History, v. 14, p. 47–61.

Gillispie, C.C., [1951] 1959, Genesis and geology: New York, Harper Torchbooks.

Gillispie, C.C., ed., 1970–1989, Dictionary of scientific biography: New York, Scribner, 18 vols.

Gohau, G., 1990, A history of geology (translated by Carrozi, A.V.): New Brunswick, New Jersey, Rutgers University Press.

Golas, P.J., 1999, Science and civilization in China, v. 5: Chemistry and chemical technology, Part XIII: Mining: Cambridge, Cambridge University Press.

Golinski, J., 1990, Humphry Davy and the 'lever of experiment', *in* Le Grand, H., ed., Experimental inquiries: Dordrecht, Kluwer Academic Publishers, p. 99–136.

Golinski, J., 1995, The nicety of experiment, *in* Wise, N.M., ed., The values of precision: Princeton, New Jersey, Princeton University Press, p. 72–91.

Gondhalekar, P., 2001, The grip of gravity: Cambridge, Cambridge University Press.

Good, G.A., 1991, Follow the needle: Seeking the magnetic poles: Earth Sciences History, v. 10, p. 154–167.

Good, G.A., ed., 1998a, Sciences of the Earth: An encyclopedia of events, people, and phenomena: v. 1: A–G; v. 2: H–Z: New York, Garland Publishing Company.

Good, G.A., 1998b, Geomagnetism, theories between 1800 and 1900, *in* Good, G.A., ed., Sciences of the Earth: An encyclopedia of events, people, and phenomena, v. 1: New York, Garland Publishing Company, p. 350–356.

Good, G.A., 1998c, Magnetometer, *in* Bud, R., and Warner, D.J., eds., Instruments of science: New York, Garland Publishing Company, p. 368–371.

Gooding, D., 1990, Experiment and the making of meaning: Dordrecht, Kluwer.

Gooding, D., Pinch, T., and Schaffer, S., 1989, The uses of experiment: Cambridge, Cambridge University Press.

Gough, J.B., 1975, René-Antoine Ferchault de Réaumur, *in* Gillispie, C.C., ed., Dictionary of scientific biography, v. 11: New York, Scribner, p. 327–335.

Gould, S.J., 1989, Response: Geological Society of America Bulletin, v. 101, no. 7, p. 998–1000.

Greenberg, A., 2000, A chemical history tour: New York, Wiley Interscience.

Greenberg, J.L., 1995, The problem of the Earth's shape from Newton to Clairaut: Cambridge, Cambridge University Press.

Griffiths, E., 1918, Methods of measuring temperature: London, Charles Griffin and Co.

Grison, E., Goupil, M., and Bret, P., eds., 1994, A scientific correspondence during the Chemical Revolution: Louis-Bernard Guyton de Morveau and Richard Kirwan, 1782–1802: Berkeley Papers in History of Science 17: Berkeley, Office for History of Science and Technology.

Guidoboni, E., 1998a, Earthquakes, theories from 1600 to 1800, *in* Good, G.A., ed., Sciences of the Earth: An encyclopedia of events, people, and phenomena, v. 1: New York, Garland Publishing Company, p. 205–214.

Guidoboni, E., 1998b, Historical seismology, *in* Good, G.A., ed., Sciences of the Earth: An encyclopedia of events, people, and phenomena, v. 2: New York, Garland Publishing Company, p. 749–754.

Guntau, M., 1984, Abraham Gottlob Werner: Leipzig, BSB B.G.Teubner Verlagsgesellschaft.

Gunther, R.T., 1931, Early science in Oxford: Roslyn, New York, W.J. Black.

Hacking, I., 1983, Representing and intervening: Cambridge, Cambridge University Press.

Hackman, W., 1998a, Electrometer, *in* Bud, R., and Warner, D.J., eds., Instruments of science: New York, Garland Publishing Company, p. 208–211.

Hackman, W., 1998b, Electroscope, *in* Bud, R., and Warner, D.J., eds., Instruments of science: New York, Garland Publishing Company, p. 219–221.

Hall, D.H., 1976, History of the earth sciences during the Scientific and Industrial Revolutions: New York, Elsevier.

Hamlin, C., 1990, A science of impurity: Water analysis in nineteenth century Britain: Berkeley, University of California Press, 341 p.

Hamm, E.P., 1997, Knowledge from underground: Leibniz mines the Enlightenment: Earth Sciences History, v. 16, p. 77–99.

Heilbron, J.L., 1979, Electricity in the 17th and 18th centuries: A study of early modern physics: Berkeley, University of California Press.

Heilbron, J.L., 1990a, Introductory essay, *in* Frängsmyr, T., Heilbron, J.L., and Rider, R.E., eds., The quantifying spirit in the eighteenth century: Berkeley, University of California Press, p. 1–23.

Heilbron, J.L., 1990b, The measure of Enlightenment, *in* Frängsmyr, T., Heilbron, J.L., and Rider, R.E., eds., The quantifying spirit in the eighteenth century: Berkeley, University of California Press, p. 207–242.

Heilbron, J.L., 1993, Weighing imponderables and other quantitative science around 1800: Supplement to Historical Studies in the Physical and Biological Science, v. 24, Part 1: Berkeley, University of California Press.

Hofmann, J.E., 1976, Ehrenfried Walther Tschirnhaus, *in* Gillispie, C.C., ed., Dictionary of scientific biography, v. 13: New York, Scribner, p. 479–481.

Hogg, J., 1886, The microscope: Its history, construction, and application: London, George Routledge and Sons.

Holmes, F.L., 1985, Lavoisier and the chemistry of life: An exploration of scientific creativity: Madison, University of Wisconsin Press.

Holmes, F.L., and Levere, T.H., eds., 2000, Instruments and experimentation in the history of chemistry: Cambridge, Massachusetts Institute of Technology Press.

Holton, G., 1973, Introduction to concepts and theories in physical science (revised by Brush, S., second edition): Menlo Park, California, Addison-Wesley Publishing Co.

Hooykas, R., 1958, The concepts of "individual" and "species" in chemistry: Centaurus, v. 5, p. 307–322.

Howarth, R., 2002, From graphical display to dynamic model: Mathematical geology in the earth sciences in the nineteenth and twentieth centuries, *in* Oldroyd, D.R., ed., The Earth inside and out: London, Geological Society of London, p. 59–98.

Howell, B.F., Jr., [1986] 1990, History of ideas on the cause of earthquakes: History of Geophysics, v. 4: Washington, D.C., American Geophysical Union, p. 118–124.

Hurlbut, C.S., and Klein, C., 1977, Manual of mineralogy (19th edition): New York, John Wiley & Sons.

Ihde, A.J., [1964] 1984, The development of modern chemistry: New York, Dover.

Jacquemart, A., 1873, History of the ceramic art: London, Sampson Low, Marston, Low, and Searle.

Jenemann, H.R., 1979, Die Waage des Chemikers/The chemists' balance: Frankfurt am Main, Dechema.

Khan, M.A., 1998, Gravimetry, *in* Good, G.A., ed., Sciences of the Earth: An encyclopedia of events, people, and phenomena, v. 2: New York, Garland Publishing Co., p. 393–398.

Klein, H.A., [1974] 1988, The science of measurement: A historical survey: New York, Dover Publications.

Klein, U., 1994, Origin of the concept of chemical compound: Science in Context, v. 7, p. 163–204.

Klein, U., 1996, The chemical workshop tradition and the experimental practice: Discontinuities within continuities: Science in Context, v. 9, p. 251–287.

Kobell, F. von, [1864] 1965, Geschichte der Mineralogie von 1650–1860: New York, Johnson Reprint Corp.

Kölbl-Ebert, M., 2003, Life, work and historical reception of alchemist and mining engineer Martine de Bertereau (d. ca 1643), *in* Pinto, M.S., ed., Geological resources and history: Aviero, Portugal, Universidade de Aveiro, p. 235–250.

Krafft, F., 1972, Otto von Guericke, *in* Gillispie, C.C., ed., Dictionary of scientific biography, v. 5: New York, Scribner, p. 574–576.

Kurlansky, M., 2002, Salt: A world history: New York, Walker and Company.

Latimer, W.M., and Hildebrand, J.H., 1951, Reference book of inorganic chemistry (third edition): New York, Macmillan Co.

Laudan, R., 1987, From mineralogy to geology: Chicago, University of Chicago Press.

Le Grand, H.E., 1989, Conflicting orientations: John Graham, Merle Tuve and paleomagnetic research at the DTM 1938–1958: Earth Sciences History, v. 8, p. 55–65.

Le Grand, H.E., ed., 1990, Experimental inquiries, Historical, philosophical and social studies of experimentation in science: Dordrecht, Kluwer Academic Publishers.

Le Grand, H.E., 1998, Paleomagnetism, *in* Good, G.A., ed., Sciences of the Earth: An encyclopedia of events, people, and phenomena, v. 2: New York, Garland Publishing Company, p. 651–655.

Leicester, H., [1956] 1971, The historical background of chemistry: New York, Dover Publications.

Lenoir, T., 1997, Instituting science: The cultural production of scientific disciplines: Stanford, California, Stanford University Press.

Lenzen, V.F., and Multhauf, R.P., 1966, Development of gravity pendulums in the 19th century: Paper 44, Contributions from the Museum of History and Technology: Washington, D.C., Smithsonian Institution, p. 301–347.

Levere, T.H., 1994, Chemists and chemistry in nature and society, 1770–1878: Aldershot, Variorum.

Lewis, C.L.E., and Knell, S.J., eds., 2009, The making of the Geological Society of London: Geological Society of London Special Publication 317 (in press).

Lin Wenzhao, 1983, Magnetism and the compass, *in* Chinese Academy of Sciences, Ancient China's technology and science: Beijing, Foreign Press, p. 152–165.

Lundgren, A., 1990, The changing role of numbers in 18th-century chemistry, *in* Frangsmyr, T., Heilbron, J.L., and Rider, R.E., eds., The quantifying spirit in the eighteenth century: Berkeley, University of California Press, p. 245–266.

Lynn, M.R., 2001, Divining the Enlightenment: Public opinion and popular science in Old Regime France: Isis, v. 92, p. 34–54.

Manten, A.A., 1966, Historical foundations of chemical geology and geochemistry: Chemical Geology, v. 1, p. 5–31.

Mathews, E.B., ed., 1927a, Fifty years' progress in geology, 1876–1926: Baltimore, Johns Hopkins Press.

Mathews, E.B., 1927b, Progress in structural geology, *in* Mathews, E.B., ed., Fifty years' progress in geology, 1876–1926: Baltimore, Johns Hopkins Press, p. 137–161.

Matthew, H.C.G., and Harrison, B., eds., 2004, Oxford dictionary of national biography: Oxford, Oxford University Press.

Mauskopf, S.H., 1976, Crystals and compounds: Transactions of the American Philosophical Society, New Series, v. 66, part 3, p. 5–82.

McDonald, E., 1970, Bayen, Johann, *in* Gillispie, C.C., ed., Dictionary of scientific biography, v. 1: New York, Scribner, p. 529–530.

Melhado, E.M., 1981, Jacob Berzelius: The emergence of his chemical system: Madison, University of Wisconsin Press.

Metayard, E., 1865, The life of Josiah Wedgwood: London, Hurst and Blackett, 2 vols.

Meyer, E.S.C. von, [1891] 1975, A history of chemistry: New York, Arno Press.

Middleton, W.E.K., 1966, A history of the thermometer and its uses in meteorology: Baltimore, Johns Hopkins Press.

Morello, N., 1981, Lazzaro Spallanzani geopaleontologo dall' origine delle sorgenti alla vulcanologia, *in* Lazzaro Spallanzani e la biologia del settecento, p. 271–281.

Morello, N., ed., 1998, Volcanoes and history: Genova, Brigati.

Morton, A.Q., and Wess, J.A., 1993, Public and private science: The King George III collections: Oxford, Oxford University Press in association with the Science Museum.

Multhauf, R.P., 1962, On the use of the balance in chemistry: Proceedings of the American Philosophical Society, v. 106, p. 210–218.

Multhauf, R.P., 1966, The origins of chemistry: London, Oldbourne.

Multhauf, R., 1978, Neptune's gift: Baltimore, Johns Hopkins University Press,

Multhauf, R.P., and Good, G., 1987, A brief history of geomagnetism and a catalog of the collections of the National Museum of American History: Smithsonian Studies in History and Technology, Paper 44: Washington D.C., Smithsonian Institution Press.

Needham, J., 1959, Science and civilisation in China: Mathematics and the science of the heavens and the Earth, v. 3: Cambridge, Cambridge University Press.

Newcomb, S., 1986, Laboratory evidence of silica solution supporting Wernerian theory: Ambix, v. 33, p. 88–93.

Newcomb, S., 1990a, Contributions of British experimentalists to the discipline of geology: 1780–1820: Proceedings of the American Philosophical Society, v. 134, p. 161–225.

Newcomb, S., 1990b, The problem of volcanic origins: Journal of Geological Education, v. 38, p. 123–131.

Newcomb, S., 1998a, Early volcanology and the laboratory, *in* Morello, N., ed., Volcanoes and history: Genova, Brigati, p. 441–458.

Newcomb, S., 1998, Laboratory variables in late eighteenth century geology, *in* Fritscher, B., and Henderson, F., eds., Toward a history of mineralogy, petrology, and geochemistry: Heft 23: München, Institut fur Geschichte der Naturwissenschaften, p. 81–100.

Newcomb, S., 2002, Characters in context, *in* Albrecht, H., and Ladwig, R., eds., Abraham Gottlob Werner and the foundation of the geological sciences: Freiberger Forschungshefte D207: Freiberg, Technische Universität Bergakademie Freiberg, p. 236–247.

Newman, W.R., 2000, Alchemy, assaying, and experiment, *in* Holmes, F.L., and Levere, T.H., eds., Instruments and experimentation in the history of chemistry: Cambridge, Massachusetts Institute of Technology Press, p. 35–54.

Nicholas, C.J., and Pearson, P.N., 2007, Robert Jameson on the Isle of Arran, 1797–1799, *in* Wyse Jackson, P.N., ed., Four centuries of geological travels: Geological Society of London Special Publication 287, p. 31–47.

Nieuwenkamp, W., 1970, Leopold von Buch, *in* Gillispie, C.C., ed., Dictionary of scientific biography, v. 2: New York, Scribner, p. 552–557.

Nobis, H.M., 1998, Der Ursprung der Steine zur Beziehung zwischen Alchemie und Mineralogie im Mittelalter, *in* Fritscher, B., and Henderson, F., eds., Toward a history of mineralogy, petrology, and geochemistry, Proceedings of the international symposium on the history of mineralogy, petrology, and geochemistry: Munich, 1996. Heft 23: München, Institut fur Geschichte der Naturwissenschaften, p. 29–52.

Oldroyd, D.R., 1972, Robert Hooke's methodology of science as exemplified in his 'Discourse on earthquakes': British Journal for the History of Science, v. 6, p. 109–130. (Reprinted *in* Oldroyd, D., 1998, Sciences of the Earth: Aldershot, Ashgate Variorum, X.)

Oldroyd, D.R., 1973a, An examination of G.E. Stahl's philosophical principles of universal chemistry: Ambix, 20, p. 36–52; 1998, Variorum V.

Oldroyd, D.R., 1973b, Some eighteenth century methods for the chemical analysis of minerals: Journal of Chemical Education, v. 50, p. 337–340; 1998, Variorum VIII.

Oldroyd, D.R., 1974a, A note on the status of A.F. Cronstedt's simple earths and his analytical methods: Isis, v. 65, p. 506–512; 1998, Variorum VI.

Oldroyd, D.R., 1974b, From Paracelsus to Haüy: The development of mineralogy in its relation to chemistry [Ph.D. diss.]: Sydney, Australia, University of New South Wales.

Oldroyd, D.R., 1974c, Some phlogistic mineralogical schemes, illustrative of the evolution of the concept of "Earth" in the 17th and 18th centuries: Annals of Science, v. 31, p. 269–305; 1998, Variorum IV.

Oldroyd, D.R., 1974d, Mechanical mineralogy: Ambix, v. 21, p. 157–78; 1998, Variorum III.

Oldroyd, D.R., 1979, Science in the silver age: Aetna, a classical theory of volcanic activity: Centaurus, v. 23, p. 1–20; 1998, Variorum I.

Oldroyd, D.R., 1986, The arch of knowledge: London, Methuen.

Oldroyd, D.R., 1996, Thinking about the Earth: A history of ideas in geology: London, T. Athlone Press.

Oldroyd, D.R., 1998, Sciences of the Earth: Aldershot, Ashgate Variorum.

Oldroyd, D.R., ed., 2002, The Earth inside and out: Some major contributions to geology in the twentieth century: London, Geological Society of London.

Oldroyd, D.R., Amador, F., Kozák, J., Carneiro, A., and Pinto, M., 2007, The study of earthquakes in the hundred years following the Lisbon Earthquake of 1755: Earth Sciences History, v. 26, no. 2, p. 321–370.

Ospovat, A.M., [1774] 1971, Translation of A.G. Werner's short classification and description of the various rocks: New York, Hafner Publishing Co.

Pancaldi, G., [2003] 2005, Volta: Science and culture in the age of Enlightenment: Princeton, New Jersey, Princeton University Press.

Partington, J.R., 1961, A history of chemistry, v. 2: London, Macmillan & Co.

Partington, J.R., 1962, A history of chemistry, v. 3: London, Macmillan & Co.

Partington, J.R., 1989, A short history of chemistry (third edition): New York, Dover.

Pinto, M.S., ed., 2003, Geological resources and history: Proceedings of the 26th Inhigeo symposium, Portugal, 2001: Aveiro, Portugal, Universidade de Aveiro.

Pintus, A., 1998, Gravity, Newton, and the eighteenth century, *in* Good, G.A., ed., Sciences of the Earth: An encyclopedia of events, people, and phenomena, v. 1: New York, Garland Publishing Company, p. 399–403.

Pitcher, W.S., 1993, The nature and origin of granite: London, Blackie Academic & Professional.

Pledge, H.T., 1939, Science since 1500: London, Science Museum.

Poirier, J.-P., 1998, Lavoisier: Chemist, biologist, economist: Philadelphia, University of Pennsylvania Press.

Polak, A., 1975, Glass its tradition and its makers: New York, G.P. Putnam's Sons.

Porter, R., 1977, The making of geology: Cambridge, Cambridge University Press.

Porter, R., 1980, The terraqueous globe, *in* Rousseau, G.S., and Porter, R., eds., The ferment of knowledge: Cambridge, Cambridge University Press, p. 285–386.

Porter, T.M., 1981, The promotion of mining and the advancement of science: The chemical revolution of mineralogy: Annals of Science, v. 38, p. 543–570.

Prestwich, J., 1895, Collected papers on some controverted questions of geology: London, Macmillan and Co.

Principe, L.M., 1998, The aspiring adept: Robert Boyle and his alchemical quest: Princeton, New Jersey, Princeton University Press.

Principe, L.M., 2000, Apparatus and reproducibility in alchemy, *in* Holmes, F.L., and Levere, T.H., eds., Instruments and experiments in the history of chemistry: Cambridge, Massachusetts Institute of Technology Press, p. 55–74.

Pumfrey, S., 1998, Geomagnetism, theories before 1800, *in* Good, G.A., ed., Sciences of the Earth: An encyclopedia of events, people, and phenomena, v. 1: New York, Garland Publishing Company p. 345–350.

Read, H.H., 1957, The granite controversy: London, Thomas Murby & Co.

Read, J., [1957] 1961, Through alchemy to chemistry: New York, Harper Torchbooks.

Reilly, R., 1989, Wedgwood: New York, Stockton Press, 2 vols.

Richet, P., 2007, A natural history of time: Chicago, University of Chicago Press.

Roberts, L., 1999, Science becomes electric: Isis, v. 90, p. 680–714.

Roger, J., translator, 1962, Les Époques de la Nature (G.L. Leclerc de Buffon), *in* Mémoires du Muséum national d'histoire naturelle: série C, Sciences de la Terre, tome X: Paris, Editions du Museum CLII.

Roger, J., 1997, Buffon: A life in natural history: Ithaca, New York, Cornell University Press.

Rossi, P., [1984] 1987, The dark abyss of time: Chicago, University of Chicago Press.

Rousseau, G.S., and Porter, R., eds., 1980, The ferment of knowledge: Cambridge, Cambridge University Press.

Rudler, F.W., 1889, Experimental geology: Proceedings of the Geologists' Association, v. 11: London, University College, Edward Stanford, p. 69–103.

Rudwick, M.J.S. [1972] 1976, The meaning of fossils (second edition): New York, Science History Publications.

Rudwick, M.J.S., 1985, The great Devonian controversy: Chicago, University of Chicago Press.

Rudwick, M.J.S., 2005, Bursting the limits of time: The reconstruction of geohistory in the age of revolution: Chicago, University of Chicago Press.

Russell, C.A., ed., 1985, Recent developments in the history of chemistry: London, Royal Society of Chemistry.

Russell, C.A., 2000, Chemical techniques in a preelectronic age: The remarkable apparatus of Edward Frankland, in Holmes, F.L., and Levere, T.H., eds., Instruments and experimentation in the history of chemistry: Cambridge, Massachusetts Institute of Technology Press, p. 311–334.

Schimkat, P., 1998, Gümbel on Bischof, or: Two different ways of losing out, in Fritscher, B., and Henderson, F., eds., Toward a history of mineralogy, petrology, and geochemistry: Heft 23: München, Institut fur Geschichte der Natuwissenschaften.

Schneer, C.J., ed., 1969, Toward a history of geology: Cambridge, Massachusetts Institute of Technology Press.

Schneer, C.J., [1969] 1988, Mind and matter: Man's changing concepts of the material world: Ames, Iowa State University Press.

Schofield, R.E., 1959, Josiah Wedgwood, industrial chemist: Chymia, v. 5, p. 180–192.

Schofield, R.E., 1963, The Lunar Society of Birmingham: Oxford, Clarendon Press.

Schuster, J.A., and Watchirs, G., 1990, Natural philosophy, experiment and discourse: Beyond the Kuhn/Bachelard problematic, in Le Grand, H.E., ed., Experimental inquiries: Historical, philosophical and social studies of experimentation in science: Dordrecht, Kluwer Academic Publishers, p. 1–47.

Scott, E.L., 1975, James Keir, in Gillispie, C.C., ed., Dictionary of scientific biography, v. 12: New York, Scribner, p. 277–278.

Şengör, A.M.C., 2001, Is the present the key to the past or the past the key to the present?: Geological Society of America Special Paper 355.

Shannon, J.M., and Shannon, G.C., 1999, The assay balance: Its evolution and the histories of the companies that made them: Privately printed JMS and GCS, 7319 W. Cedar Circle, Lakewood, Colorado, 80226.

Shapin, S., and Schaffer, S., 1985, Leviathon and the air-pump: Princeton, New Jersey, Princeton University Press.

Siegfried, R., 2002, From elements to atoms: A history of chemical composition: Transactions of the American Philosophical Society, v. 92, part 4: Philadelphia, American Philosophical Society.

Siegfried, R., and Dott, R.H., 1980, Humphry Davy on geology: The 1805 lectures for the general audience: Madison, University of Wisconsin Press.

Sigrist, R., 2001, H.-B. de Saussure (1740–1799): Un regard sur la terre: Geneva, Bibliothèque d'Histoire des Sciences.

Sigurdsson, H., ed., 2000, Encyclopedia of volcanoes: New York, Academic Press.

Simpson, A.D.C., ed., 1982, Joseph Black, 1728–1799: A commemorative symposium: Edinburgh, The Royal Scottish Museum.

Singer, C.J., ed., 1954–1959, History of technology: Oxford, Clarendon Press, 8 vols.

Smeaton, W.A., 1966, The portable chemical laboratories of Guyton de Morveau: Cronstedt and Göttling: Ambix, v. 13, p. 84–91.

Smeaton, W.A., 1987, Some large burning lenses and their use by eighteenth-century French and British chemists: Annals of Science, v. 44, p. 265–276.

Smeaton, W.A., 2000, Platinum and ground glass, in Holmes, F.L., and Levere, T.H., eds., Instruments and experimentation in the history of chemistry: Cambridge, Massachusetts Institute of Technology Press, p. 211–238.

Smith, C.S., 1956, Metallurgy and assaying, in Singer, C.J., ed., History of technology: Oxford, Clarendon Press, v. 3, p. 27–71.

Smith, C.S., 1967, Metallurgy in the seventeenth and eighteenth centuries, in Kranzberg, E.M., and Pursell, C.W., eds., Technology in western civilization, v. 1: New York, Oxford University Press, p. 142–152.

Smith, C.S., 1969, Porcelain and plutonism, in Schneer, C.S., ed., Toward a history of geology: Cambridge, Massachusetts Institute of Technology Press, p. 317–338.

Smith, D., 1998, Hardness testing instruments, in Bud, R., and Warner, D.J., eds., Instruments of science: An historical encyclopedia: New York, Garland Publishing Co., p. 303–304.

Smith, J., 1994, Fact and feeling: Baconian science and the nineteenth-century literary imagination: Madison, University of Wisconsin Press.

Smith, P.H., 1994, The business of alchemy: Science and culture in the holy Roman empire: Princeton, New Jersey, Princeton University Press.

Sorrenson, R.J., and Burnett, J., 1998, Pyrometer, in Bud, R., and Warner, D.J., eds., Instruments of science: An historical encyclopedia: New York, Garland Publishing Co., p. 497–499.

Spies, B.R., 1998, Earth conductivity measurements, in Bud, R., and Warner, D.J., eds., Instruments of science: New York, Garland Publishing Company, p. 199–201.

Stock, J.T., 1964, Development of the chemical balance: London, Her Majesty's Stationery Office.

Sweet, J.M., 1967, Robert Jameson's Irish journal, 1797: Annals of Science, v. 23, p. 97–126.

Swinbank, P., 1982, Experimental science in the University of Glasgow at the time of Joseph Black, in Simpson, A.D.C., ed., Joseph Black, 1728–1799: A commemorative symposium: Edinburgh, The Royal Scottish Museum, p. 23–35.

Szabadváry, F., 1966, A history of analytical chemistry: Oxford, Pergamon Press.

Taylor, K.L., 1971, Desmarest, Nicolas, in Gillispie, C.C., ed., Dictionary of scientific biography, v. 4: New York, Scribner, p. 70–73.

Taylor, K.L., 1973, Jean-Claude de Lamétherie, in Gillispie, C.C., ed., Dictionary of scientific biography, v. 7: New York, Scribner, p. 602–604.

Thomasian, R., 1974, Moro, Antonio-Lazzaro, in Gillispie, C.C., ed., Dictionary of scientific biography, v. 9: New York, Scribner, p. 531–534.

Torge, W., 1989, Gravimetry: Berlin, Walter de Gruyter.

Torrens, H., 1997a, Some thoughts on the complex and forgotten history of mineral exploration: Journal of the Open University Geological Society 17, Chandler's Ford, 121; and in Torrens, H., 2002, The practice of British geology, 1750–1850: Aldershot, Ashgate Variorum, I.

Torrens, H., 1997b, James Ryan (c. 1770–1847) and the problems of introducing Irish 'New Technology' to British mines in the early nineteenth century, in Bowler, P.J., and Whyte, N., eds., Science and society in Ireland: The social context of science and technology in Ireland, 1800–1950: Belfast, Queen's University of Belfast, p. 67–83, and in Torrens, H., 2002, The practice of British geology, 1750–1850: Aldershot, Ashgate Variorum, VIII.

Torrens, H., 2002, The practice of British geology, 1750–1850: Aldershot, Ashgate Variorum, I.

Torrens, H., 2004, Thomson, William: Oxford dictionary of national biography: Oxford, Oxford University Press, p. 562–563.

Touret, J.L.R., and Visser, R.P.W., eds., 2004, Dutch pioneers of the earth sciences: Amsterdam, Royal Netherlands Academy of Arts and Sciences.

Turner, A., 1987, Early scientific instruments: Europe, 1400–1800: London, Southeby's.

Turner, G.L'E., 1983, Nineteenth century scientific instruments: Berkeley, University of California Press.

Turner, G.L'E., 1998a, Scientific instruments, 1500–1900: An introduction: Berkeley, University of California Press.

Turner, G.L'E., 1998b, The compound microscope, in Bud, R., and Warner, D.J., eds., Instruments of science: An historical encyclopedia: New York, Garland Publishing Co., p. 388–390.

Tuttle, O.F., and Bowen, N.L., 1958, Origin of granite in the light of experimental studies in the system $NaAl_3O_8$-$KAlSi_3O_7$-SiO_2-H_2O: Geological Society of America Memoir 74.

Uglow, J., 2002, The lunar men: New York, Farrar, Straus and Giroux.

Vaccari, E., 1996, Lazzaro Spallanzani (1729–1799): un naturaliste italien du dix-huitième siècle et sa contribution aux sciences de la terre: COFRHIGÈO Troisième série, t.X:7, p. 78–95.

Vaccari, E., 1998a, Mineralogy and mining in Italy between the eighteenth and nineteenth centuries: The extent of Wernerian influences from Turin to Naples, in Fritscher, B., and Henderson, F., eds., Toward a history of mineralogy, petrology, and geochemistry: Heft 23: München, Institut fur Geschichte der Naturwissenschaften, p. 107–130.

Vaccari, E., 1998b, Mining academies, *in* Good, G.A., ed., Sciences of the Earth: An encyclopedia of events, people, and phenomena, v. 2: New York, Garland Publishing, p. 585–589.

Vaccari, E., and Morello, N., 1998, Mining and knowledge of the Earth, *in* Good, G.A., ed., Sciences of the Earth: An encyclopedia of events, people, and phenomena, v. 2: New York, Garland Publishing, p. 589–593.

Vai, G.B., and Caldwell, W.G., eds., 2006, The origins of geology in Italy: Geological Society of America Special Paper 411.

Vai, G.B., and Cavazza, W., 2006, Ulisse Aldrovandi and the origin of geology and science, *in* Vai, G.B., and Caldwell, W.G., eds., The origins of geology in Italy: Geological Society of America Special Paper 411, p. 43–63.

Vanity Fair, 1875, 20 February.

Veneer, L., 2009, Practical geology in the Geological Society in its early years, *in* Lewis, C.L.E., and Knell, S.J., eds., The making of the Geological Society of London: Geological Society of London Special Publication 317 (in press).

Vogel, R.M., 1966, Tunnel engineering—A museum treatment: Contributions from museum of history and technology, Paper 41: Washington, D.C., Smithsonian Institution, p. 201–239.

Whewell, W., 1857, History of the inductive sciences, v. 3: London, J.W. Parker and Son.

Whewell, W., 1858, History of scientific ideas: London, J.W. Parker and Son, 2 vols.

Wiesheu, R., and Hein, U.F., 1998, The history of fluid inclusion studies, *in* Fritscher, B., and Henderson, F., eds., Toward a history of mineralogy, petrology, and geochemistry: Heft 23: München, Institut fur Geschichte der Naturwissenschaften, p. 309–326.

Wise, M.N., ed., 1995, The values of precision: Princeton, New Jersey, Princeton University Press.

Wolf, A., [1935] 1968, A history of science, technology, and philosophy in the 16th and 17th centuries, v. 1: Gloucester, Massachusetts, Peter Smith.

Wolf, A., [1952] 1961, A history of science, technology & philosophy in the 18th century: New York, Harper Torchbooks, 2 vols.

Wyllie, P.J., 1998, Hutton and Hall on theory and experiments: The view after two centuries: Episodes, v. 21, p. 3–10.

Wyse Jackson, P.N., ed., 2000, Science and engineering in Ireland in 1798: Dublin, Royal Irish Academy.

Wyse Jackson, P.N., ed., 2007, Four centuries of geological travel: Geological Society of London Special Publication 287.

Yang Wenheng, 1983, Rocks, mineralogy, and mining, *in* Chinese Academy of Sciences, Ancient China's technology and science: Beijing, Foreign Press, p. 258–269.

Yearley, S., 1984, Proofs and reputations: Sir James Hall and the use of classification devices in scientific argument: Earth Sciences History, v. 3, p. 25–43.

Yoder, H., 1980, Experimental mineralogy: Achievements and prospects: Bulletin of Mineralogy, v. 103, p. 5–26.

Yoder, H.S., 1993, Timetable of petrology: Journal of Geological Education, v. 41, p. 447–489.

Young, D.A., 1999, The emergence of the diversity of igneous rocks as a geological problem: Earth Sciences History, v. 18, no. 1, p. 51–77.

Young, D.A., 2003, Mind over magma: The story of igneous petrology: Princeton, New Jersey, Princeton University Press.

Zittel, K.A. von, 1901, History of geology and palaeontology: London, Walter Scott.

Index

Note: Page numbers with "f" and "t" indicate material in figures and tables, respectively. The letter "n" indicates material in footnotes.